Springer Geology

The book series Springer Geology comprises a broad portfolio of scientific books, aiming at researchers, students, and everyone interested in geology. The series includes peer-reviewed monographs, edited volumes, textbooks, and conference proceedings. It covers the entire research area of geology including, but not limited to, economic geology, mineral resources, historical geology, quantitative geology, structural geology, geomorphology, paleontology, and sedimentology.

More information about this series at http://www.springer.com/series/10172

Felix V. Kaminsky

The Earth's Lower Mantle

Composition and Structure

Springer

Felix V. Kaminsky
KM Diamond Exploration Ltd.
West Vancouver, BC
Canada

ISSN 2197-9545 ISSN 2197-9553 (electronic)
Springer Geology
ISBN 978-3-319-55683-3 ISBN 978-3-319-55684-0 (eBook)
DOI 10.1007/978-3-319-55684-0

Library of Congress Control Number: 2017934461

Printed on acid-free paper

This Springer imprint is published by Springer Nature
The registered company is Springer International Publishing AG
The registered company address is: Gewerbestrasse 11, 6330 Cham, Switzerland

Contents

Abbreviations

1D	One-dimensional
3D	Three-dimensional
BLS	Brillouin light scattering
Brd, Bridg	Bridgmanite
Cal	Calcite
Carb	Carbonate
CaSiPrv	CaSi-perovskite $CaSiO_3$
CCM	Charge-coupled substitution
CF	Anhydrous aluminous phase with calcium–ferrite structure
CL	Cathodoluminescence
CMB	Core-mantle boundary
Coh	Cohenite
DAC	Diamond-anvil cell
DFT	Density function theory
Dia	Diamond
DHMS	Dense anhydrous magnesian silicates
Dol	Dolomite
EBSD	Electron backscatter diffraction
ED	Electron diffraction
EDX	Electron-dispersive X-ray spectroscopy
EELS	Electron energy loss spectroscopy
Egg	Phase 'Egg'
EHDAC	Externally-heated diamond-anvil cell
EoS	Equation of state
EPR	Electron paramagnetic resonance
$fe = Fe/(Fe + Mg)_{at}$	Iron index
f_{O_2}	Oxygen fugacity
fPer	Ferropericlase
FTIR	Fourier transform infrared spectroscopy
G	Shear modulus (modulus of rigidity)
GGA	Generalized-gradient approximation

GPa	Gigapascal (10^9 pascal = 10 kbar)
HS	High spin
HREE	Heavy rare-earth elements
Ilm	Ilmenite
IR	Infrared
IS	Intermediate spin
ISS	Impulsive stimulated light scattering
IXS	Inelastic X-ray scattering
Jfb	Jeffbenite
K_α	X-ray emission line
K_D	Partition coefficient
K_S	Isoentropic (constant-energy) bulk modulus
K_T	Isothermal bulk modulus
Ky	Kyanite
L	Liquid
LDA	Local density approximation
LHDAC	Laser-heated diamond-anvil cell
LLSVP	Large low shear velocity province
LIL	Laser-induced luminescence
LREE	Light rare-earth elements
LS	Low spin
μ	Shear elastic modulus
Maj	Majorite garnet
$mg = Mg/(Mg + Fe)_{at}$	Magnesian index
MgWus	Magnesiowüstite
MORB	Middle-ocean ridge basalt
MS	Mössbauer spectroscopy
NAL	New hexagonal aluminous phase
NMR	Nuclear magnetic resonance
OIB	Ocean island basalt
Ol	Olivine
Pilm	Picroilmenite
Per	Periclase
PREM	Preliminary Reference Earth Model
PL	Photoluminescence
QHA	Quasi-harmonic approximation
QMC	Quantum Monte Carlo method
QS	quadrupole splittings
ρ	Density in g/cm^3
REE	Rare-earth elements
σ	Electrical conductivity
Sid	Siderite
SEM	Scanning electron microscopy
ss	Solid dolution
Sti	Stishovite

TEM	Transmission electron microscopy
ULVZ	Ultra-low velocity zone
UV	Ultraviolet
V	Vacancy in crystal structure
V_p	Primary seismic compressional wave velocity
V_s	Secondary seismic shear-wave velocity
V_ϕ	Bulk sound velocity
Wus	Wüstite
XAFS	X-ray absorption spectroscopy
XANES	X-ray absorption near-edge spectroscopy
XES	X-ray emission spectroscopy
XRD	X-ray diffraction

Note: Mineral abbreviations are in accordance with Witney DL, Evans BW (2010) Abbreviations for names of rock-forming minerals. American Mineralogist 95:185–187

Chapter 1
Introduction

The Earth's lower mantle (extending from a depth of 660–2900 km) comprises more than half ($\sim$56%) of the total volume of the Earth. Most of the Earth's geodynamic processes stem from the mantle, which is reflected in the Earth's structure. However, its composition and structure are not yet well known.

Until recently, three major data sources have been used to draw conclusions regarding the composition and structure of the lower mantle: theoretical calculations; experiments conducted under high pressure, high-temperature conditions; and seismological observations. The first general models of the lower mantle were created based on these data sources. However, each of these data sources has both its advantages and disadvantages. Theoretical calculations are limited to simple compositions and may be used only as an approximation of the natural system. Experimental results depend upon the conditions of the experiments (e.g., type of compression medium and its hydrostaticity, type of heating, speed of heating and compression, duration time of high-pressure treatment and starting material) and are frequently contradictory. Depths of lower mantle scatterers established from different earthquakes, vary by hundreds of kilometres; estimation of seismic boundaries, made at different receiving stations from the same events may vary even more. The nature of the scatterers may be merely an estimation.

In recent decades, a series of lower mantle minerals and microxenoliths were discovered to be included in from diamonds in four continents. These findings support studies investigating the real composition of lower mantle material. They add significant information to our knowledge on the lower mantle but are rarely used for the lower mantle model compilations; most experimental and geophysical works ignore this data. To date, more than 2000 publications are in existence ranging from those containing results of experiments imitating lower mantle conditions, to those with ab initio calculations of mineral composition within the lower mantle, to the interpretations of seismic anomalies within the lower mantle, to other aspects of the lower mantle composition and structure. However, none of these papers used the observed geological data; hence, geological observations from experimental and theoretical data are disconnected. As a result, numerous

F.V. Kaminsky, *The Earth's Lower Mantle*, Springer Geology,
DOI 10.1007/978-3-319-55684-0_1

contradictions and discrepancies exist, which cannot be resolved if these results remain independent of each other. This represents the main reason why the author of this book set out to summarize the observed facts on the composition and structure of the material of the natural lower mantle and to compare this data with the data from areas of indirect constructions.

Some of the observed data on the natural lower mantle are in disagreement with experimental models. For example, one of the most enigmatic problems in the mineralogy of the lower mantle is the content and composition of ferropericlase. According to the theoretical and experimental constructions, ferropericlase, along with bridgmanite and CaSi-perovskite, is one of the major minerals in the lower mantle, where it should comprise up to 16–20%. However, among mineral inclusions in the lower mantle, diamond ferropericlase grains comprise 48.0–63.3% (average: 55.4%) of the total mineral content in the lower mantle, i.e., approximately three times more than the level previously suggested. Furthermore, the composition of ferropericlase, according to experiments in pyrolitic systems in the pressure range 25–60 GPa, should be highly magnesian, with the magnesium index $mg = Mg/(Mg + Fe)_{at}$ at *c*. 0.90. However, in the lower mantle per se, ferropericlase may vary locally to include more iron-rich compositions with $mg = 0.36$.

Another unexpected observation (from the theoretical and experimental perspectives) is the presence of free silica in all lower mantle associations. Silica inclusions, which were never predicted by theoretical constructions, were identified in all sets of lower mantle minerals observed in diamond from all regions and areas: 2.6–4.8% from Brazil; 8.2% from Guinea; and 10–20% from Canada.

One further feature of the real lower mantle composition that was not predicted is the existence of the carbonatitic mineral association, which was never observed in the upper mantle and comprises almost 50 mineral species, including carbonates, halides, fluorides, phosphates, sulphates, oxides, silicates, sulphides and native elements, as well as volatiles.

Kapitza (1966), in his recollections of Lord Rutherford, specifically underlined how the contradictions between theory and observation enable us to widen our knowledge of nature. Indeed, natural facts that contradict theoretical data allow regularities to be unveiled that could not be modelled by theoretical and experimental constructions alone, whilst also opening up the possibility for creating the most realistic models of the composition and structure of the deep Earth. Some discrepancies that exist between the model and the observed compositions of the lower mantle cannot be fully explained by the current hypotheses. However, I hope that even the act of just establishing such discrepancies will produce more realistic composition models of the Earth and its history.

In this work, I collected the most current geological data on the lower mantle, compared them with the available experimental, mineral physics and geophysical data, and tried to find general regularities in the composition of the lower mantle. The creation of a harmonious lower mantle model is an answer for several unknown quantities in one equation, and that would be too presumptuous to settle this problem. The main goal of the author was not to settle all the problems of the lower mantle composition and structure, but rather to summarize the current available data and to outline the major problems in this field, including its controversies. Geodynamic problems were not included in the scope of this work. I hope that the following pages will be of use for all geologists, petrologists and mineralogists studying deep Earth processes, as well as for experimental and theoretical researchers.

I would like to thank my colleagues with whom I worked in the field, collecting diamonds with lower mantle inclusions, analyzing these inclusions, studying and discussing different aspects of the lower mantle composition, and publishing results of these studies: A. Abakumov, P. Andreazza, E. Belousova, L. Bindi, F.E. Brenker, D.M. De R. Channer, P. Cartigny, E.M. Galimov, R. Davies, W.L. Griffin, B. Harte, P.C. Hayman, B. Jacobsen, G.K. Khachatryan, J.F. (Afu) Lin, S. Matsyuk, J. Matzel, C. McCammon, F. Nestola, S.Y. O'Reilly, M.P. Palot, I.D. Ryabchikov, S.M. Sablukov, A. Schreiber, A.A. Shiryaev, T. Stachel, G. Silversmit, R. Wirth, O.D. Zakharchenko and others, whose collaboration was invaluable. Images for crystal structures were created by Hong Yang (HPSTAR, Shanghai) with the use of CrystalMaker 9.2.7 software on the basis of certain observed mineral parameters obtained from geological samples or in high pressure–temperature experiments. I am particularly thankful to Yu.A. Litvin, discussions with whom inspired the writing of this monograph.

I understand that the present work may not be free of faults and will be thankful to the readers for any notes and remarks.

Reference

Kapitza, P. (1966). Recollections of Lord Rutherford. *Proceedings of the Royal Society of London. A294*, 123–137.

Chapter 2
General Physical and Chemical Models of the Earth's Lower Mantle

Abstract The most popular model of the lower mantle is the Preliminary Reference Earth Model (PREM), derived from seismic observations assuming the pyrolitic composition of the lower mantle. The uppermost part of the lower mantle ($\sim$660–770 km deep) has a steep velocity gradient, reflecting the mineral structure transformation from ringwoodite to bridgmanite and ferropericlase, after which gradual increase in both the compressional velocity (V_p) and shear velocity (V_s) reflects the near adiabatic compression of mineral phases. The adiabatic geothermal gradient within the upper mantle decreases with increasing depth without phase transitions. Subducting lithospheric slabs may significantly cool temperature profiles, particularly in the upper part of the lower mantle. However, results of experiments on the density of natural peridotite, performed within the range of entire lower-mantle pressures along the geotherm, demonstrated their significant mismatch with the PREM density model. This implies that the upper and the lower mantle must have different chemical compositions, i.e. the mantle is chemically stratified, with the inference of a non-pyrolitic composition of the lower mantle. The diapason of oxygen fugacity within the entire sequence of lower-mantle region may reach ten logarithmic units, varying from below the IW buffer to the FMQ buffer values.

2.1 General

In one of the first geophysical models of the Earth, Bullen (1942) proposed a spherically symmetric 'shell model' of the Earth's interior and distinguished eight layers labelled alphabetically from A to G, with the crust as the A layer, the upper mantle as the B layer, the transition zone as the C layer, the entire lower mantle as the D layer, the outer core as the E layer, its lower part as the F layer, and the inner core as the G layer. Later, Bullen found the D layer to be made from two different layers: the upper part of the D layer, about 1800 km thick, was renamed D′; and the lower part (the bottom 200 km) was named D″ (Bullen 1950). To date, only the

F.V. Kaminsky, *The Earth's Lower Mantle*, Springer Geology,
DOI 10.1007/978-3-319-55684-0_2

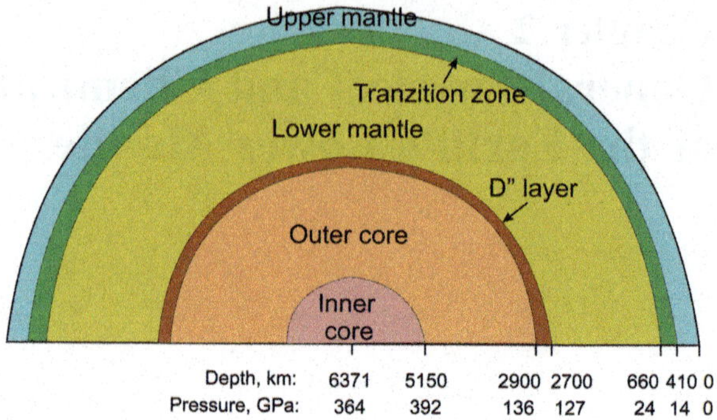

Fig. 2.1 General structure of the Earth. Borders of the major Earth's regions and corresponding pressure values, accepted in this work, are shown at the bottom

latter name is in use to refer to the lowermost 200 km of the mantle, which may be considered to be a boundary transitional layer from the lower mantle to the outer core (Fig. 2.1).

Later seismic studies revealed more complex regions of the Earth and the following principal regions were identified within the Earth, based on seismic parameters (Dziewonski and Anderson 1981).

1. Ocean layer.
2. Upper and lower crust.
3. Region above the low velocity zone (LVZ), starting at a depth of 220 km.
4. LVZ (80–220 km depth).
5. Region between the LVZ and 400 km discontinuity.
6. Transition zone (TZ), spanning the region between the 400 and 670 km discontinuities.
7. Lower mantle, subdivided into three parts by second-order discontinuities at 770, 2740 and 2740 km. The lowermost part corresponds to the D″ layer.
8. Outer core (2890–5150 km).
9. Inner core (5150–6371 km).

These depth estimates are slightly corrected in modern petrological constructions (e.g., the upper and lower borders of the TZ at 410 and 660 km, respectively, with the borders of the D″ layer at 2700 and 2900 km). The generalized structure of the Earth accepted in this work is presented in Fig. 2.1.

2.2 Seismic Velocities in the Lower Mantle

The decision to distinguish the boundary at ~660–670 km between the lower mantle and the transition zone is based on seismologic data, according to which remarkable changes (6–11%) in the velocities of seismic waves occur at this depth (Preliminary Reference Earth Model or PREM; Dziewonski and Anderson 1981). The exact position of this seismic discontinuity varies depending on the tectonic position of the area. It is located deeper than 670 km under subduction zones and shallower at 'hotspot' areas, such as Iceland, Hawaii and others (e.g., Shen et al. 1998). The increasing proportion of the basaltic component within the lower mantle in these areas may stimulate depth of this border to increase to 720 km (Deuss et al. 2006). In some areas, it bifurcates to 520 and 670 km boundaries (e.g., Deuss and Woodhouse 2001). In addition, undulation of the upper boundary of the lower mantle may be caused by the presence of high proportions of volatile components (such as water) and Mg–Fe partitioning between coexisting mineral phases in the constituent materials. The most popular 1D Earth model, PREM, is presented in Fig. 2.2.

The uppermost part of the lower mantle ($\sim$660–770 km deep) has a steep velocity gradient, reflecting the mineral structure transformation from ringwoodite to bridgmanite and ferropericlase. After that, there is a gradual increase in both the compressional velocity (V_p) and shear velocity (V_s), reflecting the near adiabatic compression of mineral phases (Dziewonski and Anderson 1981). In the normal lower mantle, three major minerals are present: bridgmanite, ferropericlase and CaSi-perovskite; stishovite can also be present. The velocities of each mineral phase vary with pressure and equilibrium composition in response to partitioning of components among phases. Comparison of the sound velocities of these phases shows that the fastest mantle phase is stishovite; however, the lowermost part of the

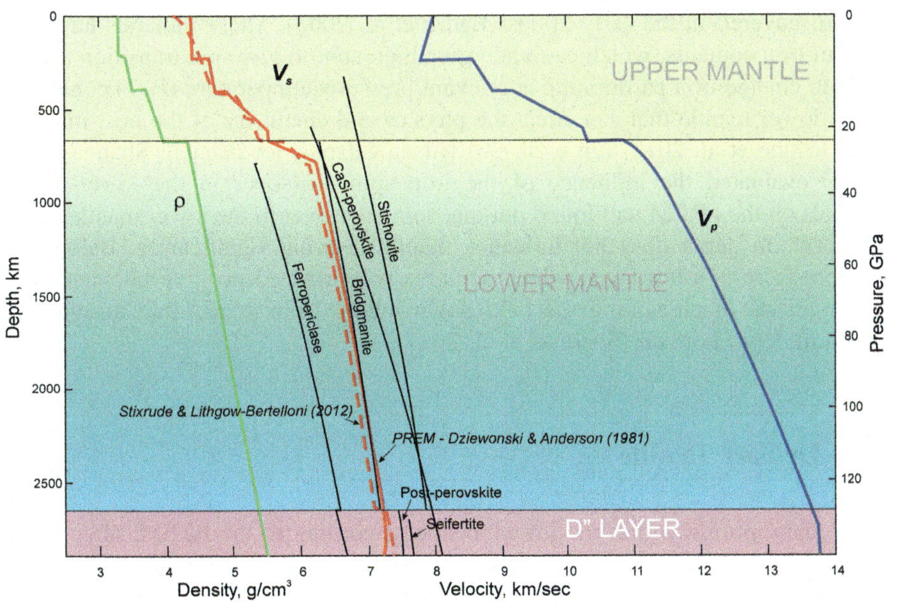

Fig. 2.2 Pyrolite mantle seismic wave velocity and density profiles as a function of depth. V_p compressional wave velocity, *blue*; V_s shear-wave velocity, *red*; velocities of the major lower-mantle minerals, *black*; ρ density of pyrolite in g/cm^3, *green*. Data from Dziewonski and Anderson (1981) and Stixrude and Litgow-Bertelloni (2012)

mantle shows CaSi-perovskite to be faster (Stixrude and Litgow-Bertelloni 2012; Fig. 1.2). Since the seismic wavelength is much greater than the size of the mineral grains, seismic waves see an averaged aggregate of constituent materials. However, lattice preferred orientations of the minerals due to the flow patterns of the mantle can cause seismic heterogeneities. The relative proportions of phases and their compositions may vary; in this case, one should expect additional variations in seismic velocities. However, in the PREM, seismic parameters are distributed homogeneously through the lower mantle, suggesting that the constituent minerals are gravitationally, almost adiabatically compressed, with the exception of the lowermost part of the lower mantle (the D″ layer), where a number of seismic anomalies have been identified (Dziewonski and Anderson 1981; Kennett et al. 1995).

Since the publication of the PREM in 1981, a number of other seismic models have been reported including *isp*91 (Kennett and England 1991), AK135 (Kennett et al. 1995), STW105 (Kustowski et al. 2008), GyPSuM (Simmons et al. 2010), and SPani (Tesoniero et al. 2015), amongst others; some of these models mainly focus on specific regions. Furthermore, some of these models deviate from PREM, showing that a re-evaluation of the PREM or a non-pyrolitic lower-mantle composition is needed.

One of most serious challenges to the one-dimensionally homogeneous seismic models such as the PREM has been the consequences of the spin crossover of iron in two major lower-mantle mineral phases, ferropericlase and bridgmanite, which were discovered in the early 2000s (Badro et al. 2003). These minerals have significant iron contents, which can undergo a high-spin to low-spin transition and, as a result, change iron partitioning at relevant pressure–temperature (*P–T*) conditions of the lower mantle that can affect the physics and chemistry of the host minerals (e.g., Irifune et al. 2010; Lin et al. 2013) (see Chap. 8 for details). Shahnas et al. (2011) evaluated the influence of the iron spin crossover in the lower-mantle minerals on the PREM and found out that spin crossover in the lower mantle above the 1800 km depth does not influence mantle layering significantly. Below this horizon, there is a hot layer, which acquires spin-induced positive buoyancy and which assists in preventing the cold downwelling from penetrating through the 1800 km depth horizon (Shahnas et al. 2011).

2.3 Density Profile

The density profile within the lower mantle, according to the PREM, after jumps related to the olivine → wadsleyite → ringwoodite → bridgmanite + ferropericlase phase transitions, becomes ~4.4 g/cm^3 at the top of the lower mantle, and then increases gradually to ~5.5 g/cm^3 at the mantle/core boundary (Fig. 2.2), mostly because of the compression of the mantle material (assuming the homogeneity of the material and adiabatic character of the process). However, even the early calculations demonstrated that the model composition of the lower mantle is at least 2.6% less dense than the observed lower-mantle density over the depth range 1000–2000 km; hence, the mantle should be stratified with the layers possibly intermixing (Jeanloz and Knittle 1989). The results of later experiments with a natural peridotite performed entirely within the range of lower-mantle pressures of up to 112 GPa along the geotherm, demonstrated the significant mismatch of density with the PREM density model, particularly in the slopes between the two profiles (Ricolleau et al. 2008, 2009). *This implies that the upper and the lower mantle must have different chemical compositions, i.e.,. the mantle is chemically stratified, if one uses the PREM as a reference. In that case, the data favour a Si-enriched, less dense lower mantle.*

Ricolleau et al. (2009) calculated that, in order to match their experimentally determined density profile of a pyrolite mantle composition to PREM, the temperature at the top of the lower mantle has to be $\sim 1500 \pm 60$ K, which is ~ 400 K lower than the estimated temperature at a depth of 660 km (Hirose 2002; Katsura et al. 2010) with a superadiabatic geotherm of gradient 0.60 K/km through the lower mantle, which is in disagreement with the adiabatic geotherm (see Sect. 2.5). The measured density for an ultramafic mantle along the geotherm shows a significant mismatch (2%) with the PREM derived from seismic observations assuming pyrolitic composition of the lower mantle. *This points to the likelihood of a non-pyrolitic composition of the lower mantle.*

2.4 Compositional Models and Chemical Homogeneity/Heterogeneity of the Lower Mantle

Knowing the composition of the lower mantle heavily depends on our understanding of the evolution and dynamics of the planet. The Earth's lower mantle is clearly layered with distinct seismic features that separate it from the overlaying upper mantle and transition zone, as well as the underlying outer core. Geochemical studies of the upper mantle materials (especially from xenoliths) show that the upper mantle is mostly made of peridotite with olivine, pyroxene and garnet being the abundant minerals. Together with petrological experiments, Ringwood (1975) proposed that the mantle is made of pyrolite, a hypothetical rock made of peridotite and basalt in a ratio of approximately 3:1. The question for the lower-mantle composition is whether or not the whole mantle remains chemically homogeneous or whether it is layered?

One key parameter in categorizing the composition of the mantle is the Mg/Si ratio that varies from approximately 1 in the chondritic composition model to 1.3 in the pyrolitic model (e.g., Ringwood 1975; Anderson 1983; Hart and Zindler 1986; Javoy 1995). As a simple reference, the Mg/Si ratio is 2 for forsterite (Mg_2SiO_4) and 1 for enstatite and bridgmanite ($MgSiO_3$). Petrological evidence obtained from the compositional trends in upper mantle peridotites suggests that the Mg/Si ratio of the bulk silicate Earth is close to 1.27, indicating that the upper and lower mantle are homogeneous in composition (Ringwood 1975; Anderson 1983). In the pyrolitic compositional model of the mantle, the lower mantle is suggested to be made of 75 vol.% bridgmanite [(Mg, Fe) (Al, Si)O_3], 17 vol.% ferropericlase [(Mg, Fe)O], and approximately 8 vol.% $CaSiO_3$-perovskite (Ringwood 1975). If the Earth was accreted from chondritic meteorites such as the CI chondrites, which have the Mg/Si ratio of approximately 1.074, the lower mantle should be chemically distinct from the upper mantle (e.g., Morgan and Anders 1980; Bass and Anderson 1984; Hart and Zindler 1986; Javoy 1995; McDonough and Sun 1995; Williams and Knittle 2005). The lower mantle in a chondritic compositional model is expected to be enriched in bridgmanite and CaSi-perovskite, with an Mg/Si ratio close to 1 (Anderson 1989).

From a geophysical prospective, a comparison of the seismic profiles with velocity and density profiles of candidate mantle minerals at relevant *P–T* conditions provides another means by which to evaluate the composition and mineralogy of the lower mantle. For example, Irifune et al. (2010) obtained the velocity profiles of pyrolite and piclogite compositions in a multi-anvil apparatus in transition zone conditions, showing that the velocity and density profiles of the region are most consistent with a pyrolite composition, instead of more pyroxene and garnet-rich piclogite. For the lower mantle, experimental results are rather more scattered because of the technical difficulties in reaching extreme *P–T* conditions and simultaneously measuring the elasticity of candidate minerals in the region (e.g., Murakami et al. 2012). Theoretical studies using first-principle methods to model the density and velocities of the lower mantle in various compositions conclude that lower mantle with a pyrolitic composition has density and velocity profiles matching the seismic PREM (Wang et al. 2015). Matas et al. (2007) studied the problem of obtaining the thermal structure and bulk chemical composition of the lower mantle from its seismologically determined velocity and density profiles, using the results on the elastic properties of the lower-mantle phases (including, of particular importance, shear moduli). They show the reverse that *these properties do not match the uniform chemical composition model of the lower mantle*. In order to satisfy the 1D seismic profiles, the averaged lower-mantle Mg/Si ratio should be lower than would be usually accepted in the pyrolite model. Instead, there should be a depth-dependent bulk composition with an Mg/Si ratio decreasing from 1.18 ± 0.14 to 1.03 ± 0.16 (i.e., the *increase of Si component to a more chondritic composition*) between 800 and 2700 km. Murakami et al. (2012) measured shear wave velocity and density profiles of bridgmanite and ferropericlase at high *P–T* using Brillouin spectroscopy and X-ray diffraction, concluding that the lower mantle is mostly made of bridgmanite, named the perovskitic lower mantle.

On the other hand, recent modelling of the shear wave velocity profile, utilizing the third-order Eulerian finite-strain equation, demonstrated that the lower mantle shear wave velocities can resolve PREM for a pyrolitic composition to within 1%. In addition, isotopic fractionation modelling of the $^{143}Nd/^{144}Nd$ and $^{176}Hf/^{177}Hf$ systems led to the conclusion that an Si-enriched lower mantle, if crystallized from the ancient magma ocean, was not preserved throughout the Earth's history and no longer exists (Hyng et al. 2016). In this model, which supports the idea that the Earth's lower mantle is pyrolitic, the mantle as a whole needs to be chemically stratified. However, the compositional heterogeneity of the mantle on the basis of modelled seismic velocity profiles were comprehensively studied by Stixrude and Lithgow-Bertelloni (2012). They demonstrated that chemical heterogeneity may survive in the mantle for as long as the age of the Earth because chemical diffusion is inefficient. Estimates of the rate of subduction and mantle processing over geologic history indicate that almost the entire mantle may be composed of lithologically heterogeneous material. The fastest lithology is silica, the presence of which causes the increase in the velocity within the lower mantle (see Sect. 4.6).

However, it should be noted that considering limited experimental constraints on the elasticity and sound velocity of lower mantle minerals at *P–T* conditions relevant to the lower mantle, recent theoretical models using new ρ and V_{Φ} profiles concluded that predictions for a pyrolitic composition of the lower mantle have seismic properties in agreement with the PREM (Wang et al. 2015; Sun et al. 2016; Zhang et al. 2016).

2.5 Geotherm

Temperature distribution in the Earth's interior reflects the energy balance between heat generation and dissipation. The heat source in the mantle is believed to be mostly primordial and radiogenic, while the dissipation is most dominant through convection, but radiative conduction can also play a certain role. However, there is no direct way to measure temperature in the Earth's deep interior, except with some drilled boreholes as deep as approximately 10 km. A combination of indirect methods was used to estimate the temperature distribution in the deep mantle (Brown and Shankland 1981; Ito and Katsura 1989; da Silva et al. 2000; Stacey and Davis 2008; Katsura et al. 2010).

In most geotherm models of the Earth, conductive and radiative heat transfer is considered to be negligible because of the small thermal conductivity of mantle minerals, and heat should be effectively transported by convection into most parts of the mantle (Katsura et al. 2010). Therefore, the temperature gradient is considered to be nearly adiabatic in the Earth's mantle. The adiabatic temperature gradient $(\partial T/\partial z)_s$ is expressed as:

$$(\partial \mathrm{T}/\partial \mathrm{z})_{\mathrm{s}} = \alpha g T/C_{\mathrm{p}};$$

where T is temperature, z is depth, g is gravitational acceleration, α and C_p are the thermal expansion coefficient and isobaric heat capacity of the constituent materials, respectively (Turcotte and Schubert 2002). Using this equation, one can calculate the temperature gradient and geotherm if the aforementioned thermodynamic parameters and a reference temperature at a given depth are known.

Based on the estimated maximal temperature of the olivine–wadsleyite transition at the 410 km seismic discontinuity of 1830 ± 48 K, Katsura et al. (2010) calculated the adiabatic temperature profile for the entire mantle using experimentally measured thermal expansion coefficients of the mantle minerals with the known gravity accelerations and the assumption that the whole mantle has a constant pyrolitic composition based on the whole mantle convection model (Stacey and Davis 2008). The thermal expansion coefficient is the most critical parameter for the adiabatic geotherm. In the lower mantle, it gradually decreases in pyrolitic composition from 3×10^{-5}/K at the topmost lower mantle to 1×10^{-5}/K to lowermost lower mantle. The adiabatic geothermal gradient within the upper mantle shows the same tendency as the thermal expansion coefficient. It decreases with increasing depth without phase transitions, but can potentially increase across phase transitions. The geothermal gradient is 0.47 K/km at the top of the lower mantle, and decreases to 0.24 K/km at 2600 km in depth (Katsura et al. 2010). In earlier works, it has been suggested to be 0.3 K/km (Brown and Shankland 1981; Turcotte and Schubert 2002), 0.4 K/km and rather constant (Stacey and Davies 2008) and 0.5–0.9 K/km and increasing with depth (Matas et al. 2007).

The temperature profile for the mantle calculated by Katsura et al. (2010) is shown in Fig. 2.3. The temperature at the top of the lower mantle is found to be 1960 ± 50 K, which coincides with earlier estimates by Hirose (2002). The temperature at 2700 km depth is 2630 ± 60 K, if the convective heat transfer still dominates in this region. This temperature profile is 70–110 K higher than that of Ito and Katsura (1989) and Brown and Shankland (1981) because the temperature

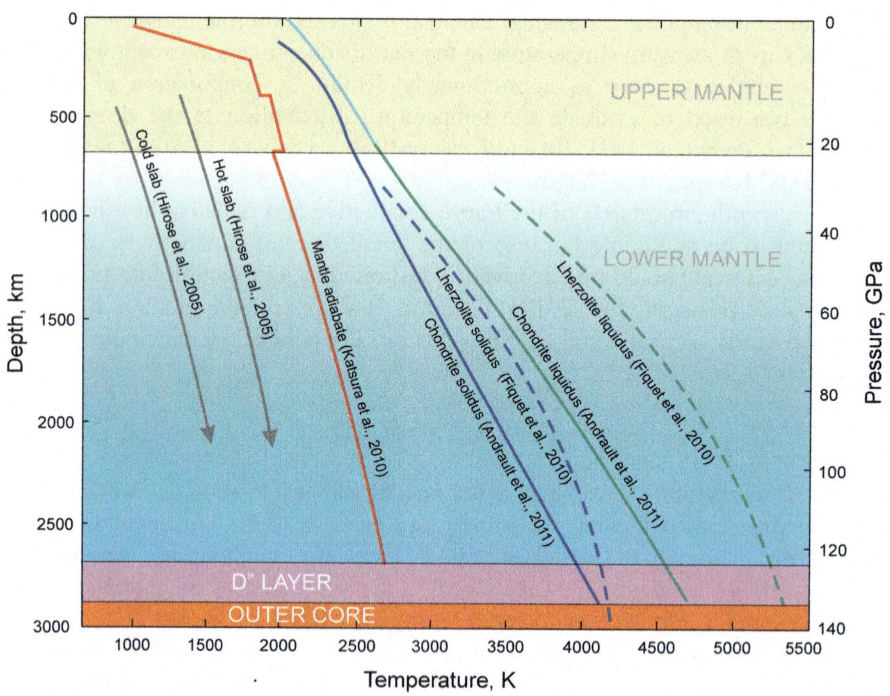

Fig. 2.3 Profiles of temperature, solidus and liquidus in the mantle. Data from Kirby et al. (1996), Katsura et al. (2010), Fiquet et al. (2010) and Andrault et al. (2011)

Table 2.1 Reference temperature at the 660 km depth used in the literature geotherm modelling

Authors	660-km phase transition temperature
Brown and Shankland (1981)	1873
Stacey and Davis (2008)	1931
Katsura et al. (2010)	1980

at the 660 km discontinuity was estimated to be lower in those studies (Table 2.1). Da Silva et al.'s (2000) temperature profiles are much steeper in the whole lower mantle because the temperature profiles in this study were obtained by matching the modelled elastic wave velocities of the lower mantle minerals with the seismic profiles, under the assumption that the heat must be mainly transferred not by convection, but by conduction.

Subducting lithospheric slabs may significantly cool temperature profiles, particularly in the upper part of the lower mantle. The profiles of both old, faster subducting 'cold' slabs and younger, slower subducting 'hot' slabs after Kirby et al. (1996) are shown in Fig. 2.3.

Experimental data demonstrated that an electronic spin transition in iron, which occurs at mid-mantle depths, results in changes to the physical properties of the ferropericlase, comprising a significant part of the bulk lower mantle. The numerical model demonstrated that as the result of this electronic transition, the lowermost region of the mantle is slightly warmed, whereas the upper mantle is slightly cooled by spin-induced effects (Shahnas et al. 2011).

2.6 Liquidus and Solidus of Mantle Material

The eutectic or peritectic melting of mantle materials consisting of phases with constant and changeable solid solution compositions is characterized by solid and liquid phase relations. Sometimes, the phase relations can be significantly affected by the addition of a small amount of volatiles. Solidus and liquidus profiles within the mantle were mostly established on the basis of high *P–T* laser-heated diamond-anvil cell (LHDAC) experiments on representative compositions using in situ synchrotron X-ray diffraction measurements and/or quenched sample analysis. Various criteria for detecting melting include diffuse X-ray scattering, melt textures and X-ray microtomographs via edge absorption have been adopted in this area of research (e.g., Fiquet et al. 2010; Andrault et al. 2011; Nomura et al. 2012).

Natural anhydrous fertile lherzolite KLB-1, a natural analogue of the upper mantle (spinel lherzolite from the Kilbourne Hole crater in New Mexico, USA; Davis et al. 2009) was studied to establish its solidus line in the lower mantle (Fiquet et al. 2010). The results show that the solidus line is approximately 900–1200 K higher than the expected adiabatic geotherms, suggested by Brown et al. (1981), within the mid-mantle region at depths of 1300–1700 km and ~1400 K higher at the bottom of the lower mantle at a depth of 2700 km. The melting temperature at the core–mantle boundary (CMB) was established to be at 4180 ± 150 K (Fiquet et al. 2010). These data are in good agreement with the mantle solidus inferred at the CMB conditions from shock wave experiments on $(Mg_{0.9}Fe_{0.1})_2SiO_4$ (Holland and Ahrens 1997) and from first principles molecular dynamics simulations (Stixrude et al. 2009).

Andrault et al. (2011) investigated the melting properties of a synthetic chondritic primitive mantle up to the CMB pressure and found a similar solidus profile; however, the solidus and liquidus chondritic temperatures of 4150 ± 150 K and 4725 ± 150 K, respectively, at the CMB at 2900 km were much lower than for lherzolite. Nomura et al. (2012) indicated that the solidus temperature of a pyrolite composition, which represents an upper boundary for the melting of the primitive lowermost mantle materials, is as low as 3570 K ± 200 K at the CMB pressures. The low solidus temperature also suggests the melting temperature of the outer core to be much lower than what may have been expected previously.

Although the studied compositions may differ from the real lower-mantle composition, the obtained results may be used as approximations.

2.7 Oxidation Potential in the Lower Mantle

2.7.1 *General*

Oxygen fugacity f_{O_2} (the chemical potential of oxygen) is a fundamental thermodynamic parameter in the growth and stability of mineral phases formed in the Earth's mantle. Oxygen fugacity can influence phase relations, the presence and speciation of a fluid phase, among other things

As an equivalent of the partial pressure of oxygen in a particular environment (melts, rocks, etc.), oxygen fugacity is measured in pascals (Pa). After recalculation in logarithmic units, it is usually compared with the value, corresponding (at given *P–T* values) to the f_{O_2} value of the fayalite-magnetite-quartz (FMQ) or iron-wüstite (IW) buffer values, corresponding to reactions $3Fe_2SiO_4 + O_2 = 2Fe_3O_4 + 3SiO_2$ and $2(1\text{-}x)Fe + O_2 = 2Fe_{1\text{-}x}O$, respectively. Usually, mantle values of f_{O_2} are located between or close to FMQ and IW buffer values.

At the top of the upper mantle, the calculated oxygen fugacity is shown to vary between −3 to +2 logarithmic units of the FMQ oxygen buffer (Frost and McCammon 2008; Stagno et al. 2013). With increasing depth, f_{O_2} decreases relative to oxygen buffers as a result of elevated pressure controlling Fe^{3+}/Fe^{2+} equilibria (Wood et al. 1990). At approximately 8 GPa (~250 km depth), f_{O_2} should be 5 log units below FMQ, and at the base of the upper mantle (~410 km; 14 GPa) f_{O_2} has been suggested to extend to ~0.8 log units below the IW buffer, assuming the precipitation of Ni in bulk silicate Earth at this depth (Frost and McCammon 2008).

This calculated data is supported by direct analyses of $Fe^{3+}/\Sigma Fe$ ratios in garnet grains hosted as inclusions within peridotite xenoliths present in South African

(Woodland and Koch 2003; Lazarov et al. 2009; Hanger et al. 2014) and Canadian (McCammon and Kopylova 2004; Creighton et al. 2010) kimberlites. Similar results were obtained from spinel and garnet peridotites in Siberia, demonstrating the decrease of f_{O_2} with increasing pressure: from +1 to −4Δlog f_{O_2} (FMQ) at a depth of 180 km (Goncharov et al. 2012) and from −2.5 to nearly −5.9Δlog f_{O_2} (FMQ) at a depth of 90–220 km (1.2–7.1 GPa pressure) (Yaxley et al. 2012). Local variations in oxygen fugacity within the upper mantle are caused by petrological processes, such as partial melting, mantle metasomatism and the recycling of oxidized material during subduction (Rohrbach and Schmidt 2011; Berry et al. 2013). Controls relating to the local tectonic environment have also been suggested (Foley 2011).

Estimates of oxygen fugacity for the deep mantle suggest that at depths in excess of 250 km (>8 GPa), f_{O_2} becomes narrowly constrained. Given equilibrium between mantle phases with molar Mg/(Mg + Fe^{2+}) (=X_{Mg}) <0.90 and $(Fe, Ni)_{metal}$, f_{O_2} can only vary from values around the IW equilibrium where the metal would be Ni-rich, to about 1.5 log units below IW where the metal would be almost pure iron (Rohrbach and Schmidt 2011). The same range of f_{O_2} variations (between IW and IW-1.5) was supposed for the entire lower mantle, suggesting that the bulk oxygen content of the lower mantle is similar to that of the upper mantle (Frost and McCammon 2008).

Direct measurements of lower-mantle redox conditions, which provide crucial constraints on the real oxidation state of the lower mantle, are scarce. Until recently, less than 20 grains of ferropericlase, bridgmanite and jeffbenite from São Luiz (Brazil) and Kankan (Guinea) placer deposits have been analysed by means of Mössbauer spectroscopy to determine their $Fe^{3+}/\Sigma Fe$ ratios (McCammon et al. 1997, 2004b). The results demonstrate the small amount of Fe^{3+} in ferropericlase (Fe^{3+}/Δ Fe = 0.1–6.0%), which is in direct contrast to the high amounts of Fe^{3+} in bridgmanite (9–75%) and, in particular, in jeffbenite (66–74%). These differences are characteristic even for coexisting phases, and have been considered to reflect differences in the crystal structures. For example, large amounts of Fe^{3+} in bridgmanite and jeffbenite were considered due to the presence of Al in their compositions, which stabilizes Fe^{3+} in the perovskite structure, independent of oxygen fugacity (McCammon 1997; McCammon et al. 2004a). Based on these results, supported by experimental data, the overall picture of redox conditions in the lower mantle was suggested to be a "generally reduced bulk mantle with possibly more oxidised regions, at least some associated with subducting slabs" (McCammon et al. 2004b).

Otsuka et al. (2013) calculated the redox conditions of earlier studies (McCammon et al. 1997, 2004b) for lower-mantle ferropericlase inclusions in diamond. The Δlog f_{O_2} (IW) values appeared to vary from 4.3–3.3 for Kankan, Guinea and from 3.2–2.6 for São Luiz, i.e. within the total Δ = 1.6 range. The estimated oxygen fugacities are close to the upper stability limit of diamond in mantle peridotite at the top of the lower mantle at adiabatic or slightly superadiabatic temperatures (Otsuka et al. 2013) and lay within a narrow range from 2.6–4.3 log units above the IW buffer (Δ = 1.7 log units), which is lower than expected by Rohrbach and Schmidt (2011), but as 'narrowly constrained' as the latter suggested.

Alternative calculations performed by Ryabchikov and Kaminsky (2013, 2014), based upon the calculated Ni/Fe partition coefficient, estimated the variation of log f_{O_2} values in the lower mantle (at the pressure conditions from 30–70 GPa) within a much wider range from −1 and −1.5 log units below the IW buffer to +3 and +5 log units (and even higher, in the presence of carbonates) above the IW buffer ($\Delta \geq 7$ log units). The proof of such great variations of f_{O_2} is the presence (among inclusions in lower-mantle diamonds) of native iron and iron carbides (Kaminsky and Wirth 2011) from one side, and magnesite and other carbonate minerals (Kaminsky et al. 2009, 2013) to the other side. The calculations below are based on Kaminsky et al. (2015).

2.7.2 *Oxygen Fugacity Conditions of the Formation of Natural Ferropericlase*

Recently, oxygen fugacity conditions in the lower mantle were estimated on the basis of the analysis of ferropericlase inclusions from lower-mantle diamonds, recovered from Brazil and Guinea (Kaminsky et al. 2015). Values of $\Delta \log f_{O_2}$ of the IW buffer vary from 1.58 to 7.76 logarithmic units (for T = 1960 K) (Fig. 2.4). Samples from Juina kimberlites and São Luiz placer deposits form a single set (except one sample from São Luiz), which is understandable, considering their close spatial proximity within the same Juina area and an identical genesis of hosting diamonds. These data demonstrate the linear correlation between $\Delta \log f_{O_2}$ and *fe* values varying in a wide range (0.150–0.641), while ferropericlase from Kankan, Guinea, vary almost within the same $\Delta \log f_{O_2}$ value range, having almost constant *fe* (0.117–0.167).

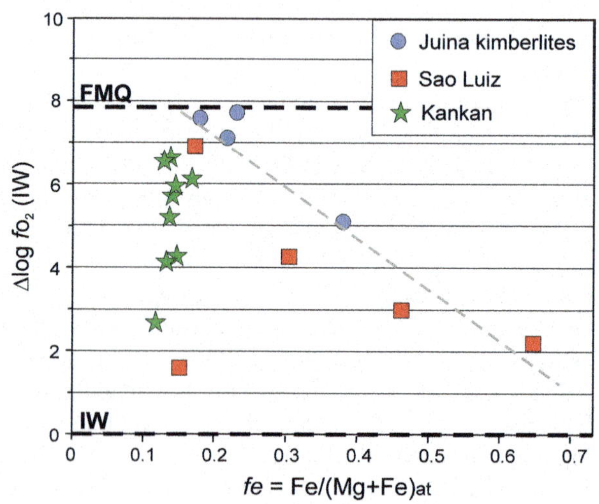

Fig. 2.4 Oxygen fugacity values versus Fe/(Mg + Fe) values in ferropericlase grains (P = 25 GPa; T = 1960 K). Trend line demonstrates linear correlation between log f_{O_2} and *fe* in Brazilian samples. FMQ buffer position (IW + 7.87) calculated for 25 GPa and 1960 K using data from Holland et al. (2013). From Kaminsky et al. (2015). ©Elsevier

2.7.3 Oxygen Fugacity Conditions in the Lower Mantle

These results demonstrate wide variations in oxygen fugacity within the lower-mantle region where diamonds are formed. The calculated variations are within a range of more than six logarithmic units ($\Delta = 6.2$), much greater than that suggested previously for the lower mantle ($\Delta = 1.6$–1.7; Rohrbach and Schmidt 2011; Otsuka et al. 2013).

The $\Delta \log f_{O_2}$ values are above the IW buffer, demonstrating unexpectedly high oxidation conditions within the lower mantle (Fig. 2.5). In general, $\Delta \log f_{O_2}$ values for the lower mantle are similar to those of the upper-mantle (Woodland and Koch 2003; McCammon and Kopylova 2004; Lazarov et al. 2009; Creighton et al. 2010;

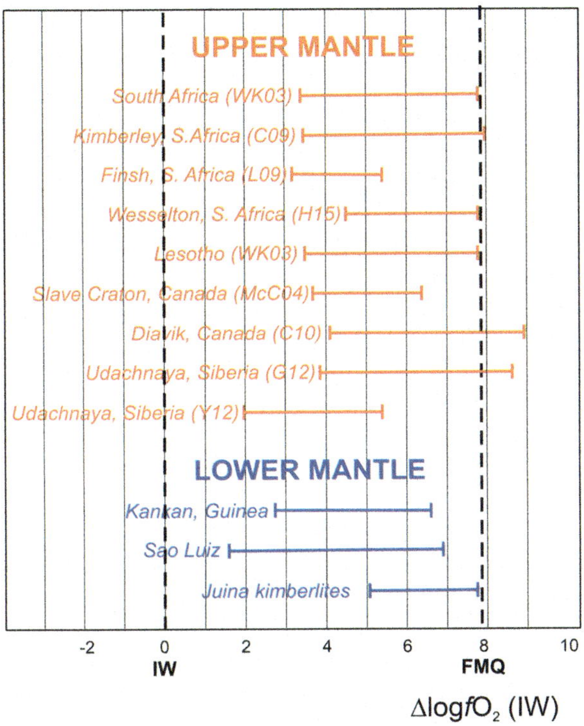

Fig. 2.5 Oxygen fugacity values for lower-mantle ferropericlase compared with the values for garnet from the upper mantle. *IW* position of the iron-wüstite buffer. *FMQ* position of the fayalite-magnetite-quartz buffer. Data on the Kaapvaal Craton in South Africa from Woodland and Koch 2003 (WC03), Lazarov et al. 2009 (L09) and Hanger et al. 2014 (H14). Data on the Slave Craton in Canada from McCandless and Kopylova 2004 (McC04) and Creighton et al. 2010 (C10). Data on the Siberian Craton in Russia from Goncharov et al. 2012 (G12) and Yaxley et al. 2012 (Y12). FMQ buffer position (IW + 7.87) calculated for 25 GPa and 1960 K using data from Holland et al. (2013). From Kaminsky et al. (2015).©Elsevier

Goncharov et al. 2012; Yaxley et al. 2012; Hanger et al. 2014), as has been suggested before (McCammon et al. 2004b).

These data are in agreement with the recent theoretic calculations based on the grounds of the calculated Ni/Fe partition coefficient (Ryabchikov and Kaminsky 2013, 2014). They demonstrate that the typical values of oxygen fugacity in zones of diamond formation in the lower mantle lie between the IW buffer and 3–5 logarithmic units above this level. The processes that operate in the lower mantle can give rise to variation of within several orders of magnitude above the elevated values, which are necessary for the formation of diamond, as compared with common oxygen fugacity levels typical of the lower mantle.

The calculated high oxidation conditions within the lower mantle are confirmed by the presence, in lower-mantle diamonds, of carbonate mineral inclusions (Kaminsky et al. 2009, 2013) and by the existence of magnesioferrite/Fe^{3+} cluster exsolution in ferropericlase with its high concentration of Fe^{3+} (Kaminsky et al. 2015). In addition to carbonates, inclusions of free silica have been identified in lower-mantle diamonds as an abundant phase for all known localities (Brazil, Guyana and Canada) (Kaminsky 2012). On the other hand, low oxygen fugacity conditions may also exist within the lower mantle. They are indicated by the presence of the metallic iron + iron carbide association found among inclusions in lower-mantle diamond (Kaminsky and Wirth 2011). Such conditions, with $\Delta\log f_{O_2}$ values at and below the IW oxygen buffer, may occur in the lowermost lower mantle and D″ layer. The performed calculations by Kaminsky et al. (2015) demonstrate that with an increase in temperature and pressure conditions, the $\Delta\log f_{O_2}$ values fall and may reach or lie below the IW oxygen buffer. Therefore, the diapason of $\Delta\log f_{O_2}$ values within the entire sequence of the lower mantle may reach ten logarithmic units, varying from below the IW buffer to the FMQ buffer values.

2.8 Conclusions

Geophysical models, based on the pyrolitic or chondritic average compositions of the Earth's mantle, demonstrate gradual changes in seismic wave velocities, density, temperature and other physical properties within the lower mantle. However, so far, none of the lower-mantle model compositions (pyrolite, CI-chondrite, 'cosmic') fully fits the data on seismic velocities. They all show a significant mismatch, particularly in the lower part of the lower mantle, observing compositional differences between the lower and upper mantle.

Of particular interest are indications of the enrichments of the lower mantle in SiO_2, in comparison with the pyrolitic upper mantle. Such indications were obtained from both re-examination of the seismic profile (Matas et al. 2007) and analysis of the density profile (Ricolleau et al. 2008, 2009) within the lower mantle.

Despite some uncertainties in the PREM, we used this model in all of our further constructions. All mismatches, in our opinion, are caused by the non-pyrolitic composition of the lower mantle, which is confirmed by data on natural mineral association from the lower mantle.

The oxygen fugacity Δlog Emphasis>o$_2$o$_2$ values for the lower mantle are similar to those of the upper mantle. The entire sequence of the lower mantle may reach ten logarithmic units, varying from below the IW buffer to the FMQ buffer values.

References

Anderson, D. L. (1983). Chemical composition of the Mantle. *Journal of Geophysical Research*, *88*, B41–B52. doi:10.1029/Jb088is01p00b41

Anderson, D. L. (1989). Theory of the Earth. Oxford: Blackwell.

Andrault, D., Bolfan-Casanova, N., Nigro, G. L., Bouhifd, M. A., Garbarino, G., & Mezouar, M. (2011). Solidus and liquidus profiles of chondritic mantle: Implication for melting of the Earth across its history. *Earth and planetary science letters*, *304*(1), 251–259.

Badro, J., Fiquet, G., Guyot, F., Rueff, J. P., Struzhkin, V. V., Vankó, G., et al. (2003). Iron partitioning in Earth's mantle: Toward a deep lower mantle discontinuity. *Science*, *300*(5620), 789–791.

Bass, J. D., & Anderson, D. L. (1984). Composition of the upper mantle: Geophysical tests of two petrological models. *Geophysical Research Letters*, *11*, 229–232.

Berry, A. J., Yaxley, G. M., Hanger, B. J., Woodland, A. B., De Jonge, M. D., Howard, D. L., et al. (2013). Quantitative mapping of the oxidative effects of mantle metasomatism. *Geology*, *41*, 683–686.

Brown, J. M., & Shankland, T. J. (1981). Thermodynamic parameters in the Earth as determined from seismic profiles. *Geophysical Journal of the Royal Astronomical Society*, *66*, 579–596.

Bullen, K. E. (1942). The density variation of the Earth's central core. *Bulletin of the Seismological Society of America*, *30*, 235–250.

Bullen, K. E. (1950). An Earth model based on compressibility-pressure hypothesis. *Monthly Notices of the Royal Astronomical Society Supplement*. *6*, 50–59.

Creighton, S., Stachel, T., Eichenberg, D., & Luth, R. (2010). Oxidation state of the lithospheric mantle beneath Diavik diamond mine, central Slave craton, NWT, Canada. *Contributions to Mineralogy and Petrology*, *159*(5), 645–657.

Da Silva, C. R. S., Wentzcovitch, R. M., Patel, A., Price, G. D., & Karato, S. I. (2000). The composition and geotherm of the lower mantle: constraints from the elasticity of silicate perovskite. *Physics of the Earth and Planetary Interiors*, *118*,103–109.

Davis, F. A., Tangeman, J. A., Tenner, T. J., & Hirschmann, M. M. (2009) The composition of KLB-1 peridotite. *American Mineralogist*, *94*, 176–180.

Deuss, A., & Woodhouse, J. (2001). Seismic observations of splitting of the mid-transition zone discontinuity in the Earth's mantle. *Science*, *294*, 354–357.

Deuss, A., Redfern, S. A. T., Chambers, K., & Woodhouse, J.H. (2006). The nature of the 660-kilometer discontinuity in Earth's mantle from global seismic observations of PP precursors. *Science*, *311* 198–201.

Dziewonski A., & Anderson D. (1981) Preliminary reference Earth model. *Physics of Earth and Planetary Interiors*, *25*, 297–356. doi:10.1016/0031-9201(81)90046-7

Fiquet, G., Auzende, A. L., Siebert, J., Corgne, A., Bureau, H., Ozawa, H., et al. (2010). Melting of peridotite to 140 Gigapascals. *Science*, *329*, 1516–1518. doi:10.1126/science. 1192448

Foley, S. F. (2011). A reappraisal of redox melting in the Earth's mantle as a function of tectonic setting and time. *Journal of Petrology*, *52*(7), 1363–1391.

Frost, D. J., & McCammon, C. A. (2008). The redox state of Earth's mantle. *Annual Review of Earth and Planetary Sciences*, *36*, 389–420.

Goncharov, A. G., Ionov, D. A., Doucet, L. S., & Pokhilenko, L. N. (2012). Thermal state, oxygen fugacity and C-O-H fluid speciation in cratonic lithospheric mantle: New data on peridotite xenoliths from the Udachnaya kimberlite, Siberia. *Earth and Planetary Science Letters, 357–358*, 99–110.

Hanger, B. J., Yaxley, G. M., Berry, A. J., & Kamenetsky, V. S. (2014). Relationships between oxygen fugacity and metasomatism in the Kaapvaal subcratonic mantle, represented by garnet peridotite xenoliths in the Wesselton kimberlite, South Africa. *Lithos, 212–215*, 443–452. doi:10.1016/j.lithos.2014.09.030

Hart, S. R., & Zindler A. (1986). In search of a bulk-Earth composition. *Chemical Geology, 57*(3–4), 247–267.

Hirose, K. (2002). Phase transitions in pyrolitic mantle around 670-km depth: Implications for upwelling of plumes from the lower mantle. *Journal of Geophysical Research, 107*(B4) 2078. doi:10.1029/2001JB000597

Holland, K. G., & Ahrens, T. J. (1997) Melting of $(Mg,Fe)_2SiO_4$ at the Core-Mantle Boundary of the Earth. *Science, 275*, 1623–1625.

Holland, T. J. B., Hudson, N. F. C., Powell, R., & Harte, B. (2013). New thermodynamic models and calculated phase equilibria in NCFMAS for basic and ultrabasic compositions through the transition zone into the uppermost lower mantle. *Journal of Petrology, 54*, 1901–1920.

Hyung, E., Huang, S., Petaev, M. I., & Jacobsen, S. B. (2016). Is the mantle chemically stratified? Insights from sound velocity modeling and isotope evolution of an early magma ocean. *Earth and Planetary Science Letters, 440*(2016) 158–168. doi:10.1016/j.epsl.2016.02.001

Irifune, T., Shinmei, T., McCammon, C. A., Miyajima, N., Rubie, D. C., & Frost D. J. (2010). Iron partitioning and density changes of pyrolite in Earth's lower mantle. *Science, 327*(5962) 193–195.

Ito, E., & Katsura, T. (1989). A temperature profile on the mantle transition zone. *Geophysical Research Letters, 16*(5), 425–428.

Javoy, M. (1995). The integral enstatite chondrite model of the Earth. *Geophysical Research Letters, 22*(16), 2219–2222.

Jeanloz, R., & Knittle, E. (1989). Density and composition of the lower mantle. *Philosophical Transactions of the Royal Society of London,* A*328*(1599), 377–389. doi:10.1098/rsta.1989.0042

Kaminsky, F. V. (2012). Mineralogy of the lower mantle: A review of 'super-deep' mineral inclusions in diamond. *Earth-Science Reviews*, 110(1–4), 127–147.

Kaminsky, F. V., & Wirth, R. (2011). Iron carbide inclusions in lower-mantle diamond from Juina, Brazil. *Canadian Mineralogist* 49(2), 555–572.

Kaminsky, F., Wirth, R., Matsyuk, S., Schreiber, A., & Thomas, R. (2009). Nyerereite and nahcolite inclusions in diamond: Evidence for lower-mantle carbonatitic magmas. *Mineralogical Magazine*, 73(5), 797–816.

Kaminsky, F. V., Wirth, R., & Schreiber, A. (2013) Carbonatitic inclusions in Deep Mantle diamond from Juina, Brazil: New minerals in the carbonate-halide association. *Canadian Mineralogist*, 51(5), 447–466.

Kaminsky, F. V., Ryabchikov, I. D., McCammon, C., Longo, M., Abakumov, A. M., Turner, S., et al. (2015). Oxidation potential in the Earth's lower mantle as recorded from ferropericlase inclusions in diamond. *Earth and Planetary Science Letters, 417*, 49–56.

Katsura, T., Yoneda, A., Yamazaki, D., Yoshino, T., & Ito E. (2010). Adiabatic temperature profile in the mantle. *Physics of the Earth and Planetary Interiors, 183*, 212–218. doi:10.1016/j.pepi.2010.07.001

Kennett, B. L. N., & Engdahl, E. R. (1991) Traveltimes for global earthquake location and phase identification. *Geophysical Journal of International, 105*, 429–465.

Kennett, B., Engdahl, E., & Buland, R. (1995). Constraints on seismic velocities in the Earth from traveltimes. *Geophysical Journal of International, 122*, 108–124.

Kirby, S. H., Stein, S., Okal, E. A., & Rubie D. C.(1996). Metastable mantle phase transformations and deep earthquakes in subducting oceanic lithosphere. *Reviews of Geophysics, 34*(2), 261–306.

Kustowski, B., Ekström, G., & Dziewoński A. M. (2008). Anisotropic shear-wave velocity structure of the Earth's mantle: A global model. *Journal of Geophysical Research, 113*, B06306. doi:10.1029/2007JB005169

Lazarov, M., Woodland, A. B., & Brey, G. P. (2009). Thermal state and redox conditions of the Kaapvaal mantle: A study of xenoliths from the Finsch mine, South Africa. *Lithos, 112*S, 913–923.

Lin, J-F., Speciale, S., Mao, Z., & Marquardt, H. (2013). Effects of the electronic spin transitions of iron in lower mantle minerals: Implications for deep mantle geophysics and geochemistry. *Reviews of Geophysics, 51*(2), 244–275.

McCammon, C. (1997). Perovskite as a possible sink for ferric iron in the lower mantle. *Nature, 387*, 694–696.

McCammon, C., & Kopylova, M. G. (2004). A redox profile of the Slave mantle and oxygen fugacity control in the cratonic mantle. *Contributions to Mineralogy and Petrology, 148*(1), 55–68.

McCammon, C., Hutchison, M. T., & Harris J. W. (1997) Ferric iron content of mineral inclusions in diamonds from São Luiz: A view into the lower mantle. *Science, 278*(5337), 434–436.

McCammon, C. A., Lauterbach, S., Seifert, F., Langenhorst, F., & van Aken, P. A. (2004a). Iron oxidation state in lower mantle mineral assemblages. I. Empirical relations derived from high-pressure experiments. *Earth and Planetary Science Letters, 222*(2), 435–449.

McCammon, C. A., Stachel, T., & Harris, J. W. (2004b). Iron oxidation state in lower mantle mineral assemblages. II. Inclusions in diamonds from Kankan, Guinea. *Earth and Planetary Science Letters, 222*(2), 423–434.

Matas, J., Bass, J. D., Ricard, Y., Mattern, E., & Bukowinsky, M. S. (2007). On the bulk composition of the lower mantle: predictions and limitations from generalized inversion of radial seismic profiles. *Geophysical Journal International 170*, 764–780.

McDonough, W. F., & Sun S. S. (1995). The composition of the Earth. *Chemical Geology, 120*(3–4), 223–253.

Morgan, J. W., & Anders E. (1980). Chemical composition of Earth, Venus, and Mercury. *Proceedings of the National Academy Science USA, 77*(12), 6973–6977. doi:10.1073/pnas.77.12.6973

Murakami, M., Ohishi,Y., Hirao, N., & Hirose K. (2012). A perovskitic lower mantle inferred from high-pressure, high-temperature sound velocity data. *Nature, 485*(7396), 90–94.

Nomura, R., Hirose, K., Uesugi, K., Ohishi, Y., Tsuchiyama, A., Miyake, A., et al. (2012). Low core-mantle boundary temperature inferred from the solidus of Pyrolite. *Science, 343*, 523–525.

Otsuka, K., Longo, M., McCammon, C. A., & Karato, S.-i. (2013). Ferric iron content of ferropericlase as a function of composition, oxygen fugacity, temperature and pressure: Implications for redox conditions during diamond formation in the lower mantle. *Earth and Planetary Science Letters, 365*, 7–16.

Ricolleau, A., Fiquet, G., Addad, A., Menguy, N., Vanni, C., Perrillat, J.-P., et al. (2008). Analytical transmission electron microscopy study of a natural MORB sample assemblage transformed at high pressure and high temperature. *American Mineralogist, 93*, 144–153. doi:10.2138/am.2008.2532

Ricolleau, A., Fei, Y., Cottrell, E., Watson, H., Deng, L., Zhang, L., et al. (2009). Density profile of pyrolite under the lower mantle conditions. *Geophysical Research Letters, 36*, L06302. doi:10.1029/2008GL036759

Ringwood, A. E. (1975). Composition and petrology of the Earth's mantle. (p. 618). New York: McGraw-Hill.

Rohrbach, A., & Schmidt, M. W. (2011). Redox freezing and melting in the Earth's deep mantle resulting from carbon–iron redox coupling. *Nature, 472* 209–214.

Ryabchikov, I. D., & Kaminsky, F. V. (2013). Redox potential of diamond formation processes in the lower mantle. *Geology of Ore Deposits, 55*(1), 1–12.

Ryabchikov, I. D., & Kaminsky, F. V. (2014). Physico-chemical parameters of material in mantle plumes: Evidence from the thermodynamic analysis of mineral inclusions in sublithospheric diamonds. *Geochemistry International, 52*(11), 963–971.

Shahnas, M. H., Peltier, W. R., Wu, Z., & Wentzcovitch, R. (2011). The high-pressure electronic spin transition in iron: Potential impacts upon mantle mixing. *Journal of Geophysical Research, 116*, B08205. DOI:10.1029/2010JB007965

Shen, Y., Solomon, S. C., Bjarnson, I. T., & Wolfe, C. J. (1998). Seismic evidence for the lower-mantle origin of the Iceland plume. *Nature, 395*, 62–65.

Simmons, N. A., Forte, A. M., Boschi, L., & Grand S. P. (2010). GyPSuM: A joint tomographic model of mantle density and seismic wave speeds. *Journal of Geophysical Research, 115*, B12310, doi:10.1029/2010JB007631

Simmons, N. A., Forte, A. M., Boschi, L., & Grand, S. P. (2010). GyPSuM: A joint tomographic model of mantle density and seismic wave speeds. *Journal of Geophysical Research, 115*, B12310. doi:10.1029/2010JB007631

Stacey, F. D., & Davis, P. M. (2008). Physics of the Earth. (4th ed., p. 532). Cambridge: Cambridge University Press.

Stagno, V., Ojwang, D. O., McCammon, C. A., & Frost, D. J. (2013). The oxidation state of the mantle and the extraction of carbon from Earth's interior. *Nature, 493*, 84–88.

Stixrude, L., & Lithgow-Bertelloni, C. (2012). Geophysics of chemical heterogeneity in the mantle. *Annual Review of Earth and Planetary Science, 40*, 569–595.

Stixrude, L., de Koker, N., Sun, N., Mookherjee, M., & Karki, B.B. (2009). Thermodynamics of silicate liquids in the deep Earth. *Earth and Planetary Science Letters, 278*, 226–232. doi:10.1016/j.epsl.2008.12.006

Sun, N., Mao, Z. Yan, S., Lin, J. F., Wu, X., & Prakapenka, V. B. (2016). Confirming a pyrolitic lower mantle using self-consistent pressure scales and new constraints on $CaSiO_3$-perovskite. *Journal of Geophysical Research, 121*(7), 4876–4892. doi:10.1002/2016JB013062

Tesoniero, A., Auer, L., Boschi, L., & Cammarano, F. (2015). Hydration of marginal basins and compositional variations within the continental lithospheric mantle inferred from a new global model of shear and compressional velocity. *Journal of Geophysical Research: Solid Earth, 120* (11), 7789–7813. doi:10.1002/2015JB012026

Turcotte, D. L., & Schubert, G. (2002). Geodynamic, (2nd ed., p. 456).Cambridge: Cambridge University Press.

Wang, X. L., Tsuchiya, T., & Hase, A. (2015). Computational support for a pyrolitic lower mantle containing ferric iron. *Nature Geoscience, 0*, 556–559. doi:10.1038/ngeo2458

Williams, Q., & Knittle, E. (2005). The uncertain major element bulk composition of Earth's mantle. In: R.D. Van der Hilst, J. Bass, J. Matas and J. Trampert (Eds.), *Earth's deep mantle: Structure, composition and evolution*. (pp. 189–202). Washington DC: AGU.

Wood, B. J., Bryndzya, L. T., & Johnson, K. E. (1990). Mantle oxidation state and its relationship to tectonic environment and fluid speciation. *Science, 248*, 337–344.

Woodland, A. B., & Koch, M. (2003). Variation in oxygen fugacity with depth in the upper mantle beneath the Kaapvaal craton, Southern Africa. *Earth and Planetary Science Letters, 214*, 295–310.

Yaxley, G. M., Berry, A. J., Kamenetsky, V. S., Woodland, A. B., & Golovin, A. V. (2012) An oxygen fugacity profile through the Siberian Craton—Fe K-edge XANES determinations of $Fe^{3+}/\Sigma Fe$ in garnets in peridotite xenoliths from the Udachnaya East kimberlite. *Lithos, 140–141*, 142–151.

Zhang, S., Cottaar, S., Liu, T., Stackhouse, S., & Militzera, B. (2016). High-pressure, temperature elasticity of Fe-and Al-bearing $MgSiO_3$: Implications for the Earth's lower mantle. *Earth and Planetary Science Letters, 434*, 264–273. doi:10.1016/j.epsl.2015.11.030

Chapter 3
Lower-Mantle Mineral Associations

Abstract There are three major sources of information about the composition of the lower mantle: high *P–T* experiments, theoretical calculations, and geological observations. Experimental data, based on the use of diamond-anvil cell technique (DAC), and theoretical calculations demonstrate that silicates, occurring in the upper mantle and the transition zone, are replaced by predominantly perovskitic assemblage in the lower mantle. Depending on the starting substrate composition, two mineral associations should occur at pressures corresponding to the lower-mantle conditions: ultramafic (bridgmanite + CaSi-perovskite + ferropericlase) and mafic (bridgmanite + CaSi-perovskite + ferropericlase + silica + Al-phase). Both iron-containing lower-mantle minerals, bridgmanite and ferropericlase, should be magnesium-rich. In recent decades, lower-mantle minerals were found as inclusions in diamonds from Brazil, Guinea, Canada, Australia and South Africa. They confirm the presence of ultramafic, mafic and carbonatitic mineral associations. Geological samples differ notably from the lower mantle compositions suggested on the basis of experimental and theoretical data for the pyrolitic composition. First, ferropericlase is the most common in the lower-mantle ultramafic association (averaging 55.6%), while bridgmanite comprises only 7.5%, about ten times lower than has been suggested (c. 70–74%) in the lower mantle. Second, silica inclusions were identified in all sets of lower-mantle minerals observed in diamond from all regions and areas. Third, wide variations in ferropericlase compositions, reaching an iron index of up to $fe = 0.64$ were observed. Minerals from the ultramafic association overwhelmingly predominate in the lower mantle samples; only two samples of mafic mineral phases, phase Egg and δ−AlOOH are found to date.

3.1 General

Historically, theoretical calculations and experiments in the 1970–1980s provided the initial information related to the development of theoretical tools and experimental techniques in this field. They produced valuable information used to

F.V. Kaminsky, *The Earth's Lower Mantle*, Springer Geology,
DOI 10.1007/978-3-319-55684-0_3

construct the first compositional models, which were in general agreement with deep seismic data. However, the results were limited by the a priori assumptions on the pyrolitic bulk composition of the lower mantle, which were the same as for the upper mantle.

Since the mid-1990s, the first data sets on real compositions of lower-mantle minerals appeared, obtained from a study of mineral inclusions in deep-seated diamonds, found in Brazil, Guinea, Canada, Australia and some other areas. They, in general, confirmed experimental and theoretical data. However, some additional discrepancies appeared as well, which remain to be interpreted with the use of petrological and geophysical materials.

3.2 Experimental Data

3.2.1 Early Techniques

According to the homogeneous ‘pyrolite model’ by Ringwood (1975), seismic discontinuities in the Earth’s mantle are caused by isochemical mineral phase transitions of major minerals, first of all olivine, which comprises 57% of the mantle. Numerous experiments confirmed phase transitions, starting within the upper mantle: (1) olivine α-(Mg, Fe)$_2$SiO$_4$ → wadsleyite β-(Mg, Fe)$_2$SiO$_4$ with orthorhombic structure (14–15 GPa or 410 km), (2) wadsleyite → ringwoodite γ-(Mg,Fe)$_2$SiO$_4$ with isometric spinel structure (17–18 GPa or 520 km), and (3) ringwoodite → bridgmanite (Mg, Fe)SiO$_3$ with a periclase structure + ferropericlase (Mg, Fe) O (23–24 GPa or 660 km) (Fig. 3.1). Each of these phase transitions is accompanied

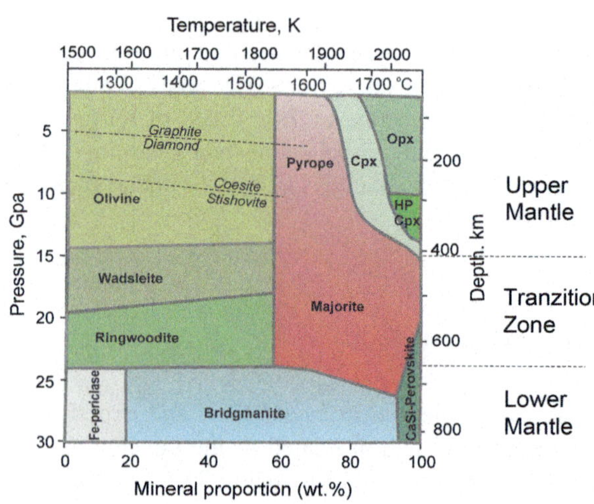

Fig. 3.1 Experimental data on mineral composition and phase transformations in the upper mantle, transition zone, and uppermost lower mantle within a pressure range of 3–30 GPa. After Akaogi (2007). © Elsevier

by an increase of density in newly formed minerals of 2–15%. Pyroxenes dissolve into garnet phase and form majorite, $Mg_3(Mg, Si)Si_4O_{12}$. At higher pressures, within the lower mantle conditions, the formation of new mineral phases with a perovskite cubic structure, which is non-characteristic for the lithosphere, was predicted (Ringwood 1967). Shortly after that the first mineral phase with a perovskite structure $CaSiO_3$ was synthesized (Ringwood and Major 1971; Liu and Ringwood 1975).

Initially, experimental works were carried out with the use of Bridgman's anvil cell, where the apparatus in a high-pressure chamber was placed into a low-pressure chamber. The first version of Bridgman anvils was constructed in 1940 and allowed the compression of various substances up to 10 GPa. The modification of the anvils using the two-step Bridgman apparatus and highly stable WC insets increased the value of pressure up to 14–15 GPa, and even to 20 GPa. Various other constructions subsequently followed: spherical, tetrahedral anvil apparatus (with four anvils), cubic apparatus (with six anvils), belt apparatus, and others.. The belt-type cell (under a pressure of 7 GPa) created the first man-made diamond and was produced in 1954 (Duffy 2005). Such equipment, however, allowed model conditions only within the upper mantle and the transition zone. The invention of multi-anvil high-pressure technology in the 1960s allowed experiments to be conducted under pressures of up to 25–28 GPa with temperatures of 2700 °C, modelling the conditions of the uppermost lower mantle at a depth of ~800 km. A good overview of multi-anvil technology is available from Ito (2009). Further development of multi-anvil equipment allowed pressures of almost 100 GPa (Liebermann 2011).

3.2.2 *Diamond Anvil Cell (DAC) and Multi-anvil Cell Apparatus with Sintered Diamond (SD) Anvil Techniques*

Principally, the new stage in studying high-pressure mineral assemblages started after the invention of the diamond anvil cell (DAC) in the National Bureau of Standards (NBS) in Washington, DC in 1958 (Bassett 2009). This allowed model pressure conditions of tens of gigapascals (up to 300 GPa), i.e. the conditions corresponding to the Earth's inner core (e.g., Mao and Bell 1976, 1978; Bassett 2001; Dewaele et al. 2007; Mao and Mao 2009). In addition to such high pressure conditions, DAC has other advantages. Diamond has a low absorption of photons in a wide range of the electromagnetic spectrum, including its visual and infrared parts; it is virtually transparent for X-rays. The latter makes the diamond cell especially suited for in situ analysis, allowing applications in a growing range of material probes. In combination with Raman spectrometers and new third-generation synchrotron X-ray emitters, DAC is now a unique equipment tool for studying mineral phases from deep Earth. In situ analysis ensures that no chemical diffusion or structural changes associated with temperature or pressure quench can influence observations of the phase assemblage. However, in typical experiments, the stability of laser heating and the availability of synchrotron X-ray diffraction beam time limits these run times to less than an hour, which, depending on temperature/pressure conditions, may be insufficient to reach equilibrium (Dorfman 2016).

In the diamond cell, high temperatures can be achieved by laser heating or by electrical resistive heating. Each of these techniques has its advantages and disadvantages. The laser heating of a DAC (LHDAC) has a large temperature gradient; it has the ability to heat samples at high pressures to temperatures up to ~5000 K. However, the major challenges in the use of laser heating are uncertainties in temperature measurement of tens to hundreds of K and great temperature gradients of ~1000 K in a cell, which can break equilibrium conditions in studied assemblages (Duffy 2005). The heterogeneous temperature during laser heating causes diffusion of elements along thermal gradients, and, as a result, in situ diffraction at only the centre of the sample may see a composition difference from the starting material and miss those phases that are segregated to the edges (Dorfman 2016).

Resistive heating methods can produce highly uniform temperatures in the pressure cell, but they have some limitations in temperature range. External heating (EHDAC) has a very small (about 5 K) temperature uncertainty (Dubrovinskaia and Dubrovinsky 2003), but it is often restricted to temperatures of ~1300 K or less, thus, limiting the direct replication of geophysical conditions to those found in cold subducting slabs. New designs for internal and external resistive heating have been developed that can reach either very high temperatures (~3000 K) at modest pressures (Zha and Bassett 2003) or modest temperatures (1200 K) at megabar pressures with a high degree of thermal stability (Dubrovinskaia and Dubrovinsky 2003). Along with the development of laser heating systems at third-generation sources, there has been continual improvement in laser heating techniques, temperature measurement and sample assemblies. With the use of DAC, major lower-mantle mineral phases (such as bridgmanite, CaSi-perovskite, ferropericlase and others) were synthesized, and phase transitions in these minerals were studied.

The main disadvantage of the diamond cell is the small volume of compressed sample (~50–200 μm^3 at P ~ 100 GPa) (Duffy 2005; Dorfman 2016). This disadvantage has been avoided with the use of multi-anvil apparatus with sintered diamond (SD) anvils (e.g., Ohtani et al. 1989; Zhai and Ito 2010). High-pressure cell assemblies for samples with volumes more than 1 mm^3 have been optimized for SD anvils, which has enabled the generation of pressures of up to 90 GPa at room temperature or up to 2273 K and stable for 1–3 h at pressures up to 60 GPa, which is equivalent to intermediate regions of the lower mantle (Tange et al. 2008; Liu et al. 2016). In some experiments, pressures up to 109 GPa in a Kawai-type multi-anvil apparatus (KMA) equipped with SD anvils by means of in situ X-ray observation with synchrotron radiation were generated (Yamazaki et al. 2014).

3.2.3 *Composition of the Lower Mantle According to Experimental Data*

Depending on the starting material, two different mineral associations can be obtained within the lower mantle pressure–temperature (*P–T*) conditions. According to experimental data, the pyrolitic (or other ultramafic source, corresponding to the juvenile mantle composition) forms the ultramafic lower-mantle association, represented by bridgmanite, CaSi-perovskite and ferropericlase (Fig. 3.2). Ringwoodite, $(Mg, Fe)_2SiO_4$, with a spinel structure forms bridgmanite, $(Mg, Fe)SiO_3$, and ferropericlase, (Mg, Fe)O; majorite, $Mg_3(Mg, Si)Si_4O_{12}$, transforms into bridgmanite and CaSi-perovskite. Both iron-containing lower-mantle minerals, bridgmanite and ferropericlase should be magnesium-rich. The magnesium index $mg = Mg/(Mg + Fe)_{at}$ for bridgmanite in a pyrolitic environment was suggested to be 0.80–0.88 (Fei et al. 1996; Lee et al. 2004; Irifune et al. 2010), and for ferropericlase it was suggested to be within the range of 0.73–0.88 (Wood 2000; Lee et al. 2004).

The basaltic source, corresponding to the subducting slab composition, approximated by the middle-ocean ridge basalt (MORB) composition, has higher contents in Si and Al. It forms the mafic lower-mantle association, in which, along with bridgmanite and CaSi-perovskite, the SiO_2 and aluminium phase(s) occur (Fig. 3.2). SiO_2 in the upper part of the lower mantle is represented by stishovite, which transforms with depth to an α-PbO_2-type phase and then to seifertite.

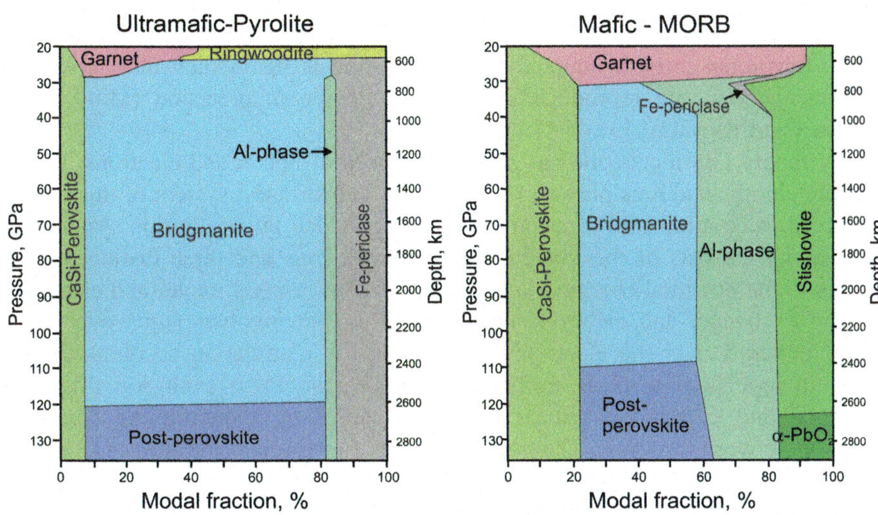

Fig. 3.2 Experimental data on mineral composition and phase transformations in the mantle. *Left* ultramafic (apo-pyrolitic) association. *Right* mafic (apo-MORB) association. After Dorfman (2016). © American Geophysical Union

Such models were obtained in experiments with both synthetic (Litasov et al. 2004; Litasov and Ohtani 2005; Irifune et al. 2010) and natural (Wood 2000; Lee et al. 2004; Ricolleau et al. 2010) starting compositions. Different experiments showed similar results, only with slight differences in the proportions of major minerals and their compositions.

3.3 Computational Simulations

3.3.1 General

Independently of experiments, theoretical computations can produce valuable data on possible mineral phase transitions and the physical properties of mineral phases under high *P*–*T* conditions. They may specify or/and explain the observed transitions; in other cases, they can even predict new phase transitions prior to the experimental studies. However, in some cases, data obtained from computer calculations and experiments may be contradictory (see Chap. 4). In other cases, even calculations carried out by different methods, e.g., density functional theory (DFT) versus molecular dynamic stimulations (Stixrude et al. 2007) or generalized-gradient approximation (GGA) versus GGA with the Hubbard correction (Mohn and Trønnes 2016) may disagree with each other. Such disagreements should be taken into account when interpreting computational simulations.

The two extremes of theoretical spectrum—semi-empirical to first principles—reflect two views of the microscopic world. In these, matter is perceived as composed of atoms and, in order to describe forces between pairs or triples of atoms, a simple functional form is adopted. Parametrization is made such that subsets of experimental data are reproduced with a fair degree of precision (Mitra 2004; Tsuchiya and Kawai 2013).

Any theory considers solids as being composed of nuclei and electrons. In this construct, atoms and ions play no primary role. The simplest system of this can be conceived to be a uniform electron gas with embedded nuclei. In this system, the total energy consists of the kinetic energy of electrons and three distinct contributions to the potential energy: Coulomb interactions among nuclei and electrons, electron exchange, and electron correlation. The first involves sums over point charge integrals over the electronic charge density. Correlation accounts for the Coulomb repulsion, whereas exchange embodies the Pauli exclusion principle. Exchange and correlation reduce the total energy by reducing the Coulomb repulsion between electrons. The pair potential is expressed as the sum of Coulomb energy (Mitra 2004).

Theoretical approaches can be categorized into three classes: first-principles approximations, ab initio methods, and semi-empirical methods based on DFT (Mitra 2004).

3.3.2 First-Principles Approximations

Computations of the structure and properties of minerals are based on the assumption that the mineral is in thermodynamic equilibrium, implying constant temperature *T*, pressure *P*, and composition *X*. First-principles methods are employed to solve the Schrodinger equation with a bare minimum of approximations. In first-principles calculations, the system of interest is represented as a collection of atomic nuclei, electrons, the total energy and the forces on all the nuclei, obtained by solving Schrodinger's equation determining the electronic ground state (Mitra 2004).

In calculations using first principles approximations, one starts from the fundamental interactions among electrons and nuclei, with little constraint of the experimental data. In the case of ferroelectrics, the first-principles methods are based on DFT, while some are based on Hartree-Fock theory. Kohn and Sham (1965) showed how to compute the charge density and energy self-consistently, using an effective exchange-correlation potential that accounts for the quantum-mechanical interactions between electrons. The local density approximation (LDA) takes the exchange-correlation potential from the uniform electron gas at the density for each point in the material. The GGA includes the effects of local gradients in density (Perdew et al. 1996). Given a form of exchange-correlation energy, one can find the self-consistent charge density and compute the energy for any arrangement of nuclei (atoms). From the energies of a zero-temperature phase diagram, phonon frequencies and elastic constants can be computed.

The calculations are based on DFT with the electronic exchange-correlation energy treated using either the LDA or the more sophisticated GGA (Wentzcovitch et al. 2010). First-principle calculations over a range of materials have been shown to give accurate predictions of both static and dynamic quantities.

In the approximation restricted only to the exchange-correlation potential, currently the most advanced method is known as the Linearized Augmented Plane Wave (LAPW). First Principles Linearized Muffin Tin Orbitals (FP-LMTO) calculations are also similarly precise. The other approximation strategy is the pseudopotential method, in which some physically motivated assumptions allow rapid computation.

First-principles theory is a complementary approach that has taken on increased significance in recent years in exploration of the properties of Earth materials under extreme conditions, such as those of the lower mantle (24–136 GPa). Elaborate electronic calculations are much more reliable than traditional atomistic calculations, which are based on interatomic potentials (usually parametrized from existing experimental information) (Karki and Stixrude 1999).

Computing power for theoretic calculations is limited to simple compositions. Computational methods are thus best applied to end members rather than to complex natural mineral compositions (Dorfman 2016). For computation, some have used the Cambridge Serial Total Energy Package (CASTEP) and Cambridge Edinburgh Total Energy Package (CETEP) codes, based on the DFT under two essential approximations (Payne et al. 1992). They first used LDA followed by pseudopotential approximation. They also optimized the lattice constant and internal structural parameters of a given structure at several pressures. For a thermodynamic computation of phase equilibria and physical properties of multiphase mantle assemblages, the HeFESTo code is the most popular (Stixrude and Lithgow-Bertelloni 2011). It was used for the recent models of the lower mantle compositions (Dorfman 2016).

3.3.3 *Ab Initio Methods*

These methods are widely used in determining the interactions between orbitals, charge density and in exploring the properties of liquids, including transport. These are more convenient to handle than the complex and elaborate first-principles calculations. In one of their modifications, such as the tight-binding method, the charge density and wave functions do not appear explicitly. The elements of the Hamiltonian and overlap matrices are approximated by simple parametric functions that are so constructed that the results of first-principles calculations are reproduced (Mitra 2004).

3.3.4 *Density Functional Theory (DFT)*

Traditionally, quantum chemistry is based on Hartree-Fock as a first approximation, followed by the inclusion of correlation effects. The correlated methods are extremely computationally intensive and cannot be fully applied to crystals because an infinite number of many-body states can be formed, even from a limited basis, by coupling different *k*-vectors (Mitra 2004). DFT in the LDA or GGA gives accurate predictions for energetic properties of closed-shell systems as well as ionic/covalent crystals. The first-principles DFT provides a powerful means for predicting the behaviour of solids that is independent of experimental data. Structural changes under pressure may be investigated theoretically using pseudopotentials.

DFT was originally established by Hohenberg and Kohn (1964) and considerably developed for lower-mantle mineral physics (Wentzcovitch et al. 2010). In principle, the DFT is an exact theory of the ground state and allows to reduces the interaction of a many-electron problem to a single-electron problem, with the nuclei being treated as an adiabatic background. A key to the application of DFT in handling the interacting electron gas was given by Kohn and Sham (1965). It may involve some approximations such as LDA, which is used for the exchange correlation energy functional and pseudopotential approximation. The latter recognises that the core electrons are essentially static (frozen) and do not vary their charge density significantly under perturbations to the lattice. The pseudopotential mimics the potential seen by the valence electrons because of the nucleus and frozen core electrons (Mitra 2004).

A good example of the successful application of the DFT method exists in the prediction and identification of the $MgSiO_3$ post-perovskite structure (Murakami et al. 2004; Oganov and Ono 2004).

However, DFT results are known to depend on approximations and corrections, particularly in the choice of exchange correlation functional. For phases with similar structures and energies, differences between the commonly used LDA and GGA functionals can lead to large differences (tens to hundreds of GPas) in calculated phase transition pressures. DFT calculations are thus more reliable for determining trends than absolute phase transition pressures and must be experimentally tested (Dorfman 2016).

3.3.5 Composition of the Lower Mantle According to Theoretical Calculations

Theoretical calculations on the composition of the lower mantle are in good agreement with experimental data (e.g., Tsuchiya and Kawai 2013; Dorfman 2016); they differ only in the proportions of lower-mantle phases (Table 3.1, Fig. 3.3). The magnesium index of ferropericlase and bridgmanite should also be similar in both theoretical calculations and experimental data (e.g., Lee et al. 2004).

The calculated MORB composite is denser than the pyrolite composite in most of the lower mantle by ~0.03–0.04 g/cm^3, corresponding to 1%, which makes negative buoyancy for subducted lithospheric slabs (Irifune and Tsuchiya 2007; Tsuchiya and Kawai 2013).

3.4 Lower-Mantle Minerals in the Terrestrial Environment

3.4.1 General

The finding of lower-mantle material on the Earth's surface is almost impossible due to the long distances in transportation from depth (more than 660 km) and the instability of some lower-mantle phases. The only possibility for such findings is the analysis of mineral inclusions in diamond. Diamond is an inert media, preserving the inclusions of interactions within the surrounding host environment. Diamond is stable in the entire range of the Earth's conditions starting from the core (e.g., Oganov et al. 2013) and keeps paragenetic inclusions encapsulated in it intact

Table 3.1 Mineralogical compositions of ultramafic and mafic associations in the upper part of the lower mantle, according to experimental and theoretical data (vol%)

Mineral	Ultramafic association (pyrolite)			Mafic association (MORB)		
	Experimental (Akaogi 2007)	Ab initio (Tsuchiya and Kawai 2013)	Thermodynamic modelling (Dorfman 2016)	Experimental (Akaogi 2007)	Ab initio (Tsuchiya and Kawai 2013)	Thermodynamic modelling (Dorfman 2016)
Bridgmanite	74	70	74	45	34	37
CaSi-perovskite	8	10	7	21	22	21
Ferropericlase	18	20	16	–	–	–
Al-phase	–	–	3	22	22	24
Silica	–	–	–	12	22	18

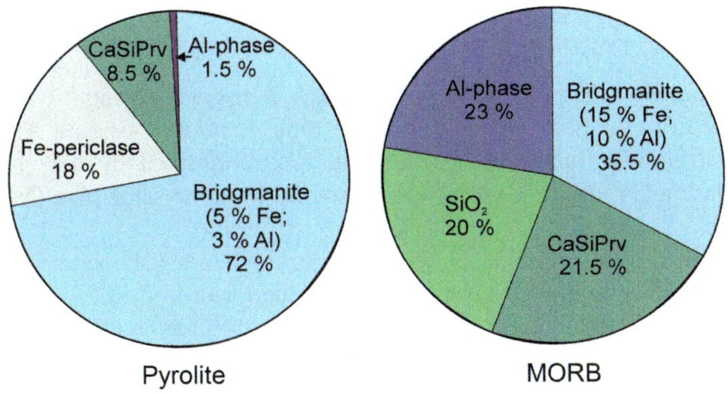

Fig. 3.3 Mineral compositions (vol%) of ultramafic and mafic associations in the lower mantle and the Fe and Al concentrations in mineral phases, according to theoretical calculations

during the ascent to the Earth's surface in kimberlitic and lamproitic rocks. To date, the Earth's lower-mantle minerals have been discovered in kimberlites from South Africa, South Australia, Brazil, Canada, and in placer deposits of Guinea, Brazil and other areas (Kaminsky 2012 and references therein) (Fig. 3.4). The number of such findings is very limited, less than lunar samples, for example. However, they have already provided valuable information about the real composition of the Earth's lower-mantle mineral phases and their associations; some of them are quite unexpected. For example, none of the earlier-known minerals were identified in the first microxenolith of the ultramafic lower-mantle rock that was found recently (Kaminsky et al. 2015).

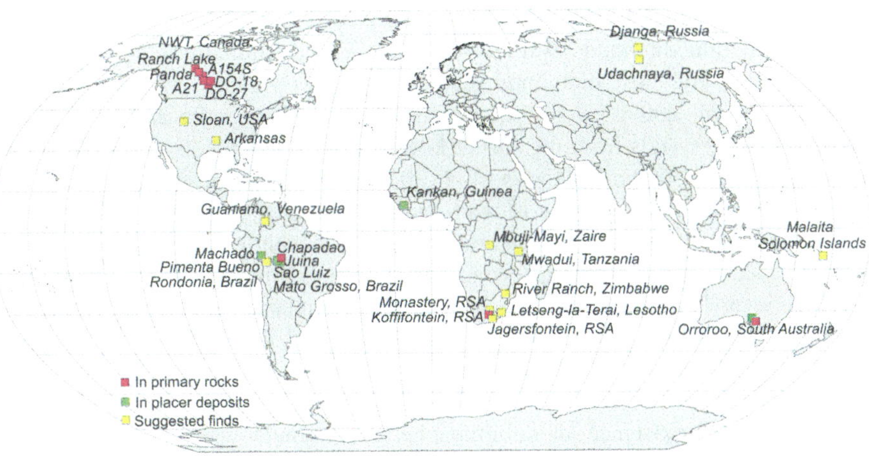

Fig. 3.4 Finds of lower-mantle minerals from inclusions in diamond. After Kaminsky (2012) with additions

3.4.2 South Africa and South Australia

The first reported occurrence of ferropericlase + retrograde enstatite was identified in diamond from the Koffifontein kimberlite pipe, Republic of South Africa (Scott Smith et al. 1984). Subsequently, this data was confirmed by new findings of ferropericlase and retrograde enstatite in diamond from the same pipe (Moore et al. 1986).

In addition, Scott Smith et al. (1984) described two ferropericlase grains in association with 'enstatite' from a diamond in the Orroroo dyke, South Australia. They were found in different diamond crystals and cannot be considered as a reliable paragenetic association. However, more recently, lower-mantle minerals (associations of ferropericlase + bridgmanite + 'olivine') were also identified in diamond from a kimberlitic dyke K7 and ancient placer deposits from the same area (Tappert et al. 2009); this confirms that the Orroroo area of South Australia is the second reliable source of lower-mantle material.

3.4.3 Brazil

In the 1990s, numerous grains of ferropericlase, bridgmanite, CaSi-perovskite, and other minerals were identified in the São Luiz placer deposit (Mato Grosso State, Brazil) and studied in detail (Wilding et al. 1991; Harte and Harris 1994; Harris et al. 1997; Harte et al. 1999; Hutchison et al. 2001). These works initiated a systematic study of lower-mantle mineral associations from the Juina area where the São Luiz placer deposit is located (Fig. 3.4). Lower-mantle mineral associations were found also in diamonds from other placer deposits within the same area: Mutum, Vermelho, Chicoria (Kaminsky et al. 2001), and Soriso (Hayman et al. 2005). A few years ago, at the sources of these rivers, primary kimberlitic sources to the lower-mantle diamonds were discovered. These kimberlites are termed the Pandrea pipes and host diamonds that contain lower-mantle mineral inclusions (Kaminsky et al. 2009a, 2010). Lower-mantle inclusions were also identified in some diamond crystals from the earlier known pipe, Collier-4, from the Juina area (Kaminsky et al. 2009a; Bulanova et al. 2010). The Juina area is now one of the major sources of data relating to lower-mantle material.

There is also information on the findings of lower-mantle inclusions in diamonds from the Machado River placer deposit, located approximately ~350 km WNW of the Juina area (Fig. 3.4) (Bulanova et al. 2008; Burnham et al. 2016). Kimberlitic pipes of Triassic age are located several kilometres upstream of the Machado alluvial deposit (Masun and Scott Smith 2008); some of them may be the sources of these diamonds.

Recently, the existence of sublithospheric diamonds was reported from the Igaparé Lajes alluvial deposit in the Pimenta Bueno kimberlite area, Rondonia State, Brazil located ~270 km east of the Juina area (Borges et al. 2016). This conclusion was based on non-destructive studies of diamonds; it did not study their inclusions and needs further confirmation.

3.4.4 *Guinea*

The fourth area, where lower-mantle inclusions in diamond were identified, is the Kankan placer deposit in south-eastern Guinea (Fig. 3.4). Lower-mantle inclusions in these diamond grains predominate over upper-mantle associations (Stachel et al. 2000). These comprise mainly ferropericlase grains, some of which coexist with bridgmanite, CaSi-perovskite and SiO_2 (initially stishovite).

3.4.5 *Canada*

The fifth area, where lower-mantle minerals have been identified within diamond, is the Slave kimberlitic province in the Lac de Gras area, Northwest Territories of Canada. Most of the superdeep inclusions are found in diamonds from kimberlite pipe DO27, where they comprise 11% of all inclusions; the remainder are represented by 'common' upper-mantle minerals (Davies et al. 2004). In addition to pipe DO27, ferropericlase and SiO_2 inclusions were found in diamonds from the nearby pipe DO18 and another pipe from the same area, Ranch Lake. CaSi-perovskite has also been found in diamond from pipe A-21 (Davies et al. 2004).

In diamond from one of the major productive kimberlite pipes, Panda, ferropericlase in association with SiO_2, CaSi-perovskite, 'olivine', and spinel were identified; they comprise $\sim$5% of all diamond inclusions in this pipe (Tappert et al. 2005a). Diamond crystals from another economic pipe, A-154S, at the Diavik mine contain numerous ferropericlase inclusions (Donnelly et al. 2007; van Rythoven and Schulze 2009).

3.4.6 *Other Suggested Finds*

In addition to the five major areas where lower-mantle inclusions have been identified, there are roughly a dozen occurrences of single mineral finds from different areas. These cannot be considered to be proven lower-mantle phases. However, since the possibility exists that they may be, these should not be discounted.

First of all, there are ferropericlase inclusions in diamonds that are not associated with other minerals. These inclusions are known in diamonds from the Guaniamo placer deposit in Venezuela (Kaminsky et al. 2000); from lamproites of Arkansas, USA (Newton et al. 1977), and kimberlites from Sloan, USA (Otter and Gurney 1989); and from the African pipes, Monastery, RSA (Moore et al. 1986),

Letseng-la-Terai, Lesotho (McDade and Harris 1999), River Ranch, Zimbabwe (Kopylova et al. 1997) and Mwadui, Tanzania (Stachel et al. 1999). In addition, ferropericlase as a single inclusion is also present in diamonds from placer deposits in north-eastern Yakutia (Sobolev et al. 1999) and within microdiamond from the Udachnaya pipe (Zedgenizov et al. 1998, 2001) in Russia. In diamonds from the Monastery pipe, in addition to ferropericlase, inclusions of majoritic garnet were also identified (Moore and Gurney 1985).

In diamonds from the Jagersfontein pipe, RSA, inclusions of high-Si majorite garnet (X_{Si} up to 3.5) and a phase with the composition of olivine were identified (Chinn et al. 1998; Tappert et al. 2005b); they can also be considered as possible lower-mantle phases.

In addition to mineral inclusions in diamond, a lower-mantle origin has been suggested for baddeleyite megacrysts from the Mbuji Mayi kimberlites. These are considered to be transformed to the present monoclinic state from a cubic ZrO_2 parent phase with an inferred lower-mantle origin (Kerschhofer et al. 2000; Schärer et al. 2011). Majorite garnet in association with Al-containing (4.57 wt% Al_2O_3) bridgmanite, CaSi-perovskite, picroilmenite, diamond, and other minerals have also been reported from xenoliths and garnet megacrysts in alnoitic pipes and sills in the Malaita Island (Ontong Java Plateau, Southwest Pacific), which led to the conclusion that they originated in the lower mantle (Collerson et al. 2000). However, subsequent analysis of this data led to some doubt concerning both the mineralogical data and the associated conclusions (Neal et al. 2001).

The deep-seated mineral phases in ophiolitic ultramafic rocks from the Swiss Alps and Tibet are also of some interest. For example, in peridotites from the Arami Massif, Switzerland, three previously unknown structures of $FeTiO_3$ were identified, indicating that the originally exsolved phase was a high-pressure perovskite polymorph of ilmenite. These data, therefore, imply a minimum depth of origin of 300 km (Dobrzhinetskaya et al. 1996).

In chromitites from the Luobusa ophiolite, southern Tibet, high-pressure nitrides and oxides were identified in association with diamond, moissanite, boron carbide, phase TiO_2-II, native iron, and other minerals. This association indicates their formation at depths of greater than 300–400 km (Dobrzhinetskaya et al. 2009). Recently, wüstite and FeNi alloy inclusions in olivine and chromite were reported from the Luobasa ophiolitic chromitites (Zhang et al. 2016). It seems possible, therefore, that some ultramafic rocks may contain lower-mantle material.

3.5 Lower-Mantle Mineral Associations

3.5.1 General

Both experimental data and geological observations indicate that, below the 660 km seismic discontinuity, new mineral associations exist. They are composed not of silicate mineral phases, like the upper mantle and the transitional zone, but of perovskite and other oxide phases. Hence, *the 'oxide Earth' replaces the 'silicate Earth' below the 660 km level.*

Geological data in general confirm the experimental and theoretical models, showing the existence of predicted mineral phases in both ultramafic and mafic associations. However, due to much more complex compositions of natural layers in comparison with model systems, geological data demonstrates compositional variabilities of these mineral phases and any additional minerals. The third mineral association, carbonatitic, was discovered among natural samples (Kaminsky et al. 2016b).

3.5.2 Ultramafic Mineral Association

Ultramafic mineral association in natural samples is composed of bridgmanite, CaSi-perovskite, ferropericlase and free silica. The compositions of major phases are discussed in Chap. 4. Here, we point to three major features of the natural association.

First, the bulk composition of the lower mantle may not correspond to the experimental and theoretical models for the pyrolite composition. In Table 3.2 and Fig. 3.5, we present frequencies of lower-mantle minerals of the juvenile ultramafic association included in diamond in three major regions, Brazil, Guinea and Canada. The variations in mineral proportions in these assemblages deviate from the average by only 10–15%, which is within an acceptable range. However, these figures differ notably from the lower mantle compositions suggested on the basis of experimental and theoretical data for the pyrolitic composition (e.g., Akaogi 2007; Tsuchiya and Kawai 2013). The estimates conclude that the most common mineral is ferropericlase, comprising 48.0–63.3% of the total number of minerals (averaging 55.6%) in the lower mantle, which is in contrast to predictions based on experimental data of *c.* 18–20% (i.e., the former estimates are around three times higher). In contrast to ferropericlase, bridgmanite comprises only 5.0–10.2% (average 7.5%) in all studied regions, which is about ten times lower than has been suggested as an average

Table 3.2 Frequency of lower-mantle mineral inclusions in diamond (in %)

Mineral	Brazil, $n = 297$	Guinea, $n = 49$	Canada, $n = 20$	Total average, $n = 366$
Ferropericlase	48.0	63.3	55.0	55.6
Bridgmanite	7.4	10.2	5.0	7.5
CaSi-perovskite	11.8	14.3	10.0	12.0
SiO_2	2.1	8.2	15.0	8.4
Others	30.7	4.0	15.0	16.5
Total	100.0	100.0	100.0	100.0

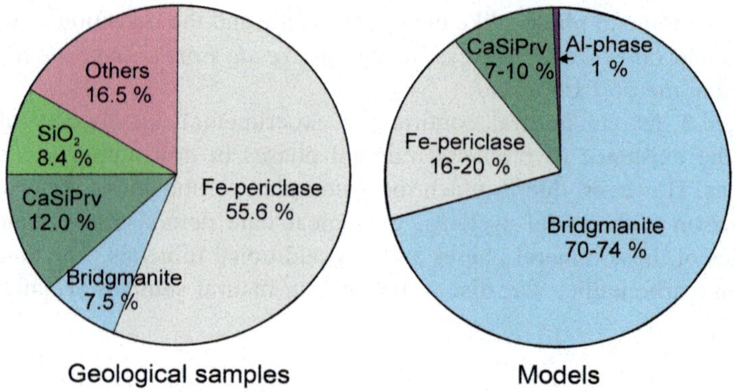

Fig. 3.5 Frequency of lower-mantle minerals included in diamond. After Kaminsky (2012) with additions

composition (c. 70–74%) in the lower mantle. According to geological data, CaSi-perovskite is more common compared with the predictions (10.0–14.3% with an average of 12.0% in experimental data compared with prediction of c. 8–10%). The frequency estimates from the geological samples cannot be considered to be necessary compositional shears of minerals for the entire lower mantle; however, they should be taken into account.

Second, free silica, which is not accounted for in experimental and theoretical models of the ultramafic lower mantle, is always present in this association where it usually coexists with ferropericlase and sometimes with native iron. This is the most principal difference between the models and the natural composition of the lower mantle. Silica inclusions were identified in all sets of lower-mantle minerals observed in diamond from all regions and areas: 1.1–4.8% from Brazilian placer deposits; 8.2% from the Kankan area, Guinea, and 10–20% from Canadian kimberlites (Harte et al. 1999; Kaminsky et al. 2001; Hayman et al. 2005; Stachel et al. 2000; Davies et al. 2004; Tappert et al. 2005a), with an average of 8.4%.

Third, unexpectedly wide variations in ferropericlase compositions, up to iron index $fe = Fe/(Mg + Fe)_{at} = 0.64$, i.e., to magnesiowüstite composition, whereas, according to experiments in natural systems in the pressure range 25–60 GPa, the iron index of ferropericlase in lower-mantle material should be $fe = 0.12–0.27$ (e.g., Wood 2000; Lee et al. 2004).

3.5.3 Mafic Mineral Association

Major minerals of the mafic mineral association, bridgmanite, CaSi-perovskite and SiO_2 are the same as in the ultramafic association, with the addition of being enriched in Al. Bridgmanite and CaSi-perovskite from the mafic association are also more iron-rich than in experiments and theoretic calculations that is their typomorphic features (see details in Sects. 5.2.1 and 5.2.2). According to experimental data, anhydrous aluminium phases should exist (22%), as well as minor amounts of hydrous aluminium phases (phase Egg and δ−AlOOH) and a series of dense anhydrous magnesium silicates (DHMS). The source of the mafic association originates from subducted lithospheric slabs, which, according to geophysical data can reach the lowermost mantle including the core–mantle boundary (CMB) (e.g., Helffrich and Wood 2001). Within the slabs, boundaries of perovskite-forming reactions should displace to higher pressures, comparing more to the juvenile lower mantle as a consequence of negative *P*–*T* slopes (e.g., Saxena 2010). However, the enrichment of mafic minerals with Al shifts the phase transition boundaries into the lower pressure field (see Chap. 5).

On the basis of a new *P*-wave tomographic model, Fukao and Obayashi (2013) showed that only a minority of the subducting slabs penetrate the 660 km discontinuity and descend well into the deep mantle; the majority of slabs stagnate in the transition zone or in the shallow lower mantle, where they turn and lay down subhorizontally (King et al. 2015). The viscosity in the more shallow 900 km of the lower mantle (in the regions surrounding settling slabs) may be about two orders of magnitude higher than previously expected, causing slabs to spread laterally and to settle very slowly through this region (Marquardt and Miyagi 2015).

The total amount of recycled crust in the lower mantle is low. Sobolev et al. (2007), based on olivine compositions in tholeiites, concluded that 2–20% of recycled crust was involved in mantle (mainly upper mantle) melting during geologic time. Ohta et al. (2008) calculated that the total amount of oceanic crust that has subducted corresponds to $\sim 11\%$ of the Earth's mantle in volume. The recycled parts of mafic oceanic crust are unlikely to be chemically homogenized on the timescales of mantle convection and will therefore persist as distinct lithologies (Hirschmann and Stolper 1996; Helffrich and Wood 2001). Numerical studies have demonstrated that both models of the Earth's dynamic regime and early to present-day Earth showed that blocks of thick basaltic crust, even after sinking to the base of the mantle, will remain stable as isolated reservoirs for billions of years (van Thienen et al. 2005). The chemical heterogeneity may survive in the mantle for as long as the age of Earth because chemical diffusion in the mantle is insignificant (Holzapfel et al. 2005; Stixrude and Lithgow-Bertelloni 2012) and decreases with increasing pressure until becoming, at 135 GPa, several magnitudes lower than that at 25 GPa (Yamazaki and Irifune 2003). This indicates that *the role of subducted lithospheric plates on the bulk composition in the Earth's mantle is minimal.*

Geological data confirm that the mafic association plays a much lesser role in the lower mantle in comparison to the ultramafic one. There are only three samples known to date, which contain typical mafic mineral phases, CAS, phase Egg and δ−AlOOH within a stishovite aggregate (Wirth et al. 2007).

3.5.4 Carbonatitic Association

Minerals of the carbonatitic association were identified in several lower-mantle diamonds from Brazil (Wirth et al. 2009; Kaminsky et al. 2009b, 2013, 2015). They belong to the natrocarbonatitic assemblage, the primary melt of which was formed as a result of partial melting in the base of the lower mantle (Kaminsky et al. 2016).

The detailed characteristics of these three lower-mantle mineral associations are presented in Chaps. 4, 5 and 6.

3.6 Conclusions

Experimental data and theoretical calculations demonstrate that silicates occurring in the upper mantle and the transition zone are replaced in the lower mantle (beneath the 660 km seismic discontinuity), predominantly by the perovskitic assemblage. Depending on the starting substrate composition, two mineral associations should occur at pressures corresponding to the lower-mantle conditions: ultramafic (bridgmanite + CaSi-perovskite + ferropericlase) and mafic (bridgmanite + CaSi-perovskite + silica + Al-phase).

In general, geological data on mineral associations in the lower mantle confirm experimental and theoretical predictions that the lower mantle is not a 'silicate Earth' (like in the upper mantle and the transition zone) and is more truly represented as an 'oxide Earth', comprising the two major mineral associations of ultramafic and mafic. However, there are some discrepancies between geological data and the experimental and theoretical predictions.

1. In addition to the two predicted major associations, ultramafic and mafic, the third carbonatitic association occurs in the lower mantle.
2. Among minerals of the ultramafic association, ferropericlase grains comprise 48.0–63.3% (averaging 55.6%), which is in contrast to the previous predictions based on theoretical and experimental data of *c.* 18% (i.e., approximately three times higher than previously suggested). By contrast, bridgmanite comprises only 7.5%, which about ten times lower than has previously been suggested (c. 70–74%) in the lower mantle.
3. In the composition of the ultramafic association, free silica permanently presents in the amount of 2.1–15.0% (average: 8.4%).
4. Theoretical and experimental predictions about the magnesium component within ferropericlase in the lower mantle (with $fe = 0.12$–0.27) are in contrast with this mineral in geological samples where it has a much wider range of compositions with $fe = 0.08$–0.64, extending to the magnesiowüstite composition.
5. Minerals of the mafic association constitute an extensive minority among natural lower-mantle minerals.

References

Akaogi, M. (2007). Phase transitions of minerals in the transition zone and upper part of the lower mantle. In: E. Ohtani (Ed.), *Advances in high-pressure mineralogy* (Vol. 421, pp. 1–13). Geological Society of America. doi:10.1130/2007_2421(01)

Bassett, W. A. (2001). The birth and development of laser heating in DACs. *Review of Scientific Instruments, 72,* 1270–1272.

Bassett, W. A. (2009). Diamond anvil cell, 50th birthday. *High Pressure Research, 29*(2), 163–186.

Borges, M.P.A.C., Moura, M.A., Lenharo, S.L.R., Smith, C.B., & Araujo, D.P. (2016). Mineralogical characterization of diamonds from Roosevelt Indigenous Reserve, Brazil, using non-destructive methods. *Lithos, 265,* 182–198. doi:10.1016/j.lithos.2016.08.003.

Bulanova, G. P., Smith, C. B., Kohn, S. C., et al. (2008). Machado river, Brazil—a newly recognised ultra deep diamond occurrence. In *9th International kimberlite Conference Extended Abstract No. 9IKC-A-00233.*

Bulanova, G. P., Walter, M. J., Smith, C. B., et al. (2010). Mineral inclusions in sublithospheric diamonds from Collier 4 kimberlite pipe, Juina, Brazil: Subducted protoliths, carbonated melts and primary kimberlite magmatism. *Contributions to Mineralogy and Petrology, 159*(4), 489–510.

Burnham, A. D., Bulanova, G. P., Smith, C. B., et al. (2016). Diamonds from the Machado river alluvial deposit, Rondônia, Brazil, derived from both lithospheric and sublithospheric mantle. *Lithos, 265,* 199–213. doi:10.1016/j.lithos.2016.05.022

Chinn, J. L., Milledge, H. J., & Gurney, J. J. (1998). Diamonds and inclusions from the Jagersfontein kimberlite. In: *Seventh international kimberlite conference extended abstracts, Cape Town* (pp. 156–157).

Collerson, K. D., Hapugoda, S., Kamber, B. S., et al. (2000). Rocks from the mantle transition zone: Majorite-bearing xenoliths from Malaita, Southwest Pacific. *Science, 288*(5469), 1215–1223.

Davies, R. M., Griffin, W. L., O'Reilly, S. Y., et al. (2004). Mineral inclusions and geochemical characteristics of microdiamonds from the DO27, A154, A21, A418, DO18, DD17 and Ranch Lake kimberlites at Lac de Gras, Slave Craton, Canada. *Lithos, 77*(1–4), 39–55.

Dewaele, A., Mezouar, M., Guignot, N., et al. (2007). Melting of lead under high pressure studied using second-scale time-resolved X-ray diffraction. *Physical Review B, 76,* 144106. doi:10.1103/PhysRevB.76.144106

Dobrzhinetskaya, L., Green, H. W., II, & Wang, S. (1996). Alpe Arami: A peridotite massif from depths of more than 300 kilometers. *Science, 271*(5257), 1841–1845.

Dobrzhinetskaya, L. F., Wirth, R., Yang, J., et al. (2009). High-pressure highly reduced nitrides and oxides from chromitite of a Tibetan ophiolite. *Proceedings of the National Academy of Sciences, 106*(46), 19233–19238.

Donnelly, C. L., Stachel, T., Creighton, S., et al. (2007). Diamonds and their mineral inclusions from the A154 South pipe, Diavik Diamond Mine, Northwest Territories, Canada. *Lithos, 98* (1–4), 160–176.

Dorfman, S. M. (2016). Phase diagrams and thermodynamics of lower mantle materials. In: H. Terasaki & R. A. Fischer (Eds.), *Deep Earth; Physics and Chemistry of the Lower Mantle and Core* (Vol. 217, pp. 241–252). Wiley: Geophysical Monograph.

Dubrovinskaia, N., & Dubrovinsky, L. (2003). Whole-cell heater for the diamond anvil cell. *Review of Scientific Instruments, 74*(7), 3433–3437.

Duffy, T. S. (2005). Synchrotron facilities and the study of the Earth's deep interior. *Reports on Progress in Physics, 68,* 1811–1859. doi:10.1088/0034-4885/68/8/R03

Fei, Y., Wang, Y., & Finger, L.W. (1996). Maximum solubility of FeO in (Mg,Fe)SiO3 perovskite as a function of temperature at 26 GPa: Implication for FeO content in the lower mantle. Journal of Geophysical Research, 101(B5), 11525-11530.

Fukao, Y., & Obayashi, M. (2013). Subducted slabs stagnant above, penetrating through, and trapped below the 660 km discontinuity. *Journal of Geophysical Research, B118,* 5920–5938. doi:10.1002/2013JB010466

Harte, B., & Harris, J. W. (1994). Lower mantle mineral association preserved in diamonds. *Mineralogical Magazine, 58A,* 384–385.

Harte, B., Harris, J. W., Hutchison, M. T. et al. (1999). Lower mantle mineral associations in diamonds from Sao Luiz, Brazil. In: Y. Fei et al. (Eds.), Mantle petrology: Field observations and high pressure experimentation: A tribute to Francis R. (Joe) Boyd (Vol. 6, pp. 125– 153). Geochemical Society Special Publication.

Harris, J. W., Hutchison, M. T., Hursthouse, M., et al. (1997). A new tetragonal silicate mineral occurring as inclusions in lower mantle diamonds. *Nature, 387*(6632), 486–488.

Hayman, P. C., Kopylova, M. G., & Kaminsky, F. V. (2005). Lower mantle diamonds from Rio Soriso (Juina, Brazil). *Contributions to Mineralogy and Petrology, 149*(4), 430–445.

Helffrich, G. R., & Wood, B. J. (2001). The Earth's mantle. *Nature, 412*(6846), 501–507.

Hirschmann, M. M., & Stolper, E. M. (1996). A possible role for garnet pyroxenite in the origin of the "garnet signature" in MORB. *Contributions to Mineralogy and Petrology, 124,* 185–208.

Hohenberg, P., & Kohn, W. (1964). Inhomogeneous electron gas. *Physical Review B, 136,* 864–871.

Holzapfel, C., Rubie, D. C., Frost, D. J., et al. (2005). Fe-Mg interdiffusion in (Mg, Fe)SiO3 perovskite and lower mantle reequilibration. *Science, 309,* 1707–1710. doi:10.1126/science.1111895

Hutchison, M. T., Hurtshouse, M. B., & Light, M. E. (2001). Mineral inclusions in diamonds: associations and chemical distinctions around the 670-km discontinuity. *Contributions to Mineralogy and Petrology, 142*(2), 119–126.

Irifune, T., & Tsuchiya, T. (2007). Mineralogy of the earth—phase transitions and mineralogy of the lower mantle. In G. D. Price (Ed.), *Treatise on Geophysics* (Vol. 2, pp. 33–62). Elsevier, Amsterdam: Mineral Physics.

Irifune, T., Shinmei, T., McCammon, C. A., et al. (2010). Iron partitioning and density changes of pyrolite in Earth's lower mantle. *Science, 327*(5962), 193–195. doi:10.1126/science.1181443

Ito, E. (2009). Theory and practice—multianvil cells and high-pressure experimental methods. In: G. D. Price & G. Schubert (Eds.), Mineral physics: *Treatise on geophysics*, (Vol. 2, pp. 197–230).

Kaminsky, F. V. (2012). Mineralogy of the lower mantle: A review of 'super-deep' mineral inclusions in diamond. *Earth-Science Reviews, 110*(1–4), 127–147.

Kaminsky, F. V., Zakharchenko, O. D., Griffin, W. L., et al. (2000). Diamond from the Guaniamo area, Venezuela. *Canadian Mineralogist, 38*(6), 1347–1370.

Kaminsky, F. V., Zakharchenko, O. D., Davies, R., et al. (2001). Superdeep diamonds from the Juina area, Mato Grosso State, Brazil. *Contributions to Mineralogy and Petrology, 140*(6), 734–753.

Kaminsky, F. V., Khachatryan, G. K., Andreazza, P., et al. (2009a). Super-deep diamonds from kimberlites in the Juina area, Mato Grosso State, Brazil. *Lithos, 112S*(2), 833–842.

Kaminsky, F., Wirth, R., Matsyuk, S., et al. (2009b). Nyerereite and nahcolite inclusions in diamond: Evidence for lower-mantle carbonatitic magmas. *Mineralogical Magazine, 73*(5), 797–816.

Kaminsky, F. V., Sablukov, S. M., Belousova, E. A., et al. (2010). Kimberlitic sources of super-deep diamonds in the Juina area, Mato Grosso State, Brazil. *Lithos, 114*(1–2), 16–29.

Kaminsky, F. V., Wirth, R., & Schreiber, A. (2013). Carbonatitic inclusions in Deep Mantle diamond from Juina, Brazil: New minerals in the carbonate-halide association. *Canadian Mineralogist, 51*(5), 669–688.

Kaminsky, F. V., Wirth, R., & Schreiber, A. (2015). A microinclusion of lower-mantle rock and some other lower-mantle inclusions in diamond. *Canadian Mineralogist, 53*(1), 83–104. doi:10.3749/canmin.1400070

Kaminsky, F. V., Ryabchikov, I. D., & Wirth, R. (2016). A primary natrocarbonatitic association in the deep earth. *Mineralogy and Petrology, 110*(2–3), 387–398. doi:10.1007/s00710-015-0368-4

Karki, B. B., & Stixrude, L. (1999). Seismic velocities of major silicate and oxide phases of the lower mantle. *Journal of Geophysical Research, 104,* 13025–13033.

Kerschhofer, L., Scharer, U., & Deutsch, A. (2000). Evidence for crystals from the lower mantle: baddeleyite megacrysts of the Mbuji Mayi kimberlite. *Earth and Planetary Science Letters, 179*(2), 219–225.

King, S. D., Frost, D. J., & Rubie, D. C. (2015). Why cold slabs stagnate in the transition zone. *Geology, 43,* 231–234.

Kohn, W., & Sham, L. J. (1965). Self-consistent equations, including exchange and correlation effects. *Physical Review A, 140,* 1133–1140.

Kopylova, M. G., Gurney, J. J., & Daniels, L. D. (1997). Mineral inclusions in diamonds from the River Ranch kimberlite, Zimbabwe. *Contributions to Mineralogy and Petrology, 129*(4), 366–384.

Lee, K. K. M., O'Neill, B., Panero, W. R., et al. (2004). Equations of state of the high-pressure phases of a natural peridotite and implications for the Earth's lower mantle. *Earth and Planetary Science Letters, 223,* 381–393.

Liebermann, R. C. (2011). Multi-anvil, high pressure apparatus: a half-century of development and progress. *High Pressure Research, 31*(4), 493–532. doi:10.1080/08957959.2011.618698

Litasov, K., Ohtani, E., Suzuki, A., et al. (2004). Absence of density crossover between basalt and peridotite in the cold slabs passing through 660 km discontinuity. *Geophysical Research Letters, 31,* L24607. doi:10.1029/2004GL021306

Litasov, K. D., & Ohtani, E. (2005). Phase relations in hydrous MORB at 18–28 GPa: Implications for heterogeneity of the lower mantle. *Physics of the Earth and Planetary Interiors, 150,* 239–263. doi:10.1016/j.pepi.2004.10.010

Liu, L.-G., & Ringwood, A. E. (1975). Synthesis of a perovskite-type polymorph of $CaSiO_3$. *Earth and Planetary Science Letters, 28,* 209–211.

Liu, Z., Irifune, T., Nishi, M., Tange, Y., et al. (2016). Phase relations in the system $MgSiO_3$–Al_2O_3 up to 52 GPa and 2000 K. *Physics of the Earth and Planetary Interiors, 257,* 18–27. doi:10.1016/j.pepi.2016.05.006

Mao, H. K., & Bell, P. M. (1976). The 1-megabar mark on the ruby R_1 static pressure scale. *Science, 191*(4229), 851–852.

Mao, H. K., & Bell, P. M. (1978). Design and varieties of the megabar cells. *Carnegie Institution of Washington Yearbook, 77,* 904–908.

Mao, H.-K., & Mao, W. L. (2009). Theory and practice—diamond-anvil cells and probes for high *P-T* mineral physics studies. In: G. D. Price & G. Schubert (Eds.), Mineral physics: Treatise on geophysics (Vol. 2, pp. 231–267).

Marquardt, H., & Miyagi, L. (2015). Slab stagnation in the shallow lower mantle linked to an increase in mantle viscosity. *Nature Geoscience, 8,* 311–314. doi:10.1038/ngeo2393

Masun, K. M., & Scott Smith, B. H. (2008). The Pimenta Bueno kimberlite field, Rondônia, Brazil: Tuffisitic kimberlite and transitional textures. *Journal of Volcanology and Geothermal Research, 174*(1–3), 81–89. doi:10.1016/j.jvolgeores.2007.12.043

McDade, P., & Harris, J. W. (1999). Syngenetic inclusion bearing diamonds from Letseng-la-Terai, Lesotho. *Proceedings of the VIIth International Kimberlite Conference* (Vol. 2, pp. 557–565). Cape Town: Red Roof Design.

Mitra, S. (2004). *High-pressure geochemistry in mineral physics* (Vol. 9, 1233 pp). Elsevier: Developments in Geochemistry.

Mohn, C. E., & Trønnes, R. G. (2016). Iron spin state and site distribution in $FeAlO_3$-bearing bridgmanite. *Earth and Planetary Science Letters, 440,* 178–186. doi:10.1016/j.epsl.2016.02.010

Moore, R. O., & Gurney, J. J. (1985). Pyroxene solid solution in garnets included in diamonds. *Nature, 318*(6046), 553–555.

Moore, R. O., Otter, M. L., Rickard, R. S. et al. (1986). The occurrence of moissanite and ferro-periclase as inclusions in diamond. In *4th International Kimberlite Conference Extended Abstracts, Perth* (Vol. 16, pp. 409–411). Geological Society of Australia Abstracts.

Murakami, M., Hirose, K., Kawamura, K., et al. (2004). Post-perovskite phase transition in $MgSiO_3$. *Science, 304,* 855–858.

Neal, C. R., Haggerty, S. E., Sautter, V. (2001). "Majorite" and "silicate perovskite" mineral compositions in xenoliths from Malaita. Science *292*(5519), 1015a.

Newton, M. G., Melton, C. E., & Giardini, A. A. (1977). Mineral inclusions in the Arkansas diamond. *American Mineralogist, 62*(5–6), 583–586.

Oganov, A. R., & Ono, S. (2004). Theoretical and experimental evidence for a post-perovskite phase of $MgSiO_3$ in Earth's D″ layer. *Nature, 430,* 445–448.

Oganov, A. R., Hemley, R. J., Hazen, R. M., et al. (2013). Structure, bonding, and mineralogy of carbon at extreme conditions. *Reviews in Mineralogy and Geochemistry, 75,* 47–77. doi:10.2138/rmg.2013.75.3

Ohta, K., Hirose, K., Lay, T., et al. (2008). Phase transitions in pyrolite and MORB at lowermost mantle conditions: Implications for a MORB-rich pile above the core–mantle boundary. *Earth and Planetary Science Letters, 267,* 107–117.

Ohtani, E., Kagawa, N., Shimomura, O., et al. (1989). High-pressure generation by a multiple anvil system with sintered diamond anvils. *Review of Scientific Instruments, 60*(5), 922. doi:10.1063/1.1140344

Otter, M. L., & Gurney, J. J. (1989). Mineral inclusions in diamond from the Sloan diatremes, Colorado-Wyoming State Line kimberlite district, North America. In J. Ross et al. (Eds.), *Kimberlites and related rocks* (Vol. 2, pp. 1042–1053)., Proceedings of the Fourth International Kimberlite Conference, Perth 1986 Carlton: Blackwell.

Payne, M. C., Teter, M. P., Allen, D. C., et al. (1992). Interactive minimization techniques for ab initio total-energy calculations: Molecular dynamics and conjugate gradient. *Reviews of Modern Physics, 64,* 1045–1097.

Perdew, J. P., Burke, K., & Ernzerhof, M. (1996). Generalized gradient approximation made simple. *Physical Review Letters, 77,* 3856–4868.

Ricolleau, A., Perrillat, J.-P., Fiquet, G., et al. (2010). Phase relations and equation of state of a natural MORB: Implications for the density profile of subducted oceanic crust in the Earth's lower mantle. *Journal of Geophysical Research, 115,* B08202. doi:10.1029/2009JB006709

Ringwood, A. E. (1967). The pyroxene-garnet transformation in the earth's mantle. *Earth and Planetary Science Letters, 2*(3), 255–263.

Ringwood, A. E. (1975). *Composition and petrology of the Earth's mantle.* New York: McGraw-Hill.

Ringwood, A. E., & Major, A. (1971). Synthesis of majorite and other high pressure garnets and perovskites. *Earth and Planetary Science Letters, 12,* 411–418.

Saxena, S. K. (2010). Thermodynamic modeling of the Earth's interior. *Elements, 6,* 321–325. doi:10.2113/gselements.6.5.321

Scott Smith, B. H., Danchin, R. V., Harris, J. W., et al. (1984). Kimberlites near Orroroo, South Australia. In J. Kornprobst (Ed.), *Kimberlites I: Kimberlites and Related Rocks* (pp. 121–142). Amsterdam: Elsevier.

Schärer, U., Berndt, J., & Deutsch, A. (2011). The genesis of deep-mantle xenocrystic zircon and baddeleyite megacrysts (Mbuji-Mayi kimberlite): Trace-element patterns. *European Journal of Mineralogy, 23,* 241–255.

Sobolev, A. V., Hofmann, A. W., Kuzmin, D. V., et al. (2007). The amount of recycled crust in sources of mantle-derived melts. *Science, 316,* 412–417. doi:10.1126/Science.1138113

Sobolev, N. V., Yefimova, E. S., & Koptil, V. I. (1999). Mineral inclusions in diamonds in the Northeast of the Yakutian diamondiferous province. *Proceedings of the VIIth International Kimberlite Conference* (Vol. 2, pp. 816–822). Cape Town: Red Roof Design.

Stachel, T., Harris, J. W., & Brey, G. P. (1999). REE patterns of peridotitic and eclogitic inclusions in diamonds from Mwadui (Tanzania). *Proceedings of the VIIth International Kimberlite Conference* (Vol. 2, pp. 829–835). Cape Town: Red Roof Design.

Stachel, T., Harris, J. W., Brey, G. P., et al. (2000). Kankan diamonds (Guinea) II: Lower mantle inclusion parageneses. *Contributions to Mineralogy and Petrology, 140*(1), 16–27.

Stixrude, L., & Lithgow-Bertelloni, C. (2007). Influence of phase transformations on lateral heterogeneity and dynamics in Earth's mantle. *Earth and Planetary Science Letters, 263*(1–2), 45–55.

Stixrude, L., & Lithgow-Bertelloni, C. (2011). Thermodynamics of mantle minerals—II, Phase equilibria. *Geophysical Journal International, 184,* 1180–1213. doi:10.1111/j.1365-246X.2010.04890.x

Stixrude, L., & Lithgow-Bertelloni, C. (2012). Geophysics of chemical heterogeneity in the mantle. *Annual Review of Earth and Planetary Sciences, 40,* 569–595.

Stixrude, L., Lithgow-Bertelloni, C., Kiefer, B., et al. (2007). Phase stability and shear softening in $CaSiO_3$ perovskite at high pressure. *Physical Review B, 75,* 024108.

Tange, Y., Irifune, T., & Funakoshi, K. (2008). Pressure generation to 80 GPa using multianvil apparatus with sintered diamond anvils. *High Pressure Research, 28*(3), 245–254. doi:10.1080/08957950802208936

Tappert, R., Stachel, T., Harris, J. W., et al. (2005a). Mineral inclusions in diamonds from the Slave Province, Canada. *European Journal of Mineralogy, 17*(3), 423–440.

Tappert, R., Stachel, T., Harris, J. W., et al. (2005b). Diamonds from Jagersfontein (South Africa): Messengers from the sublithospheric mantle. *Contributions to Mineralogy and Petrology, 150* (5), 505–522.

Tappert, R., Foden, J., Stachel, T., et al. (2009). The diamonds of South Australia. *Lithos, 112S,* 806–821.

Tsuchiya, T., & Kawai, K. (2013). Ab initio mineralogical model of the Earth's lower mantle. In: Karato S.-I. (Ed.), Physics and chemistry of the deep earth (pp. 213–243). Wiley.

Van Rythoven, A. D., & Schulze, D. J. (2009). In-situ analysis of diamonds and their inclusions from the Diavik Mine, Northwest Territories, Canada: Mapping diamond growth. *Lithos, 112S,* 870–879.

Van Thienen, P., van Summeren, J., van der Hilst, R. D. et al. (2005). Numerical study of the origin and stability of chemically distinct reservoirs deep in Earth's mantle. In: R. D. van der Hilst et al. (Eds.), Earth's deep mantle: Structure, composition, and evolution (pp. 117–136). Washington, D.C: Geophysical Monograph 160, American Geophysical Union.

Wentzcovitch, R. M., Wu, Z., & Carrier, P. (2010). First principles quasiharmonic thermoelasticity of mantle minerals. *Reviews in Mineralogy and Geochemistry, 71*(1), 99–128. doi:10.2138/rmg.2010.71.5

Wilding, M. C., Harte, B., & Harris, J.W. (1991). Evidence for a deep origin for the Sao Luiz diamonds. In *Fifth International Kimberlite Conference Extended Abstracts, Araxa* (pp. 456–458).

Wirth, R., Vollmer, C., Brenker, F., Matsyuk, S., et al. (2007). Nanocrystalline hydrous aluminum silicate in superdeep diamonds from Juina (Mato Grosso State, Brazil). *Earth and Planetary Science Letters, 259*(3–4), 384–399.

Wirth, R., Kaminsky, F., Matsyuk, S., et al. (2009). Unusual micro- and nano-inclusions in diamonds from the Juina Area, Brazil. *Earth and Planetary Science Letters, 286*(1–2), 292–303.

Wood, B. J. (2000). Phase transformations and partitioning relations in peridotite under lower mantle conditions. *Earth and Planetary Science Letters, 174,* 341–354.

Yamazaki, D., & Irifune, T. (2003). Fe-Mg interdiffusion in magnesiowustite up to 35 GPa. *Earth and Planetary Science Letters, 216,* 301–311. doi:10.1016/S0012-821X(03)00534-X

Yamazaki, D., Ito, E., Yoshino, T., Tsujino, N., Yoneda, A., Guo, X., et al. (2014). Over 1 Mbar generation in the Kawai-type multianvil apparatus and its application to compression of (Mg 0.92 Fe 0.08)SiO3 perovskite and stishovite. *Physics of the Earth and Planetary Interiors, 228,* 262–267. doi:10.1016/j.pepi.2014.01.013

Zedgenizov, D. A., Logviniva, A. M., Shatsky, V. S., et al. (1998). Inclusions in microdiamonds from some Yakutian kimberlite diatremes. *Dokladi Akademii Nauk, 359*(2), 204–208.

Zedgenizov, D. A., Yefimova, E. S., Logvinova, A. M., et al. (2001). Ferropericlase inclusions in a diamond microcrystal from the Udachnaya kimberlite pipe, Yakutia. *Doklady Akademii Nauk, 377*(3), 319–321.

Zha, C. S., & Bassett, W. A. (2003). Internal resistive heating in diamond anvil cell for in situ X-ray diffraction and Raman scattering. *Review of Scientific Instruments, 74,* 1255–1262.

Zhai, S., & Ito, E. (2010). Recent advances of high-pressure generation in a multianvil apparatus using sintered diamond anvils. *Geoscience Frontiers, 2*(1), 101–106.

Zhang, R. Y., Yang, J.-S., Ernst, W. G., et al. (2016). Discovery of in situ super-reducing, ultrahigh-pressure phases in the Luobusa ophiolitic chromitites, Tibet: New insights into the deep upper mantle and mantle transition zone. *American Mineralogist, 101*(6), 1285–1294. doi:10.2138/am-2016-5436

Chapter 4
Ultramafic Lower-Mantle Mineral Association

Abstract The juvenile ultramafic lower mantle is composed of the following mineral association: bridgmanite + ferropericlase + CaSi-perovskite + free silica. Bridgmanite, with $mg = 0.84–0.96$ forms two compositional groups: low-Al and high-Al. High-Al bridgmanite is richer in Fe and infers the characteristic of deeper layers in the lower mantle. The crystal structure of bridgmanite is orthorhombic through the entire lower mantle down to the D″ layer. The chemical composition of ferropericlase differs from the predicted composition, with the magnesium index *mg* varying from 0.36 to 0.90. Low-Fe ferropericlase has a cubic rocksalt structure, which is stable throughout the entire lower mantle. Iron contents in both ferropericlase and bridgmanite and ferropericlase increase with increasing pressure, indicating higher Fe concentration in the lower mantle with increasing depth. CaSi-perovskite is remarkably clean in its chemical composition with only minor admixtures of Ti, Al and Fe, but is enriched in trace elements. CaSi-perovskite within the lower mantle has a cubic structure, which, at low temperatures (in subsolidus conditions), may transfer into a tetragonal or orthorhombic structure. The presence of free silica in the lower mantle was identified in geological samples from all areas. In the upper part of the lower mantle, it is represented by stishovite; at a depth of 1600–1800 km, stishovite transforms into the $CaCl_2$-structured polymorph; and at the core–mantle boundary, into a α-PbO_2 phase seifertite. In addition to the major minerals, a variety of other mineral phases occur in the lower mantle: Mg–Cr–Fe, Ca–Cr and other orthorhombic oxides, jeffbenite, ilmenite, native Ni and Fe, moissanite and some others.

4.1 General

According to experimental data, three major minerals comprise the juvenile, ultramafic lower mantle: bridgmanite, CaSi-perovskite, and ferropericlase. However, significant differences occur between the model experimental composition and observations made in natural ultramafic lower-mantle associations.

First, the proportion of the major minerals that occur in natural samples differs significantly the proportion of major minerals suggested by experimental data.

F.V. Kaminsky, *The Earth's Lower Mantle*, Springer Geology,
DOI 10.1007/978-3-319-55684-0_4

Bridgmanite, which is supposed to comprise 77% of the lower mantle volume, comprises (in different regions) only 5–10% of all lower-mantle mineral inclusions in diamond; while ferropericlase, which is expected to comprise 18%, accounts for the majority of the lower-mantle inclusions at 48–63%.

Second, the fourth major mineral, free silica, SiO_2, occurs in the lower-mantle ultramafic association; it comprises (in different areas) between 2 and 20% of all lower-mantle inclusions in diamond (average: 8.4%). At depth, it exists as stishovite and post-stishovite phases; however, because of their instability it occurs in samples as a form of quartz. However, its usual association with ferropericlase and CaSi-perovskite proves its primary origin within the lower mantle.

Third, the composition of major minerals from the calculated ultramafic association differs from the theoretical and experimental data. Bridgmanite in natural samples has a magnesium index *mg* = 0.84–0.96 (average: 0.916), conversely, the predicted values of magnesium index were *mg* = 0.80–0.88. Differences in the composition of ferropericlase are more drastic: instead of the expected *mg* = 0.73–0.88, in the natural mineral, *mg* varies widely from 0.36 to 0.90, forming a series of periclase–wüstite solid solution compositions.

Fourth, other minerals exist in the lower-mantle ultramafic association. For example, the recently found lower-mantle microxenolith is composed of ferropericlase; iron carbide; and two new orthorhombic oxides, $Mg(Cr,Fe)_2O_4$ and $CaCr_2O_4$. Some other minerals are known in association with major lower-mantle minerals, such as jeffbenite; ilmenite; perovskite $CaTiO_3$; minerals with a composition of olivine and spinels; titanite; native nickel and iron; sulphides; and some other phases, the genesis of which remains unclear.

4.2 Bridgmanite $(Mg,Fe)SiO_3$

4.2.1 General

Bridgmanite (Brd) is the most abundant mineral in the lower mantle (Tschauner et al. 2014) and was previously known as magnesium silicate perovskite. The name 'perovskite' was initially derived from the $CaTiO_3$ analog in the perovskite family, which actually has a cubic crystal structure. It is a member of the perovskite family with an orthorhombic distortion of the ideal cubic perovskite structure and has the general formula of $^{VIII/XII}A^{2+}\,^{VI}B^{4+}O^{2-}{}_3$, where the A site is a larger pseudo-dodecahedral site (with 8/12-fold coordination), mainly occupied by divalent cations, such as Mg, Fe, Mn, Ni, Ca; and the B site is a smaller six-fold octahedral site occupied by Si^{4+}, Al^{3+} and Fe^{3+} cations. The simplified formula that most represents lower-mantle bridgmanite is $(Mg,Fe)(Si,Al,Fe)O_3$, as other cations are much less abundant, but it should be noted that Fe^{3+} may also be present in the A site (McCammon et al. 1997; Lin et al. 2013). The crystal has an orthorhombic structure with space group *Pbnm* (Liu 1974; Fiquet et al. 2000).

Bridgmanite is formed as a result of the dissociation of ringwoodite $(Mg,Fe)_2SiO_4$ into ferropericlase and bridgmanite and is responsible for the 660 km seismic discontinuity. Due to the similarity in Fe–Mg partitioning across the transformation, it should take place over a narrow pressure interval, equivalent to <2 km in depth (Ito and Takahashi 1989; Ito et al. 1990), which is consistent with seismic observations of short-period reflected and converted phases (Kind and Li 2007).

Ringwood (1962) first suggested that $MgSiO_3$ might have a perovskite structure in the deep mantle and pointed out that such a structure would be a few percent denser than an isochemical mixture of MgO and SiO_2. This MgSi-perovskite was first synthesised by Liu (1974) from almandine-pyrope garnet, which was laser heated to about 1400–1800 °C at approximately 27 GPa in a diamond-anvil cell (DAC). Subsequently, Liu (1975) synthesised silicate perovskite by transformation of forsterite and enstatite at high pressure and temperature. Ever since then, bridgmanite, in various chemical compositions of geological relevance, had been synthesized in laser-heated DACs (LHDACs) and in multi-anvil apparatus (Ito and Matsui 1978). It is interesting to note that the discovery of the silicate perovskite, amongst many other discoveries in mineral physics, was made possible as a result of the invention of the LHDAC technique by Bassett (2001).

Bridgmanite has also been found in nature, including in meteorites and diamond inclusions (Scott Smith et al. 1984; Tomioka and Fujino 1997; Tschauner et al. 2014). In diamond inclusions, it has been found to be associated with ferropericlase, CaSi-perovskite, jeffbenite, a phase with the composition of olivine, chrome spinel, native nickel and sulphide. It forms colorless grains, 15–150 μm in size, which tend not to exhibit crystal faces. It may have an elongated ellipsoid, moderately well-formed euhedral shape, but its overall form is either indistinguishable or it has no discernable form at all. There is evidence for imposed 'negative' octahedral shape on a few grains (Fig. 4.1).

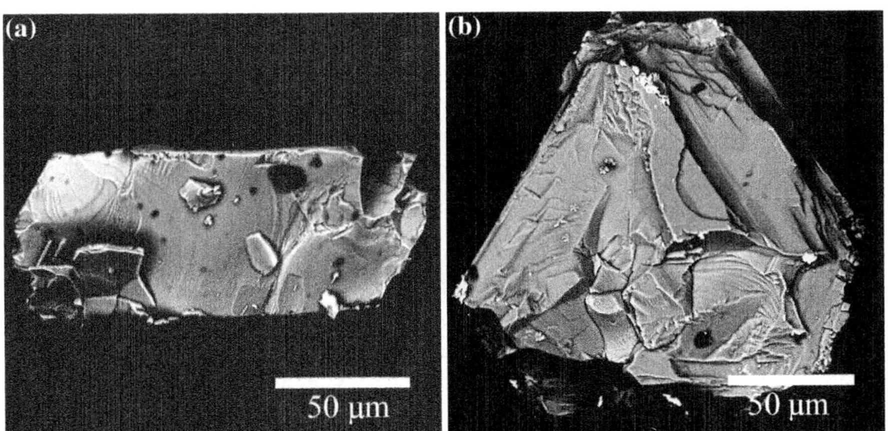

Fig. 4.1 Grains of bridgmanite extracted from diamonds of the Rio Soriso area, Brazil. SEM image. From Hayman et al. (2005)

The retrograde phase fully inherits its initial bridgmanite chemical composition, including a high Al content, which is not characteristic for enstatite. In its natural environment, it was found first as an inclusion in diamond from the Koffifontein kimberlite pipe, South Africa, where it was initially identified as 'enstatite' (Scott Smith et al. 1984). Subsequently, it was confirmed to be in association with ferropericlase in another diamond from the same pipe (Moore et al. 1986). Scott Smith et al. (1984) described it also as 'enstatite' in association with ferropericlase from a diamond in the Orroroo dyke, South Australia. Later 'MgSi-perovskite' was identified as inclusions in diamonds from placer deposits and kimberlite pipes in the Juina and Machado areas in Brazil (Wilding et al. 1991; Hutchison 1997; Harte et al. 1999; Kaminsky et al. 2001, 2009a, b; Hayman et al. 2005; Bulanova et al. 2008, 2010; Thomson et al. 2014; Zedgenizov et al. 2014a, 2015), from the Kankan placer deposit in Guinea (Stachel et al. 2000), from kimberlitic pipes in the Northwest Territories of Canada (Davies et al. 2004; Tappert et al. 2005a), and from the kimberlitic dyke K7 and ancient placer deposits from the Orroroo area in South Australia (Tappert et al. 2009b).

Due to its unstable nature under ambient conditions, this phase undergoes retrograde transformation to a pyroxene-type structure after extraction from diamond inclusions, inheriting the initial chemical composition. The identification of the retrograde phase as 'perovskite' was based on the analogy with experimental data and the association with other high-pressure (supposedly lower-mantle) minerals.

In meteorites, $(Mg,Fe)SiO_3$-perovskite was identified in the shock-induced veins in the Tenham L6 chondritic meteorite (Tomioka and Fujino 1997) using XRD and TEM; the perovskite grains showed an orthorhombic diffraction pattern, but became amorphous after the electron beam irradiation. This observed phase was inferred to have transformed from pyroxene at high pressures and temperatures by shock metamorphism. The structural data was sufficient for the International Mineralogical Association to approve bridgmanite as a new mineral (specimen IMA 2014-017). Bridgmanite was named after Percy W. Bridgman (1882–1961), the 1946 Nobel laureate, for his fundamental contributions to high-pressure physics (Tschauner et al. 2014).

4.2.2 Chemical Composition of Natural Bridgmanite

Analysis of the chemical composition of natural bridgmanite samples is mainly based on the full empirical formula of the ABO_3 oxide, where the pseudo-dodecahedral A site is occupied by large-size cations, Mg, Fe, Mn and Ca; and the octahedral B site contains smaller cations, Si and Al, with various cation substitutions. Its full formula is (Mg, Fe, Mn, Ca)(Si,Al)O_3. Ti (0–0.23 wt% TiO_2), Ni (0–0.09 wt% NiO) and K (0–0.26 wt% K_2O) are considered to be minor impurities (Table 4.1). All analyzed samples show almost ideal stoichiometry in accordance with experimental and theoretical data, with the exception of Al-rich samples, in which the B site shows excess cations, implying the presence of Al in

Table 4.1 Representative compositions of bridgmanite inclusions in diamond from different areas (wt%)

Country	Brazil				Canada	Guinea		Australia	Tenham L6 meteorite
Area	Juina			Machado	Northwest Territories	Kankan		Orroroo	
Location	Sao Luiz placer		Rio Vermelho placer	Placer	Pipe DO27	Placer		Dyke Eurelia K7	
Sample No.	BZ207C	BZ242B	4-104	P33-1a	DO2700100	KK-103b	USNM 7703	FBS5-11	KK-16c
Mineral association	Brd + fPer + Jbn	Brd + fPer	Brd + Jbn	Brd + Jbn + 'Ol'	Brd + fPer + CaSiPrv	Brd + fPer	Brd + Akm + Rng + Maj + Msk	Brd + fPer	Brd + fPer + Sid
SiO_2	55.43	54.96	55.96	56.97	57.40	57.38	55.6	60.02	57.50
TiO_2	0.20	0.03	0.23	0.14	0.00	0.02	0.19	0.03	0.01
Al_2O_3	2.66	12.58	2.16	2.24	0.69	0.61	0.2	0.25	0.55
Cr_2O3	0.15	1.75	0.17	0.03	0.28	0.17	0.13	0.38	0.28
FeO	9.02	3.95	8.57	4.71	2.85	3.39	13.7	2.93	3.34
NiO	0.03	0.01	n.a.	0.02	0.00	0.02	n.a.	0.09	0.01
MnO	0.30	1.29	0.20	0.04	0.07	0.10	0.49	0.15	0.11
MgO	32.03	16.35	32.26	33.68	37.90	35.92	27.9	36.58	35.94
CaO	0.05	4.38	0.03	0.73	0.39	0.07	0.9	0.05	0.07
Na_2O	0.07	4.63	0.10	0.42	n.a.	0.03	0.9	0.00	0.05
K_2O	0.00	0.01	0	0.18	n.a.	0.01	0.04	0.00	0.01
P_2O_5	n.a.	n.a.	n.a.	n.a.	n.a.	0.00	n.a.	0.00	0.00
Total	99.94	99.94	99.68	99.16	99.58	97.71	100.05	100.48	97.87
fe	0.136	0.119	0.130	0.073	0.040	0.050	0.216	0.043	0.050
mg	0.864	0.881	0.870	0.927	0.960	0.950	0.784	0.957	0.950
References	Hutchison (1997)		Kaminsky et al. (2001)	Burnham et al. (2016)	Davies et al. (2004)	Stachel et al. (2000)		Tappert et al. (2005a)	Tschauner et al. (2014)

Notes: *n.a.* not analyzed, *Akm* akimotoite, *Brd* bridgmanite, *CaSiPrv* CaSi-perovskite, *fPer* ferropericlase, *Jbn* jeffbenite, *Maj* majorite, *Msk* maskelynite, *'Ol'* phase with a composition of olivine, *Rng* ringwoodite; *Sid* siderite

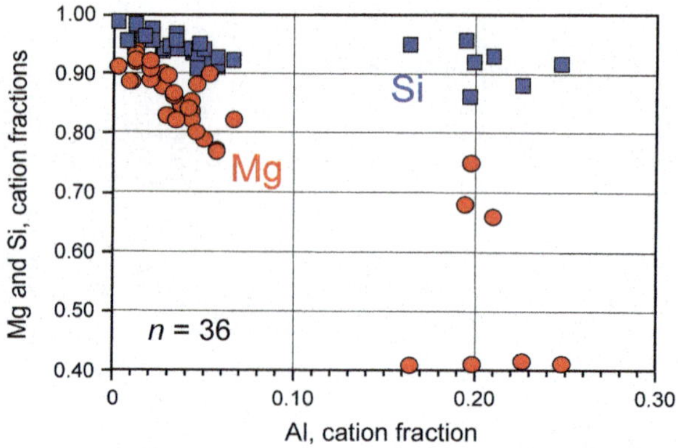

Fig. 4.2 Cation fractions of Mg (*red circles*), Si (*blue squares*) and Al in natural bridgmanite from diamond inclusions. The formula is on the basis of three oxygens

both A and B sites (Fig. 4.2). A permanent small deficit in the A site points to the presence of Fe, independent of its valence state, only in this site. An experimental single-crystal XRD study confirmed the presence of both Fe^{2+} and Fe^{3+} only in the A site (Vanpeteghem et al. 2006). Representative chemical compositions of natural bridgmanite are shown in Table 4.1.

The composition of the bridgmanite from the Tenham L6 meteorite is also included in the table. It significantly differs from the terrestrial bridgmanite (low Al and high *fe* first of all). It also differs from the 'mafic' bridgmanite (see Sect. 5.2.1). The specific composition of the meteoritic bridgmanite is most likely to have been caused by its occurrence in a shock-melt vein in the meteorite and its impact origin under non-equilibrium conditions.

Concentration of iron and magnesium index in bridgmanite. Iron, mainly in a divalent ferrous form of Fe^{2+}, enters into the pseudo-dodecahedral site A, as a substitute for Mg^{2+} because of their similarity in ionic radii (Shannon and Prewitt 1969; Fei et al. 1996). Although the modeled solubility of iron in bridgmanite reaches $fe = Fe/(Mg + Fe)_{at.} = 0.68$ (Dorfman et al. 2013), its concentration in natural bridgmanite is 2.85–10.40 wt% in total as FeO; and $fe = 0.04$–0.16. It was demonstrated experimentally that even great concentrations of Fe (Fe^{2+} or Fe^{3+}) do not occur in the octahedral B site; Fe replaces Mg only in the A site: ${Fe_A}^{2+} \leftrightarrow {Mg_A}^{2+}$ (Vanpeteghem et al. 2006).

The magnesium index in bridgmanite, $mg = Mg/(Mg + Fe)_{at.}$, is relatively uniform, varying within a range of 0.84–0.96 (average: 0.917 ± 0.066 (2σ)) (Fig. 4.3). The magnesium index in bridgmanite appears to show regional differences: for example, bridgmanites in Guinea and Canada are more magnesium rich ($mg = 0.92$–0.96) than bridgmanites in Brazil, where samples show lower Mg contents, with $mg = 0.84$–0.94. This possibly reflects different depths and/or sources of origin of bridgmanite in different areas. On the other hand, experimental mineral physics data suggest a lower magnesium index for bridgmanite, at 0.80–0.88 (Fei 1996; Lee et al. 2004; Irifune et al. 2010; Lin et al. 2013). In the

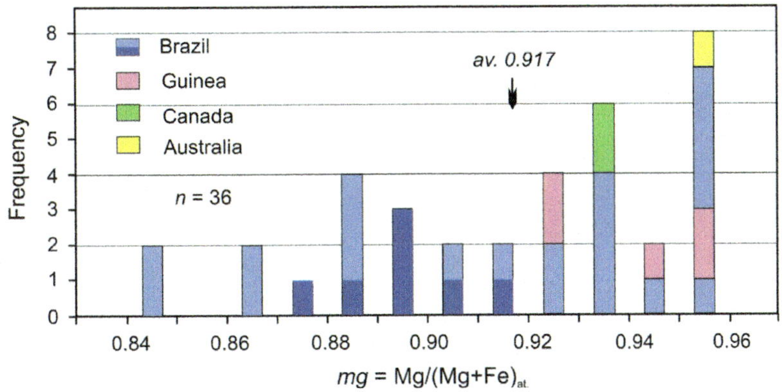

Fig. 4.3 Distribution of the magnesium index *mg* in natural bridgmanite from different areas. Data from Hutchison (1997), Stachel et al. (2000), Kaminsky et al. (2001), Davies et al. (2004), Hayman et al. (2005), Tappert et al. (2009b), Zedgenizov et al. (2014a, 2015), Burnham et al. (2016). *Deep blue color* indicates high-Al samples from Brazil

lowermost mantle, more Mg-rich bridgmanite was expected owing to the substantial decrease of the suggested Fe–Mg partition coefficient between bridgmanite and melt $K_D^{Brd-melt} = (Fe/Mg)^{Brd}/(Fe/Mg)^{melt}{}_{at.}$ from 0.25–36 GPa to 0.06–0.08 above 76 GPa, corresponding to a depth of ~1800 km (Irifune et al. 2010; Nomura et al. 2011). Similar values of $K_D^{Brd-melt}$ were experimentally established more recently: 0.329 at 34 GPa, 0.113 at 68 GPa, and 0.085 at 88 GPa (Tateno et al. 2014). Much higher values of $K_D^{Brd-melt}$ were found in experiments with Al-rich, pyrolite-like glass at pressures from 40 to 120 GPa (from 0.6 to 0.45) (Andrault et al. 2012), but they also show the same tendency of decrease of Fe–Mg partition, leading to crystallization of bridgmanite with higher magnesium index than the sourcing liquid.

In one of the diamonds from pipe Pandrea-2 in the Juina area, Brazil, two grains of a phase with Mg–Ca–Si composition were identified (Kaminsky et al. 2009a). Conventionally, they may be attributed as a perovskite phase only, by analogy with other perovskite phases met as inclusions in diamonds from the Juina area (Harte et al. 1999; Kaminsky et al. 2001; Hayman et al. 2005). The *mg* numbers of these grains are higher than usual in bridgmanite (0.966 and 0.969). The only noticeable admixture is Al; in one of the grains, Al_2O_3 = 2.04 wt% (Table 4.2). Since the grains are opaque, and the two analyses show a spread, especially in Al and Ca, it is possible that the grains were originally a mixture of bridgmanite and CaSi-perovskite, and are now a fine-grained mixture due to decompression breakdown, as described by Harte and Cayzer (2007) for majoritic garnets from the Sao Luiz diamonds.

Concentration of ferric iron Fe^{3+} in bridgmanite. According to mineral physics and geochemical results, which suggest relatively low oxygen fugacity in the lower mantle, one should expect a very low (if any) concentration of ferric iron Fe^{3+} in bridgmanite. However, it has been shown that the Al admixture in

Table 4.2 Chemical compositions of Mg–Ca–Si phase from a diamond in pipe Pandrea-2, Juina area, Brazil (wt%) (Kaminsky et al. 2009a)

Sample No.	014-2a	014-2a-1
SiO_2	55.70	57.65
TiO_2	0.04	0.03
Al_2O_3	2.04	0.10
Cr_2O_3	0.01	0.02
FeO	1.01	0.90
NiO	0.09	0.07
MnO	0.06	0.08
MgO	22.50	21.89
CaO	12.64	17.56
Na_2O	0.15	0.04
K_2O	0.17	0.02
Total	94.40	98.37
mg	0.966	0.969
Reference	Kaminsky et al. (2009a)	

bridgmanite enhances the concentration of the ferric ion in the mineral even under reducing conditions. The valence states of iron in mantle minerals are typically analyzed using Mössbauer spectroscopy and/or electron energy-loss spectroscopy (EELS). It was experimentally demonstrated that the $Fe^{3+}/\Sigma Fe$ ratio may exceed 50% with the increased substitution of Al content in bridgmanite synthesized at 26 GPa (Lauterbach et al. 2000). Incorporation of ferric iron via the oxygen vacancy mechanism is not expected in bridgmanite and post-perovskite at lower-mantle conditions (Brodholt 2000). The high Fe^{3+} content in bridgmanite is enhanced by an increased Al^{3+} substitution via the charge-coupled substitution of Al for Mg and Si mechanism (McCammon 1997, 2005; Andrault et al. 2007). This means that the $Fe^{3+}/\Sigma Fe$ ratio in bridgmanite may not be caused by the oxygen fugacity conditions. However, this conclusion was challenged by the results of later experiments with Al-free bridgmanite, showing a strong dependence of Fe^{3+} incorporation in bridgmanite with f_{O_2} (Nakajima et al. 2012). Recent experiments at moderate pressure (25 GPa) and reducing oxygen fugacities (between IW-1 and IW) also demonstrated a weak correlation of Fe^{3+} content with Al content (Boujibar et al. 2016). A self-disproportionation mechanism has also been proposed to explain the co-existence of metallic iron with bridgmanite, in which Fe^{2+} disproportionates into Fe^{3+} and Fe^0, drastically increasing the amount of Fe^{3+} in the bridgmanite lattice (Frost et al. 2004).

The issue of site occupancy and valence states of iron in bridgmanite is still much debated experimentally (Shukla et al. 2015). For example, Lin et al. (2012) propose that 50% of Fe^{3+} is substituted through charge-coupled substitution in an Mg and Si site, while the remaining 50% Fe^{3+} goes in the Si site through oxygen vacancy. Hummer and Fei (2012) suggest that the ratio of Fe^{3+} in Mg and Si sites depends on total iron concentration. Hsu et al. (2012) and Sinmyo et al. (2014) concluded that all Fe^{3+} would be in an Mg site. At the same time, evidence from geological data shows that all Fe in bridgmanite exists in site A replacing Mg (Kaminsky and Lin 2017).

This was confirmed in recent high-pressure XRD experiments with single-crystal samples (Mao et al. 2017), Fe^{3+} comprises only a small share ($Fe^{3+}/\Sigma Fe$ = 9–12 at.%; Kaminsky et al. 2015b). The thermodynamic model, based on ab initio calculations, also demonstrates a very low amount of $Fe^{3+}/\Sigma Fe$ = 1–7% in Al-free bridgmanite under low-mantle conditions (Xu et al. 2015).

Concentration of Al in bridgmanite is very important. In contrast to upper-mantle enstatite, which usually shows Al_2O_3 contents of 0.3–0.9 wt% but does not exceed 1 wt%, natural bridgmanite in diamond inclusions commonly contains 1–3 wt% Al_2O_3, and in some cases this value may be up to 12.58 wt%, as observed in some samples from São Luiz in Brazil. Using the concentration of Al in bridgmanite, the studied natural samples can be clearly subdivided into two groups: low-Al and high-Al species (Fig. 4.4). Low-Al bridgmanite has 0.25–3.10 wt% Al_2O_3 (average: 1.63 wt%), while the high-Al group has 8.34–12.58 wt% Al_2O_3 (average: 10.46 wt%).

Both low-Al and particularly high-Al groups of bridgmanite show the correlation with *fe* and Fe speciation with Al substitution (Fig. 4.5a). High-pressure–temperature experiments show that significant amounts of the Fe in bridgmanite may be incorporated as Fe^{3+} depending on Al-content and, in contrast to Al^{3+}, Fe^{3+} apparently has an affinity for the 8/12-fold A site over the six-fold B site (McCammon 1997; Frost et al. 2004). The existing graph demonstrates the correlation of Al in natural bridgmanite with the total concentration of Fe.

The presence of Al enhances the ability of a perovskite structure to accept various minor and trace elements (Andrault et al. 2007; Righter et al. 2014). This explains much higher Cr concentrations in high-Al bridgmanite compared with the low-Al variety (1.16–3.14 wt% Cr_2O_3 vs. 0.04–0.45 wt% Cr_2O_3; Fig. 4.5b); some high-Al samples of bridgmanite contain high concentrations of Ca and Na as well (4.38–5.07 wt% CaO; 0.82–6.21 wt% Na_2O; Fig. 4.5c, d).

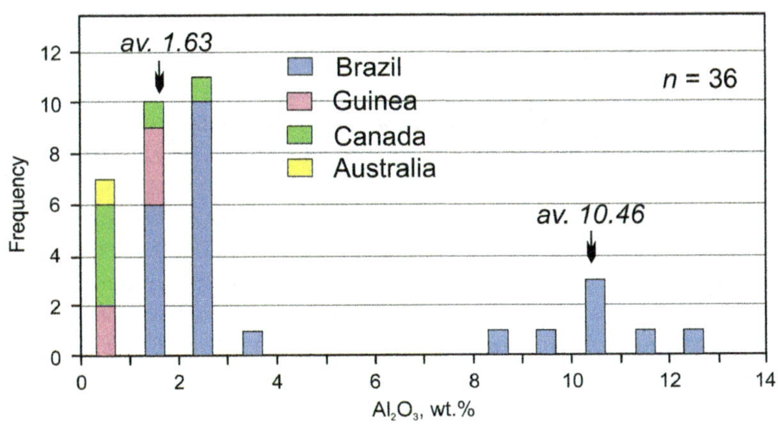

Fig. 4.4 Distribution of Al_2O_3 (wt%) in natural bridgmanite. Data from Hutchison (1997), Stachel et al. (2000), Kaminsky et al. (2001), Davies et al. (2004), Hayman et al. (2005), Tappert et al. (2009b), Zedgenizov et al. (2014a, 2015), Burnham et al. (2016)

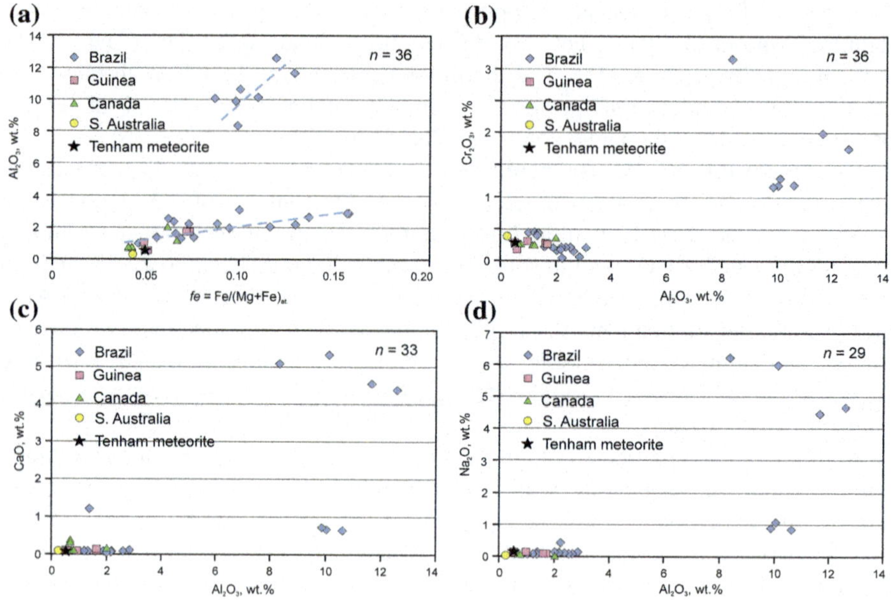

Fig. 4.5 Chemical characteristics of natural bridgmanite. **a** Al_2O_3 versus iron index *fe*; **b** Cr_2O_3 versus Al_2O_3; **c** CaO versus Al_2O_3; **d** Na_2O versus Al_2O_3. Data from Hutchison (1997), Stachel et al. (2000), Kaminsky et al. (2001), Davies et al. (2004), Hayman et al. (2005), Tappert et al. (2009b), Tschauner et al. (2014), Zedgenizov et al. (2014a, 2015), Burnham et al. (2016)

According to experimental garnet-to-perovskite transitions at c. 24–26.5 GPa pressure, in the $MgSiO_3$–Al_2O_3 system, mineral physics results indicate an increase of Al_2O_3 in bridgmanite with increasing pressure (Irifune et al. 1996; Wood 2000). For Al-free bridgmanite, the transition pressure occurs at ~23 GPa, whereas dissolution of majoritic garnet into bridgmanite can extend the transition pressure to approximately 27–28 GPa (~800 km in depth). Recently, an experimental study of the system $MgSiO_3$–Al_2O_3 under pressures of up to 52 GPa and 2000 K, with the use of sintered diamond anvils combined with in situ synchrotron X-ray diffraction observations in a multi-anvil apparatus were performed (Liu et al. 2016). This study demonstrated that the Al_2O_3 content in bridgmanite increases from 12 mol% at 27 GPa to 29 mol% at 52 GPa. These results suggest the Al_2O_3 content in bridgmanite is a good pressure indicator at pressures above 30 GPa (Liu et al. 2016). It is possible, in such a case, to find transitional compositions of natural bridgmanite.

The incorporation of Al^{3+} in bridgmanite may occur by two competing mechanisms (Richmond and Brodholt 1998; Vanpeteghem et al. 2006). The first is a Tschermak-like charge-coupled substitution (CCM) $2Al^{3+} \rightarrow Si^{4+} + Mg^{2+}$, where Al enters both pseudo-dodecahedral (A) and octahedral (B) cationic sites. The effective ionic radius of Al^{3+} lies between those of Si^{4+} and Mg^{2+} (Shannon and Prewitt 1969), and Al^{3+} substitution reduces charges of the octahedra (Al^{3+} substitutes for Si^{4+}) and increases charges in the dodecahedra (Al^{3+} substitutes for

Mg^{2+}), making local charge neutrality easier to attain. As a result, this mechanism does not require the creation of oxygen vacancies for the charge balance. In the second mechanism, Al replaces Si in the octahedral site only while oxygen vacancies are created to balance the charge: $2Si^{3+} \rightarrow 2Al^{4+} + V_O$ (OVM). In this mechanism, it is necessary to remove an oxygen ion in order to balance the charge.

Experiments performed under 26 GPa and 1973 K with the use of X-ray absorption spectroscopy (XAFS) demonstrated that the CCM substitution is preferable (Andrault et al. 1998a). Later, Stebbins et al. (2001) reported the results obtained by the use of nuclear magnetic resonance (NMR) spectra (according to which only CCM substitution was detected for the substitution mechanism of Al^{3+} into bridgmanite) and theoretical calculations also demonstrated that the CCM substitution mechanism is energetically favoured (e.g., Navrotsky et al. 2003; Akber-Knutson and Bukowinski 2004).

For high-Al bridgmanite (starting from 7.7 mol% Al_2O_3), the CCM substitution was suggested to evolve rapidly as the Al content in bridgmanite increases (Daniel et al. 2001). It was concluded that the oxygen vacancy mechanism may dominate over the coupled substitution mechanism in Al-rich bridgmanite (Daniel et al. 2004). This would be consistent with ab initio calculations by Brodholt (2000), which showed that within the experimentally investigated pressure range, oxygen vacancies are energetically favoured for a bridgmanite structure with a high Al content. Yamamoto et al. (2003) carried out first-principles density functional theory (DFT) calculations with large supercells consisting of 80 atoms and corresponding to 6.25 mol% Al_2O_3 incorporation and found that the charge-coupled substitution in bridgmanite is more favorable than the oxygen vacancies mechanism at all pressures up to 100 GPa.

The data obtained for natural assemblages demonstrate that it is possible that both mechanisms are realised: oxygen vacancy mechanism with Al replacement of Si only in site B is characteristic for low-Al bridgmanite; while, according to the mineral stoichiometry, Al-rich samples show Al^{3+} replacement in both sites, A (for Mg^{2+}) and B (for Si^{4+}).

Admixtures of minor elements in bridgmanite. In addition to Fe and Al, there are admixtures of minor elements in bridgmanite such as of Cr and Ni; Cr_2O_3 reaches 3.14 wt% (Table 4.3; Fig. 4.5b). The admixture of Na correlates with Al: high-Al bridgmanite contains 0.82–6.21 wt% Na_2O (Fig. 4.5d). High concentrations of Na in bridgmanite (increasing its unit-cell volume) was observed experimentally in the $MgSiO_3$–Na_2CO_3–Al_2O_3 system at 24 GPa and 1700 °C (Bindi et al. 2016). The mechanism of such incorporation remains unknown because of the lack of sufficient crystallographic data. However, these experimental results demonstrate that bridgmanite can incorporate significant concentrations of Na and possibly other alkalis in the lower mantle.

The total trace element concentrations in bridgmanite are low, usually one to two orders of magnitude below chondritic. The REE pattern is chondrite-like, but lower than in chondrite concentrations, only showing the enrichment in heavy REE and mainly in Lu (Fig. 4.6). Considering their generally low mineral/melt partition coefficient under high-pressure conditions (from c. 10^{-3} for La to c. 1 for Lu;

Table 4.3 Minor elements in bridgmanite (wt%) (from Kaminsky 2012)

Oxides	Brazil			Guinea	Canada
	Rio São Luiz (low-Al/high-Al)	Rio Vermelho	Rio Soriso	Kankan	DO27 pipe
Al_2O_3	1.23–2.86/8.34–12.58	2.16	1.60–3.10	0.55–1.68	0.69–2.03
Na_2O	0.03–0.11/0.82–6.21	0.10	0.07–0.08	0.03–0.09	0.04–0.07
Cr_2O_3	0.04–0.45/1.16–3.14	0.17	0.20–0.22	0.17–0.29	0.23–0.36
NiO	0.01–0.03/0–0.02	<0.06	n.d.	0.01–0.02	n.d.
mg	0.843–0.945/0.871–0.913	0.870	0.900–0.938	0.927–0.951	0.934–0.960
References	Wilding (1990), Harte et al. (1999)	Kaminsky et al. (2001)	Hayman et al. (2005)	Stachel et al. (2000)	Davies et al. (2004)

Note: *n.d.* not detected

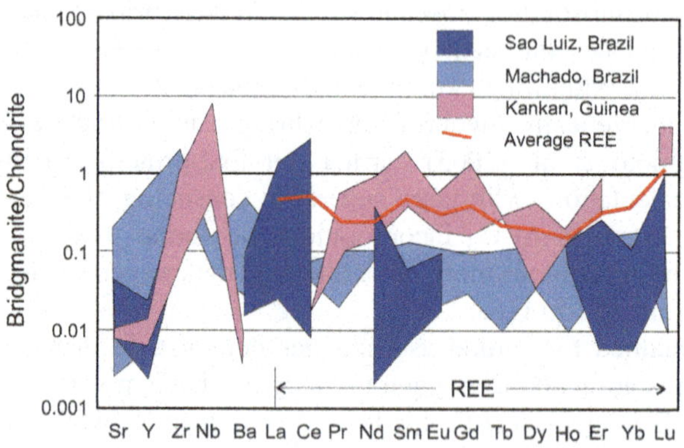

Fig. 4.6 Chondrite-normalized trace element patterns in bridgmanite. Data from Harte et al. (1999), Stachel et al. (2000) and Burnham et al. (2016). Chondrite composition here and in the other spider-diagrams (Figs. 3.10, 3.22, 3.32, 3.34, 3.37 and 3.39) are taken from McDonough and Sun (1995). Modified after Kaminsky (2012), with additions

Thomson et al. 2016), such concentrations indicate the primary origin of the analyzed bridgmanite grains, with the subsequent trapping of them from pristine ambient mantle by host diamonds. One should expect the increase of trivalent and tetravalent cations of trace elements entering the A-cite in Al-rich bridgmanite because of their high crystal-liquid partition coefficients (Wood and Corgne 2009). However, there is currently not enough data on this subject.

Water in bridgmanite. Bridgmanite can also accommodate a certain amount of water (hydroxyl), which may play a role in affecting the physical and chemical properties of the mineral in the lower mantle. From analysis of FTIR spectra at ambient conditions taken from bridgmanite that was synthesized at 25–26 GPa in a multi-anvil apparatus, the solubility of water in pure bridgmanite is $\sim$100 ppm

H_2O, while in Al-rich bridgmanite it increases up to 1000–1500 ppm (Litasov et al. 2003). In LHDAC experiments, bridgmanite was suggested to contain about 0.2 wt % H_2O (Murakami et al. 2002). With the use of first-principles methods, Hernández et al. (2013) suggested that pure bridgmanite can contain as much water as 1000 ppm.

It has been proposed that water can enter into the bridgmanite structure with oxygen vacancies originating from the substitution of Si by Al, potentially making this mineral the major storage for water in deep mantle (Daniel et al. 2001; Merli et al. 2016). $V_{si}(H_4)$ cation-vacancy complexes, in which four protons are placed at the site of a Si vacancy, were found most stable in the perovskite structure. Other polyvalent substitutions (e.g., $Na^{1+} \rightarrow Mg^{2+}$) can also favour the origin of $V_{Mg}(H_2)$ oxygen vacancies and subsequently water incorporation in bridgmanite (Hernández et al. 2013). Townsend et al. (2016) suggested a possibility of hydrogen incorporation for Al-free bridgmanite without oxygen-vacancy formation with the so-called “Mg–2H mechanism”: $Mg^{2+} \leftrightarrow 2H^+$.

During the course of bridgmanite → post-perovskite structural transformation, water favours bridgmanite over post-perovskite by a factor of about 5:1 for Al-free system; in the Al-bearing system, in reverse, water favours post-perovskite over bridgmanite by a factor of about 3:1 at ambient mantle conditions (see Sect. 9.2.2).

4.2.3 Crystal Structure of Bridgmanite

Structure. Initially, the crystal structure of bridgmanite was suggested to be cubic with the lattice parameter $a = 3.426 \pm 0.006$ Å (Reid and Ringwood 1975). The *d*-spacing of the first synthesized ‘MgSi-perovskite’, however, was indexed as an orthorhombic perovskite structure (space group *Pbnm*) with $a = 4.818 \pm 0.005$ Å, $b = 4.869 \pm 0.005$ Å, and $c = 6.906 \pm 0.007$ Å with $Z = 4$ (Liu 1974) (Fig. 4.7). The bridgmanite sample from the Tenham meteorite has the orthorhombic *Pnma* perovskite structure with unit cell parameters $a = 5.02 \pm 0.03$ Å, $b = 6.90 \pm 0.03$ Å, $c = 4.81 \pm 0.02$ Å, and a unit cell volume of 167 ± 2 Å^3 (Tschauner et al. 2014). The space group *Pnma* is equivalent to *Pbnm*, with the following conversion: $(a, b, c)_{Pnma} \rightarrow (b, c, a)_{Pbnm}$ (Wicks and Duffy 2016). Deviation from the ideal cubic arrangement is achieved through both rotation and tilting of the SiO_6 octahedra and the offset in position of the central Mg atom (Horiuchi et al. 1987). These structural distortions increase with increasing pressure (Fiquet et al. 2000). Addition of Al in the bridgmanite structure also increases the orthorhombic distortion (Kojitani et al. 2007; Walter et al. 2004a), while substitution of Fe^{2+} for Mg^{2+} in the A site of perovskite expands the structure and decreases the degree of distortion (Kudoh et al. 1990).

Phase transitions in bridgmanite. The possibility of phase transitions in bridgmanite has been debated for decades. Early high-pressure XRD measurements on temperature-quenched samples by Knittle and Jeanloz (1987) demonstrated the stability of bridgmanite up to 127 GPa. However, later studies reported its

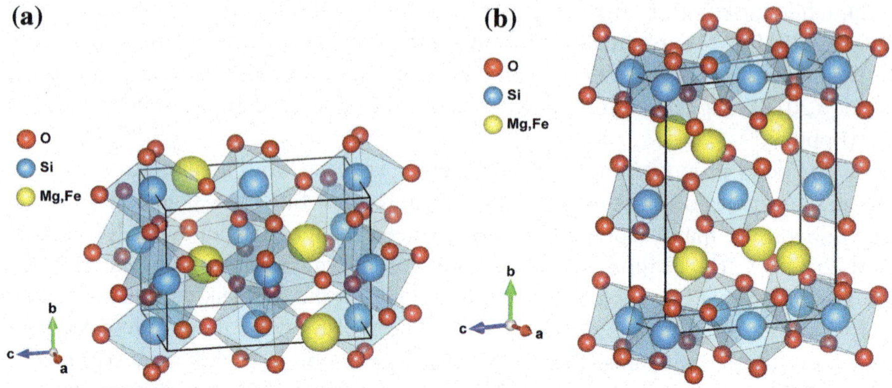

Fig. 4.7 Orthorhombic crystal structures of bridgmanite in the *Pbnm* space group (*left*) and post-perovskite in the *Cmcm* space group (*right*). Tilting of the SiO_6 octahedra results in the shortening of crystallographic axis and produces a displacement of Mg,Fe cations reducing the co-ordination from 12 (characteristic for the ideal cubic perovskite structure) to 8. *Thick lines* outline the unit cell. Lattice parameters for bridgmanite taken from Sugahara et al. (2006); for post-perovskite from Zhang et al. (2014)

dissociation into MgO and SiO_2 during the course of laser heating (temperature was not measured) at 60–70 GPa (Meade et al. 1995), at 58–100 GPa and 1900–3000 K (Saxena et al. 1996), or at 82 GPa and 1140–1250 K (Saxena et al. 1998). These conclusions appeared to be a result of technical issues, possibly because of extreme temperature gradients during the experiments and reactions between the pressure medium and samples. The several experimental studies that followed used inert pressure mediums and in situ XRD (Fiquet et al. 2000; Andrault et al. 2001; Shim et al. 2001; Ono et al. 2004). Raman measurements (Serghiou et al. 1998) have indicated the stability of orthorhombic *Pbnm* $MgSiO_3$ to pressures of 100–106 GPa. Shim et al. (2001) reported a subtle phase change of bridgmanite from *Pbnm* to one of three possible space groups, $P2_1/m$, *Pmmn*, or $P4_2/nmc$ above 83 GPa and 1700 K, but this was not confirmed by later studies. Recently, the dissociation of Fe-bearing perovskite was reported by Zhang et al. (2014). They found that in LHDAC experiments within the pressure interval of 95–101 GPa and temperatures of 2200–2400 K bridgmanite with the $mg = 85$ disproportionates into almost iron-free bridgmanite and an iron-rich phase with a hexagonal structure, named H phase, using XRD analysis. It should be cautiously noted that such a dissociation has not yet been confirmed by other groups who have tried to reproduce the study, although an analogous dissociation of ferropericlase into almost pure wüstite and an Mg-rich phase at 86 GPa and 100 K was obtained experimentally (Dubrovinsky et al. 2000, 2001a) and was observed in lower-mantle inclusion in diamond from the Soriso area, Brazil (Kaminsky et al. 2015a).

Since most scientists had believed that perovskite is stable at lower-mantle pressure–temperature (*P–T*) conditions, by the early 2000s most groups did not pursue higher-pressure experiments on the stability of perovskite toward the core–

mantle boundary conditions (the D″ layer up to 136 GPa and ~3500 K). Interestingly, a first-order solid phase transition was proposed by a group of seismologists (Sidorin et al. 1999) to exist approximately 200 km above the core–mantle boundary (the D″ region) with a Clapeyron slope of approximately 6 MPa/K. The occurrence of such a structural transition can explain most of the observed seismic variations in travel times and the spatial intermittance of the seismic triplication (Sidorin et al. 1999). It was not until 2004 that the transition of bridgmanite to silicate post-perovskite was discovered experimentally and theoretically to occur at pressure–temperature conditions of the D″ region (Murakami et al. 2004a) (see Sect. 9.2.2). In 2004, after experiments reaching higher *P*–*T* parameters, corresponding to the D″ layer conditions, the post-perovskite crystal structure was observed at 113–125 GPa and 2200–2500 K with the use of LHDAC and in situ XRD (Murakami et al. 2004a, 2005; Oganov and Ono 2004; Tsuchiya et al. 2004b). According to first-principle computations, it has another orthorhombic, the $CaIrO_3$ structure with the *Cmcm* space group (Tsuchiya et al. 2004b).

In post-perovskite, Si^{4+} and Mg^{2+} have the same coordination environment as in bridgmanite, but there are profound differences in their structural arrangement. In contrast to the corner-sharing network in bridgmanite, the SiO_6 octahedra in post-perovskite share edges along the *a* axis and corners along c. The structure is thus sheet-like with layers of SiO_6 octahedra alternating with MgO_8 layers along the *b* axis (Fig. 4.7b). The Mg sites in post-perovskite are smaller and less distorted than those in bridgmanite, resulting in a volume reduction of ~1–1.5% across the transition (Wicks and Duffy 2016).

The phase boundary between *Pbnm* and *Cmcm* phases has a Clapeyron slope of ~7.5 ± 0.3 MPa/K at ~2750 K and ~125 GPa (Tsuchiya et al. 2004b) or 13–16 MPa/K at 3750 K (Li et al. 2015 and references therein) (see Fig. 9.2).

The concentration of iron suppresses the stability of bridgmanite, which was supported in early experiments (e.g., Fei et al. 1996). Dorfman et al. (2013), using double-sided LHDAC with in situ XRD, demonstrated that iron-rich bridgmanite (with $fe > 0.35$–0.40) decomposes into low-Fe Brd + fPer + SiO_2 at pressures below 70 GPa. In natural samples analyzed to date, maximal concentrations of FeO in bridgmanite reach 10.4 wt% ($fe = 0.157$); they do not experience such decomposition.

4.2.4 Physical Properties

The role of the chemical composition of bridgmanite. A series of experiments and theoretical calculations have been conducted on the role of the Fe and/or Al incorporation in the bridgmanite on physical properties (Zhang and Weidner 1999; Brodholt 2000; Kubo et al. 2000; Daniel et al. 2001, 2004; Andrault et al. 2001, 2007; Yamamoto et al. 2003; Jackson et al. 2004; Walter et al. 2004a; Yagi et al. 2004; Akber-Knutson and Bukowinski 2004; Gu et al. 2016), but these results have been extensively debated and remain reconciled. For example, the substitution of Fe in bridgmanite inherently increases its density significantly because the iron ion is much heavier than the magnesium ion; however, the lattice parameters are not expected to change substantially because of their similar ionic radii. Substitution of Fe^{3+} at the A site would be preferred at a high iron content coupled with Mg vacancies and would lead to a significant distortion of the dodecahedral site. Thus, substitution of Fe^{3+} would diminish the density increase compared with the same amount of Fe^{2+} substitution. Analogically, Al incorporation into the bridgmanite structure increases the cell volume (Daniel et al. 2004; Yagi et al. 2004)

Recently, Fukui et al. (2016) studied the role of cation substitution on bridgmanite elasticity at ambient conditions and concluded that substitution of Mg with Fe^{2+} and Al influences the anti-correlation between bulk sound velocity and shear velocity. At the same time, it was shown that the effect of Fe^{3+} concentration on the bulk modulus of bridgmanite is not significant for lower-mantle pressures (Gu et al. 2016).

Volume. Independently of the composition of bridgmanite, its volume significantly (by $\sim 13\%$) decreases with increasing pressure from 30 to 90 GPa, at room temperature (Gu et al. 2016).

Elastic properties. In general, the addition of Fe in bridgmanite can cause an increase in the density, bulk modulus, and bulk sound velocity independently on the Al concentration at lower-mantle pressures. The analysis of the EoS parameters of bridgmanite at lower-mantle pressures, obtained in low-Fe (*fe* = 0.06–0.11) single-crystal samples in a hydrostatic He medium, indicates that the substitution of iron in bridgmanite will cause an anti-correlation between the enhanced density and reduced bulk sound velocity (V_Φ), whereas the combined Al and Fe substitution has a reduced effect on density and a negligible effect on V_Φ (Mao et al. 2017). In experiments with high-Fe bridgmanite the increase in concentrations of iron (*fe* = 0.25) increases the density and bulk sound velocity (V_Φ) of bridgmanite by 4–6 % and 6–7 % respectively at lower-mantle pressures (Mao et al. 2011). The presence of Fe^{3+} in the B-site of bridgmanite (if happens) can further enhance this increase (Mao et al. 2015). However, this data may not be related to natural bridgmanite because in natural samples the concentrations of Fe are significantly smaller ($fe = Fe/(Mg + Fe)_{at} = 0.04–0.16$).

The role of Al seems more complex; the experimental results are controversial. Based on experiments in cubic-anvil apparatus, Zhang and Weidner (1999) reported that bridgmanite with only 5 mol% Al_2O_3 might have a bulk modulus K_T 10% lower than pure bridgmanite, showing that the addition of aluminium can drastically decrease the incompressibility of bridgmanite. Such a large effect on the K_T of bridgmanite, resulting from a relatively low concentration of Al, has also been reported in DAC experiments with higher Al contents (up to 7.7–10.3 mol% Al_2O_3) (Kubo et al. 2000; Daniel et al. 2001). However, later XRD experimental studies performed at lower-mantle pressure conditions in a DAC showed that the K_T of aluminous bridgmanite is, in reverse, higher than that of pure bridgmanite, and hence, dissolution of Al makes it even more incompressible than pure $MgSiO_3$ (Andrault et al. 2001). In another publication, Daniel et al. (2004) demonstrated, on the basis of monochromatic synchrotron XRD from bridgmanite with 5 mol% Al_2O_3, that its bulk modulus K_T is decreased under pressure conditions of up to 40 GPa only by 1.8% at most, i.e., it is affected little if at all by the presence of Al^{3+}. Daniel et al. (2004) suggested that the relative proportions of the minerals characteristic of the lower mantle, as inferred from seismological models, are unlikely to be significantly altered by the introduction of Al in the system.

However, the experiments performed in DAC by Yagi et al. (2004) again confirmed the results made by Zhang and Weidner (1999), Kubo et al. (2000) and Daniel et al. (2001), and were opposite to the results obtained by Andrault et al. (2001). The results of experiments performed by Yagi et al. (2004) in DAC at pressures of up to 33 GPa have shown that the dissolution of Al_2O_3 into $MgSiO_3$ decreases the bulk modulus considerably. This effect is highly nonlinear with Al_2O_3 content; the bulk modulus decreases by as much as 7% with only 1% Al_2O_3. Walter et al. (2004a) found in their experiments a decrease in bulk modulus of bridgmanite with the addition of Al in its structure; however, they found this decrease to be "mild" and established its rate at -67 ± 35 GPa/X_{Al}. The general results of major experiments on the elastic properties of bridgmanite under high-pressure conditions are summarized in Table 4.4.

The discrepancies between the different series of experiments may be caused by sample chemistries and/or technical problems, including experimental *P*–*T* conditions, pressure medium and calibrant, starting sample conditions, *P*–*T* effects on solubility of cations, and types of stable defects at high *P*–*T* conditions (Yagi et al. 2004; Walter et al. 2004a). For example, the amorphization of bridgmanite can occur in response to grinding the sample by mortar and pestle under atmospheric pressure, or by heating to above 400 K at atmospheric pressure (Yagi et al. 2004). Use of such a sample would cause problematic results at high pressures. The experiments performed by Walter et al. (2004a) demonstrated that Al^{3+} substitution increases orthorhombic distortion and unit cell volume, which causes a decrease in K_{0T} at a rate of $\sim 67 \pm 35$ GPa/X_{Al} and, as a result, a mild increase in compressibility. This relatively mild increase in compressibility is in good agreement with theoretical calculations if Al^{3+} substitutes into the bridgmanite structure by means of a charge-coupled substitution mechanism ($2Al^{3+} \rightarrow Si^{4+} + Mg^{2+}$). By contrast, the much lower incompressibility for aluminous bridgmanite measured in several other studies is consistent with Al^{3+} substitution by oxygen-vacancy formation ($2Al^{3+} \rightarrow 2Si^{4+} + O^{2-}{}_V$). The formation of vacancies in the structure of bridgmanite caused by the latter mechanism makes the concentration of vacancies increase in proportion to the Al concentration. At a low concentration, the vacancies are randomly distributed; above a certain concentration threshold, they are ordered, as shown by Becerro et al. (1999), for Fe in solid solution between $CaTiO_3$ (perovskite) and $Ca_2Fe_2O_5$ (brownmillerite). The formation of vacancies can decrease the incompressibility considerably. Brownmillerite, for example, is approximately 25% more compressible than expected from the systematics of unit cell volume versus bulk modulus relation in bridgmanite (Ross et al. 2002). Brodholt (2000) made theoretical calculations for $(Mg_3Al)(AlSi_3)O_{12}$ perovskites and $Mg_2Al_2O_5$

Table 4.4 Results of major experiments on elastic properties of bridgmanite under high pressure conditions

References	Method	Conditions	Al_2O_3 mol%	Results
Zhang and Weidner (1999)	Multi-anvil No medium	1–10 GPa	5.0	Brd with 5 mol% Al_2O_3 has a bulk modulus K_T 10% lower than pure Brd
Kubo et al. (2000)	DAC Medium: Alcohol mixture	1–9 GPa	10.3	Same
Daniel et al. (2001)	DAC Medium: Ar, He Synchrotron XRD	1–32 GPa	7.7	Same Al^{3+} is likely to be incorporated in place of Si^{4+} in the octahedral site of Brd, hence requiring the creation of oxygen vacancies for charge balance. As a consequence, Al-Brd may have a high affinity for water
Andrault et al. (2001)	LHDAC Medium: Ar Monochromatic synchrotron XRD	29–57; 29–69 GPa	5.0; 11.0	Opposite Al increases bulk modulus K_T of Brd in comparison to that of pure Brd; i.e., dissolution of Al makes Brd even more incompressible than pure Brd
Daniel et al. (2004)	DAC Medium: Ne Monochromatic synchrotron XRD	Up to 40 GPa Ambient temperature	5.0	K_T is affected little if at all by the presence of Al^{3+}. Hence, seismological models should not be significantly altered by the introduction of Al in the system
Jackson et al. (2004)	Multi-anvil Brillouin spectroscopy	~25 GPa, 1873 K	5.1 ± 0.2	The presence of 5.1 wt% does not affect bulk modulus and decreases the shear elastic modulus (μ) by 5.6%
Yagi et al. (2004)	LHDAC Medium: He Synchrotron XRD	2–33 GPa	1–10	The dissolution of Al_2O_3 into Brd decreases the bulk modulus considerably. The effect is highly nonlinear with Al_2O_3 content; the bulk modulus decreases by 7% with only 1% Al_2O_3
	Multi-anvil cell No medium	17–23 GPa.	0–5	The dissolution of Al_2O_3 does not decrease the bulk modulus of Brd
Walter et al. (2004a)	LHDAC Medium: KBr In situ angle-dispersive XRD	30–100 GPa 2000–2600 K	0–25	Al^{3+} substitution causes a mild decrease in bulk modulus K_0 at a rate of −67 ± 35 Ga/X_{Al}
Andrault et al. (2007)	LHDAC Medium: Ar	Up to 60 GPa. 2300 K	0–5	Al–Brd synthesized in the SiO_2-rich system is more compressible than Al–Brd synthesized in the MgO-rich system

(continued)

Table 4.4 (continued)

References	Method	Conditions	Al_2O_3 mol%	Results
Mao et al. (2011)	LHDAC Medium: Ne Synchrotron XRD	Up to 135 GPa. 1300–1600 K	0; 0–25 Fe	The isothermal bulk modulus of Brd with 25% $FeSiO_3$ is ~20% larger at high pressures than that of the pure Brd
Mao et al. (2015)	LHDAC Medium: Ne In situ synchrotron XRD	30–125 GPa	10 Fe	The addition of Fe causes an increase in the density, bulk modulus, and bulk sound velocity in both Al-free and Al-bearing bridgmanite at lower-mantle pressures. The presence of Fe^{3+} in the B-site of bridgmanite can further enhance this increase
Mao et al. (2017)	LHDAC Medium: Ne Synchrotron XRD and Mossbauer spectroscopy	Up to 130 GPa	1–11; 6–11 Fe	The substitution of Fe in Brd causes an anti-correlation between the enhanced density and reduced bulk sound velocity (V_Φ), whereas the combined Al and Fe substitution has a reduced effect on density and a negligible effect on V_Φ

Notes: *Brd* bridgmanite, *DAC* diamond-anvil cell, K_T bulk modulus, *XRD* X-ray diffraction

with the brownmillerite structure. He showed that the bulk modulus of $(Mg_3Al)(AlSi_3)O_{12}$, which forms by the charge-coupled mechanism and contains no vacancies, is approximately 6% smaller than that of pure bridgmanite. On the other hand, in $Mg_2Al_2O_5$, in which 1/6 of the oxygen sites are vacant, the bulk modulus is approximately 40% lower than that of pure $MgSiO_3$. With regards to aluminous bridgmanite, there have been extensive considerations concerning the dominant substitution mechanism, while experimental results have suggested that at least some vacancies are formed by the dissolution of alumina. However, on the basis of this idea alone, it is difficult to explain the large drop in the bulk modulus (7%) for the sample with only 1% Al_2O_3 because the maximum concentration of vacancies, expected from this composition, is very small (Yagi et al. 2004).

In addition, most of the experimental studies of bridgmanite elasticity have been performed in static compression experiments using XRD in a DAC, which yield the isothermal bulk modulus (K_T) through modeling the pressure–volume relationship. The EoS, using Birch-Murgnahan EoS Brillouin Light Scattering (BLS), is a powerful technique to measure the sound velocities of the mantle samples and to derive elasticity, including single-crystal elastic constants at high pressures in a DAC (Bass 2008). This method, using the measured V_p and V_s velocities (from Brillouin) and density (from diffraction and chemical analysis), provides a direct means of determining the adiabatic bulk modulus (K_S) that is quite different in methodology from the static compression pressure–volume methods that give the K_T based on EoS modeling (Bass 2008). The sound velocity of the same sample of bridgmanite used by Daniel et al. (2004), which had almost the same Al content (5.1 mol% Al_2O_3), was investigated using Brillouin spectroscopy at ambient conditions by Jackson et al. (2004). The performed Brillouin scattering measurements in bridgmanite under room *P–T* conditions demonstrated the lack of a reduction in bulk modulus K_S caused by the presence of 5.1 wt% of Al_2O_3; that is, they did not support an increase of the bulk modulus with increasing Al content. The presence of 5.1 wt% of Al_2O_3 is also shown to decrease the shear elastic modulus (μ) by 5.6% compared with Al-free bridgmanite samples (Jackson et al. 2004). Based on the Brillouin results, the end-member polycrystalline pure $MgSiO_3$ bridgmanite exhibits $K_S = 253(3)$ GPa and $\mu = 175(2)$ GPa, which are 4 and 1% lower, respectively, than the commonly accepted values ($K_S = 264$ GPa, $\mu = 177$ GPa) (Sinogeikin et al. 2004).

Static first-principles DFT, as well as atomistic calculations, in general, confirm the experimental data on the decrease of bulk modulus with the incorporation of Al in bridgmanite (Brodholt 2000; Yamamoto et al. 2003; Akber-Knutson and Bukowinski 2004). They have demonstrated that the dissolution of Al^{3+} into bridgmanite should increase its compressibility. Later lattice dynamic methods found no effect of the Al incorporation on thermodynamic properties of bridgmanite (Tsuchiya and Wang 2013).

This means that, despite some contradictions in high-pressure experimental results, one may suggest a decrease in bulk modulus of bridgmanite with the incorporation of Al in its structure, i.e. a decrease in incompressibility of the mineral. This effect seems weak, and thus should be not significant for seismologic interpretations. However, this problem needs further study.

Ferroelasticity. It is known that many distorted perovskite samples exhibit ferroelastic behavior, in which acoustic velocity, thermal expansio, and heat capacity show large anomalies near the phase transitions to higher symmetric structures (Carpenter and Salje 1998). However, bridgmanite has been shown to be far from a ferroelastic phase transformation by both theory (Karki et al. 2001; Oganov et al. 2001; Shim et al. 2001) and experiments (Stixrude and Cohen 1993; Wolf and Jeanloz 1985; Wentzcovitch et al. 1995).

Thermoelasicity. Experimental data on the thermoelasticity of bridgmanite is scarce (Murakami et al. 2012) and the uncertainty could be largely due to technical issues relating to the extreme conditions. Currently, there is only theoretical elastic constants data available for the extended *P–T* range of the lower mantle. In recent years, there have been several theoretical studies on the effects of ferrous and ferric iron on the thermoelastic properties of bridgmanite and post-perovskite (Metsue and Tsuchiya 2011; Tsuchiya and Wang 2013; Shukla et al. 2015); however, an agreement with seismological models of the lower mantle has not been reached. One potential reason is that the influence of Al^{3+} has not yet been included in calculations, another reason could be related to the possible inadequacy of using the quasi-harmonic approximation (QHA) in computing thermoelasticity at high *P–T* levels. There has been one study that used density functional molecular dynamics (DFT-MD) to calculate the properties of bridgmanite and post-perovskite enriched with ferrous iron (Muir and Brodholt 2015); however, it calculated values at only one pressure, exceeding the bridgmanite stability field (136 GPa). Recently, calculations with the use of ab initio molecular dynamics elastic and seismic properties of pure, Fe^{3+}-, Fe^{2+}-, and Al^{3+}-bearing bridgmanite and post-perovskite over a wide range of pressures (up to 137 GPa), temperatures and Fe/Al compositions were performed, in which the high-spin Fe^{3+} and Fe^{2+} in various possible spin states and compositions were considered (Zhang et al. 2016a). The results demonstrated that a mineral assemblage resembling pyrolite (neglecting some phases, such as $CaSiO_3$) fits a 1D seismological model well, down to at least the top of the D″ layer (Zhang et al. 2016a).

Thermal and electrical conductivity. Ohta et al. (2017) experimentally demonstrated that bridgmanite (and post-perovskite) is the best heat conductor in the Earth's lower mantle, while the electronic thermal conduction in ferropericlase is negligible.

Electrical conductivity of bridgmanite has the general increase with temperature (Deschamps and Khan 2016). The increase of the Fe^{3+} iron content in aluminous bridgmanite makes an electrical conductivity greater than that of an Al-free bridgmanite. In experiments at 25 GPa and 1400–1600 °C, bridgmanite with *fe* = 0.085 containing 2.89 wt% Al_2O_3 has an electrical conductivity approximately 3.5 times greater than that of Al-free bridgmanite with the same iron index (Xu et al. 1998). Further experiments on Al-bridgmanite with *fe* = 0.010 under 26–28 GPa pressure and up to 2000 K temperature confirmed the significant increase in its electrical conductivity with respect to the quantity of Al and established the increase of the electrical conductivity with the increase of iron content in bridgmanite (Yoshino et al. 2016). The resulting conductivity of Al-bearing bridgmanite first increases up to 0.06 Al atoms per formula unit and then remains constant or increases with increasing Al content at higher temperatures. These results are in agreement with the observed 1D conductivity-depth profiles from the Earth's surface to the deep mantle, which have shown that conductivity (σ) increases from 10^{-3} to 10^{0} S/m with increasing depth from 500 up to 1000 km and becomes constant at greater depths (Civet et al. 2015).

4.3 Periclase–Wüstite Solid Solution Phases (Mg,Fe)O

4.3.1 General

According to experimental and theoretical data based on the pyrolite composition, another abundant mineral in the lower mantle is ferropericlase, which should comprise approximately 16–20 wt% of the lower mantle (Irifune 1994; Fei and Bertka 1999; Wood 2000; Irifune et al. 2010) (Figs. 2.2 and 2.3). However, ferropericlase grains included in deep-mantle diamond inclusions account for 50–56% of all identified lower-mantle inclusions and are more frequently found than bridgmanite, which is believed to be the most abundant mineral, accounting for

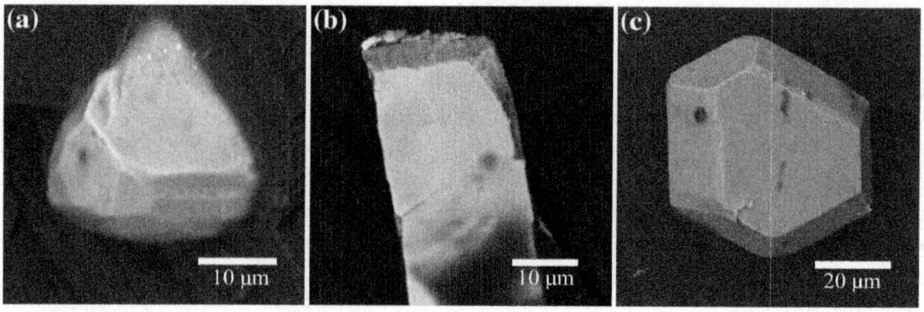

Fig. 4.8 Grains of ferropericlase extracted from diamonds of the Rio Soriso area, Brazil. SEM image. From Hayman et al. (2005)

75% of the region. Compositionally, these ferropericlase grains occupy a much wider range in FeO content than what is expected for ferropericlase, with approximately 20% FeO in the pyrolite model; some of these grains even contain more than 50% FeO and should be referred to as magnesiowüstite. These yellow-orange small (<1 mm) grains are usually euhedral with octahedral, cubic or cubic-octahedral crystal habit that reflects the rocksalt crystal structure of the lattice (Fig. 4.8). Their occurrences are either as single inclusions in diamond or form associations with bridgmanite, $CaSiO_3$ perovskite, stishovite (SiO_2), jeffbenite (formerly called TAPP), a phase with the composition of olivine, chrome spinel, ilmenite and native nickel (Kaminsky 2012 and references therein).

4.3.2 *Chemical Composition of Natural (Mg,Fe)O*

Representative chemical analyses of the chemical composition of ferropericlase–magnesiowüstite seria (which, for short, we refer to as 'ferropericlase') are presented in Table 4.5.

Magnesium index. According to high *P–T* experimental results in pyrolitic systems in the pressure range 25–60 GPa, the magnesium index of ferropericlase in lower-mantle material with $mg = \text{Mg}/(\text{Mg} + \text{Fe})_{\text{at.}} = 0.80–0.95$ should be 0.73–0.88 (e.g. Wood 2000; Lee et al. 2004, Irifune et al. 2010). In the lower mantle with the most likely lower-mantle magnesium index of $mg = 0.89–0.92$, the magnesium index of ferropericlase would lie within a much narrower range, at c. 0.90 (Kesson and Fitz Gerald, 1991). However, lower-mantle geological samples of ferropericlase grains that originated from different regions of the lower mantle varied in iron contents from a 'normal' iron index to more iron-rich compositions with $mg = 0.36–0.90$ (Fig. 4.9); one of the analyzed samples had a magnesium index as low as 0.16 (Zedgenizov et al. 2014a), which is close to the end-member wüstite stoichiometry. Interestingly, this range is very similar to the compositional ranges of olivine and low-Ca pyroxene in anhydrous interplanetary dust particles ($mg = 0.46–1.00$), which represents the entire solar protoplanetary disk (Flynn et al. 2016).

Table 4.5 Representative compositions of ferropericlase-magnesiowüstite inclusions in diamond from different areas (wt%)

Country	Brazil					Canada	Guinea	Australia
Area	Juina				Machado	Northwest Territories	Placer	Orroroo
Location	Collier-4 pipe	Sao Luiz placer		Rio Soriso placer	Machado placer	DO27 pipe	Placer	Dyke Eurelia K7
Sample No.	C-18	BZ207B	BZ66	3-2G	P15-1	DO2700100	KK-16a	FBS5-11
Mineral association	fPer	fPer + Brd + Jbn	fPer	fPer + Brd + CaSiPrv + 'Ol' + Jbn	2fPer + Brd + Dol	fPer + Brd + CaSiPrv + SiMg	2fPer + Brd + Sid	6fPer + Brd
SiO_2	0	0.03	0.01	0.08	0.07	0	0.02	0.02
TiO_2	0	0.02	0.01	n.a.	<0.02	n.a.	0.00	0.02
Al_2O_3	0.01	0.11	0.13	n.a.	0.06	n.a.	0.01	0.00
Cr_2O_3	0.25	0.51	1.06	0.35	0.69	0.67	0.49	0.14
FeO	22.91	43.14	73.56	24.67	19.44	16.80	22.47	23.78
NiO	1.48	0.90	0.10	1.36	0.16	1.13	1.42	1.16
MnO	0.16	0.43	0.80	0.22	1.11	0.19	0.18	0.14
MgO	75.55	54.84	23.13	72.41	79.07	81.60	75.20	73.96
CaO	0	0.01	0.01	n.a.	0.01	n.a.	0.00	0.01
Na_2O	0.03	0.14	1.05	0.18	0.03	0.26	0.12	0.00
K_2O	0	0.01	0	n.a.	<0.02	n.a.	0.00	0.00
P_2O_5	n.a.	n.a.	n.a.	n.a.	n.a.	n.a.	0.01	0.00
Total	100.39	100.14	99.86	99.27	100.64	100.65	99.93	99.23
fe	0.145	0.306	0.641	0.160	0.121	0.103	0.143	0.153
mg	0.855	0.694	0.359	0.840	0.879	0.897	0.857	0.847
References	Kaminsky et al. (2009b)	Hutchison (1997)		Hayman et al. (2005)	Burnham et al. (2016)	Davies et al. (2004)	Stachel et al. (2000)	Tappert et al. (2005a)

Notes: *n.a.* not analysed, *Brd* bridgmanite, *CaSiPrv* CaSi-perovskite, *Dol* dolomite, *fPer* ferropericlase, *Jbn* jeffbenite, '*Ol*' phase with a composition of olivine, *Sid* siderite, *SiMg* phase with a Si–Mg composition

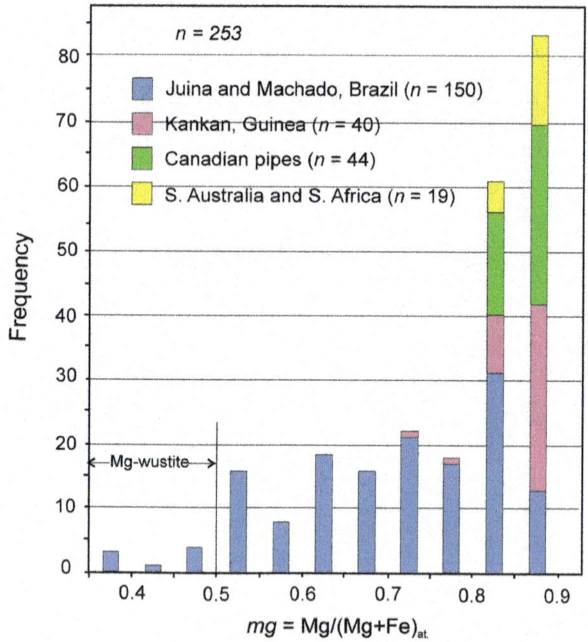

Fig. 4.9 Distribution of magnesium index *mg* in ferropericlase-magnesiowüstite inclusions in diamond. Data from Scott Smith et al. (1984), Moore et al. (1986), Wilding (1990), Hutchison (1997), Stachel et al. (2000), Hutchison et al. (2001), Kaminsky et al. (2001, 2009b), Davies et al. (2004), Hayman et al. (2005), Tappert et al. (2005a, 2009a), van Rythoven and Schulze (2009), Thomson et al. (2014), Zedgenizov et al. (2014a), Burham et al. (2016). Modified after Kaminsky (2012), with additions

There are systematic differences in ferropericlase *mg* values from different regions. Ferropericlase grains from Canada, South Australia and South Africa have *mg* in the range of 0.80–0.90, close to the expected values, while ferropericlase from Guinea, along with similar *mg* values, may show *mg* values as low as 0.75. On the other hand, almost half of all ferropericlase grains from Brazil (c. 46.5%) are iron-rich; their compositions may be magnesiowüstite. It has been suggested that such great deviations may reflect the compositional variations of bridgmanite, from which iron-rich ferropericlase and SiO_2 were formed as a result of decomposition (Fei et al. 1996; Harte et al. 1999). However, this model, while potentially explaining one discrepancy, actually raises another, that is: how can highly iron-rich bridgmanite be in equilibrium in the lower mantle whose bulk *mg* = 0.80–0.95? Liu (2002) proposed a model that involves the decarbonation of ferromagnesite (with variable iron content), to result in the formation of ferropericlase with a correspondingly variable magnesium index, along with carbon (in the form of diamond) and oxygen. Considering the complexity of this model and even assuming the presence of ferromagnesite in the lower mantle, the notion of variability in its iron content remains enigmatic.

The value of the magnesium index mg = Mg/(Mg + Fe)$_{at.}$ in ferropericlase–magnesiowüstite strongly correlates with the content of Ni in the mineral, varying from 0.08 wt% in most iron-rich magnesiowüstite to more than 1 wt% in ferropericlase (Fig. 4.10). The concentration of Ni in ferropericlase with $mg > 0.8$ is close to 1%, which is significantly higher than that estimated for lherzolite with 1% of the metal alloy.

Several explanations can be offered for such a high variability of mg values in lower-mantle ferropericlase. First, variations in mg may be caused by the occurrence and concentration of the metallic phase in the lower mantle, which should result in an increase of mg in both ferropericlase and bridgmanite (Ryabchikov and Kaminsky 2013a, 2014). The content of the iron–nickel alloy is suggested to be close to 1% under lower mantle conditions (Frost and McCammon 2008). The concentration of Ni in iron-rich ferropericlase should decrease remarkably due to its transition into the metal phase, as was demonstrated in Ryabchikov (2011) and Ryabchikov and Kaminsky (2013b). Second, different bulk compositions of host rocks are another factor that may control the compositional variations of sub-lithospheric ferropericlase. The abundance of the lower mantle minerals included in diamonds (ferropericlase, 56%; bridgmanite, 8%; CaSi-perovskite, 12%; SiO_2, 8%; other minerals, 17%) differs significantly from the calculated values for pyrolite (ferropericlase, 18–20%; bridgmanite, 70–74%; CaSi-perovskite, 8–10%), which points to the difference between the compositions of the lower mantle and the upper mantle. In particular, the presence of more high-temperature rocks, such as

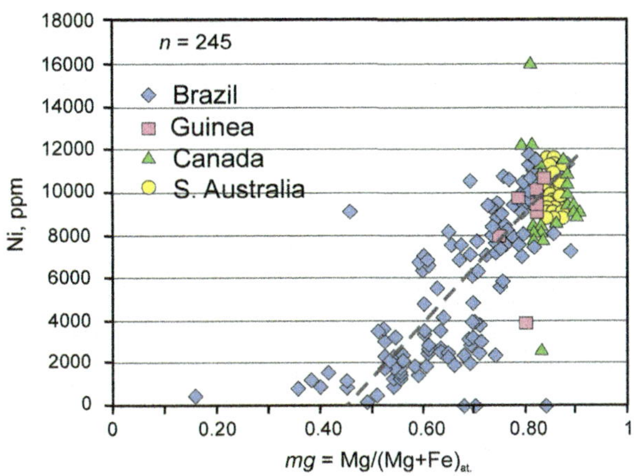

Fig. 4.10 Dependence of the nickel concentration on mg for ferropericlase-magnesiowüstite in diamond inclusions from various regions worldwide. Data from Scott Smith et al. (1984), Moore et al. (1986), Wilding (1990), Hutchison (1997), Stachel et al. (2000), Hutchison et al. (2001), Kaminsky et al. (2001, 2009b), Davies et al. (2004), Hayman et al. (2005), Tappert et al. (2005a, 2009a), van Rythoven and Schulze (2009), Thomson et al. (2014), Zedgenizov et al. (2014a), Burham et al. (2016). Modified after Ryabchikov and Kaminsky (2013a), with additions

harzburgite in addition to primitive lherzolite in the lower mantle, may result in the appearance of ferropericlase with different compositions compared to the mineral in the pyrolite bulk environment. The third scenario is that iron-rich ferropericlase may be formed as a result of magma fractional crystallisation directly in the lower mantle. Such a possiblility is supported by the experimental data. High *P–T* DAC results from Nomura et al. (2011) and Andrault et al. (2012) demonstrated that the partitioning coefficient of Mg/Fe between lower mantle phases and melt $K_D = (Mg/Fe)^{crystal}/(Mg/Fe)^{melt}$ are always >1, which is similar to those for olivine and pyroxenes under crustal conditions; consequently, the Fe/Mg ratio increases during fractional crystallization.

It should be noted that some researchers assume the presence of cumulates of iron-rich magnesiowüstite near the core–mantle boundary, the origin of which results from the solidification of the basal magmatic ocean at the early stages of the Earth's history (e.g., Nomura et al. 2011). As a whole, comparisons from the results of model calculations within the database of the composition of ferropericlase inclusions in diamonds demonstrates that the lower mantle is mainly composed of rocks of peridotitic composition, most likely including fertile varieties close to pyrolite, as well as more refractory rocks of a harzburgitic composition (Irifune et al. 2010).

Distribution of ferrous and ferric ions in ferropericlase. The distribution of ferrous and ferric ions in the structure of ferropericlase was revealed recently in studies by A. Abakumov involving ferropericlase grains from inclusions in diamonds from the Juina kimberlites (Kaminsky et al. 2015b). High-angle annular dark-field scanning transmission electron microscopy (HAADF-STEM) images demonstrated the presence of iron Fe^{3+}-rich octahedral-shaped clusters, varying in size from 5 to 10–20 nm and numerous smaller shapeless clusters with a size of 1–2 nm, within natural ferropericlase grains (Fig. 4.11a, b). The clusters are faceted by the {111} crystal planes delimiting truncated octahedra. High-resolution HAADF-STEM images demonstrated that the clusters have an Mg/Fe ordered structure, which is responsible for the superlattice reflections causing doubled $a \approx 8.4$ Å supercell (Fig. 4.11c). Both EDX analysis and STEM–EELS compositional mapping showed that the clusters are Fe-enriched and Mg-depleted (Fig. 4.11d, e).

The observed Fe/(Fe + Mg) ratio in the octahedral clusters corresponds to non-stoichiometric $Mg_{1+x}Fe_{2-x}O_{4-x/2}$ compounds with the structure intermediate between those of the B1 (rock salt) and $H1_1$ (spinel) structure types (Moran et al. 2002). The end member wüstite ($Fe_{1-x}O$) is well known to be highly nonstoichiometric. Commercially available wüstite typically has a composition of $Fe_{0.88}O$; there is relatively high ferric iron content in its lattice (Fei 1996). Relatively high stoichiometric wüstite, with a composition as high as $Fe_{0.98}O$ has been synthesized at high *P–T* conditions near the iron-wüstite buffer (McCammon 1993).

High-energy resolution EEL spectra demonstrated that signals from the non-stoichiometric clusters clearly showed that both the L_3 and L_2 peaks shift to higher energies, characteristic of Fe^{3+} ions, which are octahedrally coordinated (van

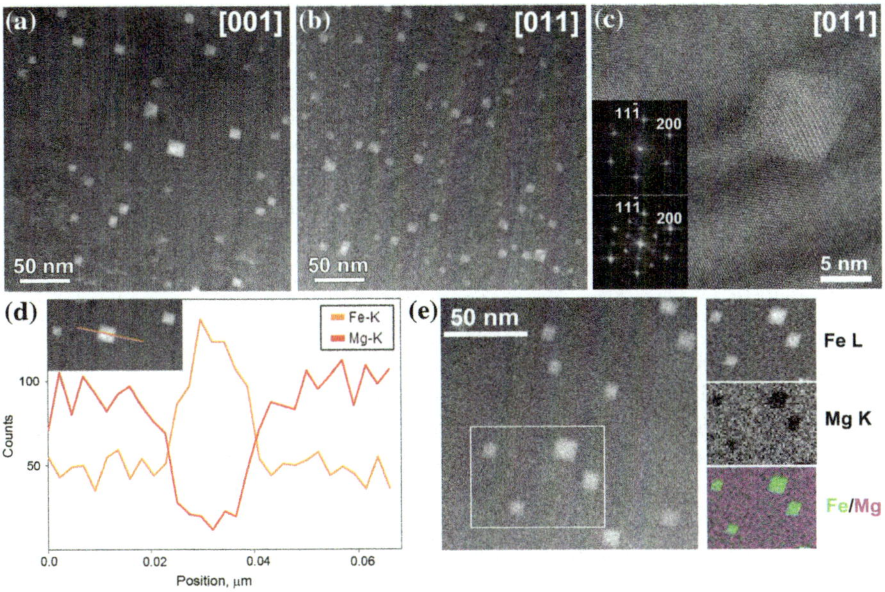

Fig. 4.11 Low magnification HAADF-STEM image of ferropericlase #C-143 in [001] pattern (**a**) and low magnification HAADF-STEM image of sample #7-100 in [011] pattern (**b**). The octahedral clusters on these images look prominently brighter due to a higher Fe content than in the surrounding ferropericlase matrix. High resolution HAADF-STEM image of a cluster in sample #7-100 with [011] pattern (**c**). Fourier transforms from the areas of the ferropericlase matrix and the cluster are shown as *top* and *bottom inserts*, respectively. The superlattice reflections originate exclusively from the cluster. The intensity profiles of the EDX Fe K and Mg K lines across the octahedral cluster in sample #C-143 showing Mg-depletion and Fe-enrichment (**d**). The insert shows the corresponding HAADF-STEM image with the line selected for acquiring the STEM EDX signal. HAADF-STEM image of the sample #7-100 with the area selected for EELS compositional mapping (**e**); the signal maps of the Mg K, Fe L edges and their color-coded mixed image. From Kaminsky et al. (2015b). © Elsevier

Aken and Liebscher 2002). According to the same EEL spectra collected from the ferropericlase matrix, the iron cations in the matrix are in the ferrous Fe^{2+} state.

Weak superstructure reflections were observed in both the [001] and [011] electron diffraction patterns. These reflections require a doubling of the unit cell parameter, resulting in a face-centered cubic unit cell with $a \approx 8.4$ Å. With the exception of reflection conditions imposed by F-centering, the $hk0$: $h + k = 4n$ reflection condition is deduced from the [001] ED pattern, suggesting that the most symmetric space group is $Fd\bar{3}m$. This unit cell parameter and the space group symmetry are characteristic of the $H1_1$ magnesioferrite structure. In other fPer inclusions in diamonds, collected from the same Juina area in Brazil, crystals of magnesioferrite, 10–50 nm in size, were observed to exhibit cubic and octahedral forms along dislocation lines (Wirth et al. 2014; Kaminsky et al. 2015a; Palot et al. 2016).

These results demonstrate the inhomogeneous distribution of ferric ions in ferropericlase and its concentrations in iron-rich, magnesioferrite-like clusters.

In early works, which used Mössbauer spectroscopy, the ferric iron proportion ($Fe^{3+}/\Sigma Fe$) in MgO-rich ferropericlase was low, approximately 0–7% of the total iron content (McCammon et al. 1997; Jacobsen et al. 2002), which is much lower than in the FeO-rich magnesiowüstite and wüstite samples (Fei 1996). Recent use of synchrotron radiation facility to collect energy domain Mössbauer spectra of ferropericlase inclusions in diamond gained similar results ($Fe^{3+}/\Sigma Fe$ = 3–9%; Nestola et al. 2016a). However, Mössbauer spectra of Fe^{3+}-rich magnesioferrite clusters within ferropericlase are magnetically split, with absorption divided over many lines with intensities similar to the baseline scatter. Hence, the signal from the magnesioferrite/Fe^{3+} cluster is not included in the Mössbauer spectrum of ferropericlase, and the concentration of Fe^{3+} in ferropericlase, established with the use of Mössbauer spectroscopy, is less than the total by 2–3% (according to the share of magnesioferrite/Fe^{3+} clusters in the matrix of ferropericlase; Kaminsky et al. 2015b).

Flank method was used for the determination of $Fe^{3+}/\Sigma Fe$ in ferropericlase with the use of an electron microprobe. This method has an advantage over Mössbauer spectroscopy because it has better spatial resolution ($\sim$1–10 μm against $\sim$100 μm) and a possibility to measure single ferropericlase samples containing exsolved clusters ('magnesioferrite'), even when these materials are too small to be resolved by electron microprobe measurements (Longo et al. 2011). The results demonstrated that values of $Fe^{3+}/\Sigma Fe$ in ferropericlase varied from 8 to 12%, and samples analyzed earlier, corrected for Fe^{3+} in clusters, comprised 3–10% (Kaminsky et al. 2015b).

Minor elements in ferropericlase. The most abundant minor elements in ferropericlase, in addition to Ni, are Cr, Mn and Na, each comprising up to 1–2.5 wt% of the total (Table 4.6). This corresponds to the experimentally established high values of $K_D^{\mathrm{fPer-Melt}}$ for Cr and Mn, according to which ferropericlase may be a host phase for these elements (Righter et al. 2014 and references therein). Mn has a negative covariance with *mg*. Na is positively related to Cr.

Trace element contents in ferropericlase show an irregular distribution that vary considerably in abundance and are usually below chondrite level; only Li, Ta, Th, and U are consistently enriched. REE patterns show the enrichment in heavy REE around the chondrite level (Fig. 4.12). Walter et al. (2004b) earlier determined ferropericlase-melt partition coefficients for a series of trace elements in ferropericlase. Despite large measurement uncertainties, most minor and trace elements, with the exceptions of Cr and Ni, were considered to be incompatible in ferropericlase. In particular, REE have experimental K_D values below 0.1; however, natural samples do not confirm this finding (Fig. 4.12). U and Th were not measured but were expected to be highly incompatible based on the results obtained for other tetravalent cations (Si, Ti, Zr, and Hf). However, natural samples show the reverse.

Table 4.6 Minor elements in ferropericlase (wt%) (from Kaminsky 2012)

Oxides	Brazil					Gunea	Canada	
	São Luiz	Rio Mutum. Vermelho. Chicoria	Rio Soriso	Pandrea pipes	Collier-4 pipe	Kankan	DO27, DO18, A21, Ranch Lake pipes	Panda pipe
Cr_2O_3	0.19–2.56	0.11–2.26	0–1.30	0.10–0.65	0.09–0.27	0.12–1.04	0.41–2.56	0.69–0.75
NiO	0.10–1.49	0.02–1.37	0.11–1.46	0.59–1.34	1.02–1.48	0.51–1.46	0.33–1.55	1.34–2.14
MnO	0.07–0.81	0.19–0.94	0.16–1.46	0.15–0.42	0.14–0.18	0.16–0.26	0.13–0.34	0.12–0.24
Na_2O	0.02–1.25	0.04–1.35	0–1.20	0.01–0.24	0–0.03	0.07–0.79	0.04–1.11	0.07–0.19
mg	0.36–0.87	0.49–0.83	0.45–0.89	0.75–0.81	0.76–0.86	0.75–0.92	0.84–0.90	0.80–0.87
References	Wilding (1990), Harte et al. (1999)	Kaminsky et al. (2001)	Hayman et al. (2005)	Kaminsky et al. (2009a)		Stachel et al. (2000)	Davies et al. (2004)	Tappert et al. (2009a)

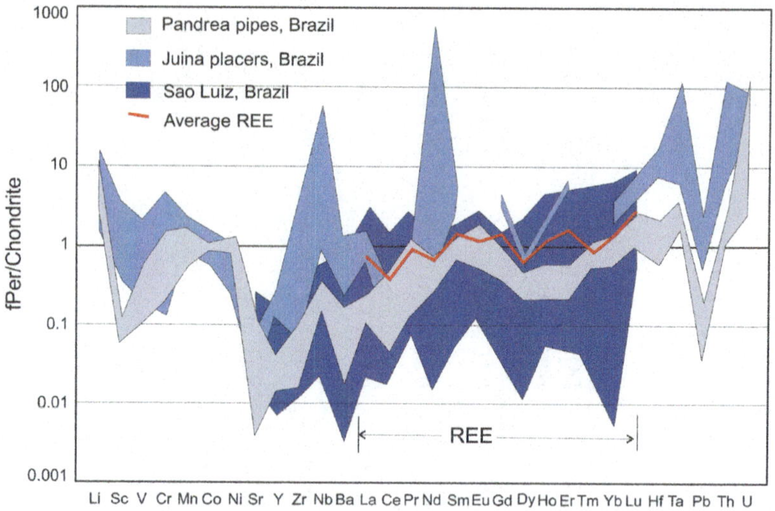

Fig. 4.12 Chondrite-normalised trace element patterns in ferropericlase-magnesiowüstite. Data from Harte et al. (1999), Kaminsky et al. (2001, 2009a). Chondrite composition from McDonough and Sun (1995). Modified after Kaminsky (2012), with additions

4.3.3 Crystal Structure and Phase Transitions of Ferropericlase–Magnesiowüstite at High Pressures and Temperatures

Cubic crystal structure. MgO–FeO forms a complete solid solution series with a gradual compositional transition between the end-member periclase and wüstite in the cubic crystal structure B1 with space group $Fm\bar{3}m$ (NaCl type with $Z = 4$) (Fig. 4.13), in which end-members differ only in crystal parameters: $a = 4.203$–4.212 Å for pure periclase and $a = 4.296$ Å for pure wüstite. For intermediate compositions, the structure consists of interpenetrating face-centered-cubic lattices producing alternating Mg^{2+}/Fe^{2+} cations and O^{2-} anions. The structure can also be described as an edge-share array of $(Mg,Fe)O_6$ octahedra in which all edges are shared with neighbouring octahedra (Wicks and Duffy 2016).

In natural ferropericlase–magnesiowüstite samples found in diamond inclusions, the series occupies a composition range, almost reaching both end-members. The magnesium end-member periclase MgO was found to be stable as the cubic rocksalt (B1) structure ($Fm\bar{3}m$ space group) to at least 227–250 GPa along the Hugoniot curve in early shock wave experiments or to 250 GPa in static DAC experiments, i.e., to pressures in the bottom of the lower mantle and in the outer core (Duffy et al. 1995; Dorfman et al. 2012). Its transition to the cubic CsCl-type (B2) phase (space group $Pm\bar{3}m$) is observed only at ~600 GPa, and the B2 phase is found to be stable up to 900 GPa (Coppari et al. 2013). Theoretical calculations, performed with the use of an evolutionary algorithm demonstrated that at pressure 116 GPa,

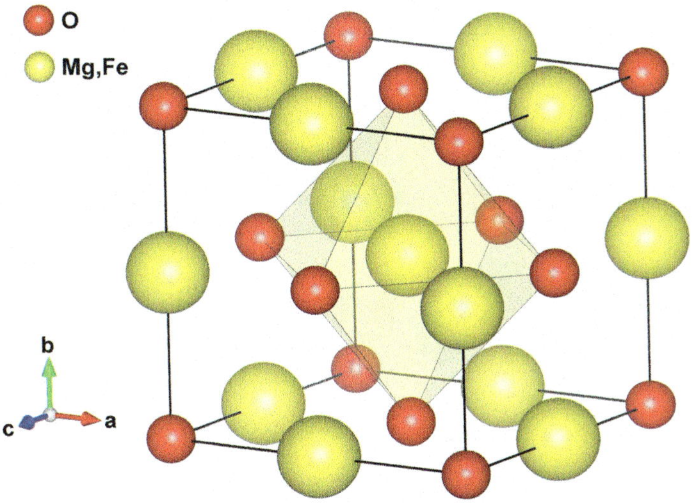

Fig. 4.13 Crystal structure of ferropericlase with cubic rock-salt structure (space group *Fm–3m*). Lattice parameters from Mao et al. (1996)

magnesium peroxide MgO_2 becomes stable, and contains transitional ions $[O–O]^{2-}$ (Zhu et al. 2013). Recently these calculations were confirmed experimentally with the use of DAC and synchrotron XRD (Lobanov et al. 2015). Raman spectroscopy detected the presence of a peroxide ion ($O_2{}^{2-}$) in the synthesised material as well as in the recovered specimen. MgO_2 forms as a result of the reaction for $2MgO + O_2 \rightarrow 2MgO_2$ in the presence of free oxygen at a pressure 94 GPa and temperature 2150 K. The resulting magnesium peroxide MgO_2 has a tetragonal structure with space group *I4/mcm*, which is significantly denser than the periclase with the *Fm*3*m* structure (Lobanov et al. 2015). The formation of magnesium peroxide may happen only under high oxygen fugacity conditions.

Rhombohedral structure. By contrast, the iron end-member wüstite FeO experiences various phase transitions under high-pressure conditions. FeO transition to the rhombohedral $R\bar{3}m$ structure (Fig. 4.14a–c) occurs at a pressure above 9–18 GPa at 300 K (Zou et al. 1980; Yagi et al. 1985; Mao et al. 1996; Shu et al. 1998; Kantor et al. 2004). The B1 → rhombohedral transition has also been observed at approximately 16 GPa not only in pure wüstite, but also in iron-rich Fe–Mg solid solutions (*fe* = Fe/(Mg + Fe) = 0.80–0.95) using in situ XRD in DAC experiments (Mao et al. 2002; Kondo et al. 2004; Kaercher et al. 2012). At the initial stage of the B1 → rhombohedral transition (still in cubic structure), a magnetic ordering transition was observed by Mössbauer spectroscopy and neutron powder diffraction at approximately 4.5 GPa, which is associated with an unusual discontinuity in the elastic constants as a function of pressure (Kantor et al. 2004). The B1 → rhombohedral structural transition in FeO can be viewed as an elongated distortion along <111> direction of the B1 lattice (Fig. 4.14b, c) such that it is also

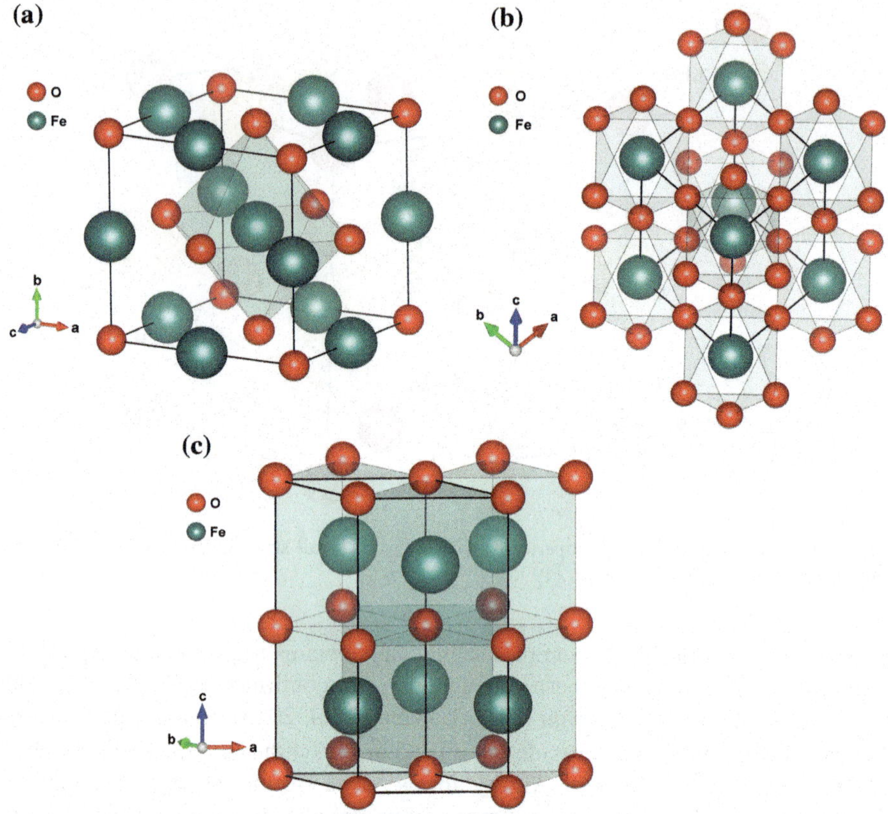

Fig. 4.14 Crystal structures of FeO phases under high pressure: cubic B1 with space group *Fm–3m* (*left*), B-rhombohedral with space group $R\bar{3}m$ (*center*) and hexagonal B8 with space group $P6_2/mmc$ (*right*). Lattice parameters from Mao et al. (1996)

found to be fairly sensitive to the pressure medium and thus hydrostaticity in the sample chamber as well as stoichiometry of the sample.

Hexagonal structure. At higher pressures (~70–75 GPa and 600–1600 K temperatures), the B1 wüstite transforms to the hexagonal NiAs (B8) structure (space group $P6_3/mmc$) (Fei and Mao 1994; Sherman and Jansen 1995; Kondo et al. 2004; Murakami et al. 2004b; Fischer et al. 2011b; Kaercher et al. 2012) and the lattice parameters of the high-pressure hexagonal B8-structured FeO, established at 96 GPa and 800 K in the first synthetic crystals with the use of in situ synchrotron XRD, were: $a = 2.574$ Å and $c = 5.172$ Å (Fei and Mao 1994) (Fig. 4.15). However, the location and the Clapeyron slope of the phase boundary at high *P–T* remain extensively debated: earlier results suggested the transition from a vertical slope near 70 GPa (Murakami et al. 2004b) to a strongly positive slope in *P–T* space (Kondo et al. 2004; Ozawa et al. 2010). Some experiments have failed to obtain the transition to the B8 phase at all (Mao et al. 1996; Seagle et al. 2008) and

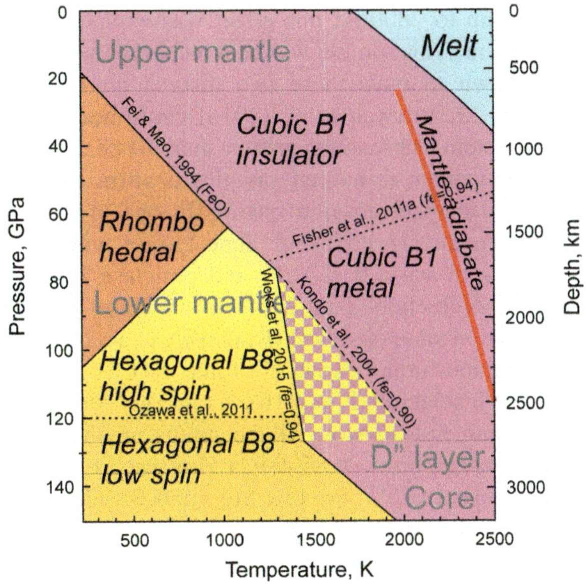

Fig. 4.15 Phase diagram for wüstite at high pressures and temperatures. Data from Fei and Mao (1994), Mao et al. (1996), Kondo et al. (2004), Fischer and Campbell (2010), Fischer et al. (2011a, b), Ozawa et al. (2010), Ohta et al. (2014), Wicks et al. (2015). Mantle adiabate after Katsura et al. (2010)

have attributed this to kinetic barriers at room temperature (Mao et al. 1996) or to differences in stoichiometry (Seagle et al. 2008).

The transformation from the B1 to the B8 structure has been observed only for the Fe end-member, and was not observed to occur in less iron-rich samples (*fe* = 0.61–0.75) up to 86–102 GPa and 2500 K (Lin et al. 2003; Kondo et al. 2004). Recent experiments with polycrystalline magnesiowüstite with *fe* = 0.94 obtained from XRD under heating up to 122 GPa and 2100 K demonstrated that the addition of Mg to FeO stabilizes the B1 phase with respect to the B8 phase at these conditions (Wicks et al. 2015).

In the deep Earth, wüstite can undergo the B1 → B8 phase transition at temperatures well below the geotherm and may occur only in the subsolidus conditions of the lower mantle (Fig. 4.15).

Based on the thermodynamic analysis of the MgO–FeO solid solution at high pressures and temperatures, an immiscibility gap has been proposed to exist between the B8 and B1 structures (McCammon et al. 1983). XRD experiments demonstrated that intermediate solid solutions (*fe* = 0.20–0.50) dissociate, at least partially, at conditions near 83–86 GPa and 1000 K (corresponding to a depth of 1900–2000 km) to an Mg-enriched composition and an almost pure wüstite (FeO) composition (Dubrovinsky et al. 2000, 2001a). However, this finding was initially not confirmed by other groups. Experiments on peridotite compositions of up to 124 GPa have reported no evidence of the dissociation (Murakami et al. 2005) and diffraction experiments on more Fe-rich compositions (*fe* = 0.61–0.75) at 86–102 GPa and 2500 K found no evidence of such a dissociation (Lin et al. 2003). More recently, an iron-rich composition (*fe* = 0.96) was found to decompose into hexagonal B8-FeO and $(Mg_{0.06}Fe_{0.94})O$ in a diffraction pattern at 123 GPa after heating up to 2800 K; this effect was explained by the metallization of the FeO component. This sample was also recovered to 1 bar for ATEM measurements and was shown to have these two distinct phases (Ohta et al. 2014). In addition, recently, Martirosyan et al. (2016) confirmed the immiscibility gap in the FeO–MgO system at pressures greater than 70 GPa.

Observations in natural samples confirm decomposition of ferropericlase into iron- and magnesium-rich phases. In one of the samples from the Juina area in Brazil, coexisting grains of almost pure wüstite and periclase were identified: wüstite with *mg* = 0.02–0.15 and periclase with *mg* = 0.85–0.92 (confirmed by electron diffraction data); no ferropericlase or magnesiowüstite of intermediate composition were observed (Kaminsky et al. 2009b). The analyzed grains exhibited some compositional variation (Table 4.7).

On a histogram (Fig. 4.16), inclusions form two groups with respect to their magnesium index. Most of the grains are highly enriched in Mg with low Fe concentrations (*mg* = 0.85–0.92). However, a significant part of the grains analyzed (approximately 25%) are low-Mg, high-Fe wüstite with *mg* = 0.02–0.15. A subset of these have extremely high ferrogenosity (*fe* = 0.96–0.98), close to stoichiometric wüstite.

In another sample from the same Juina area in Brazil, a single crystal of almost pure wüstite with composition $(Mg_{0.07}Fe_{0.93})O$ completely encapsulated by

Table 4.7 Chemical compositions of coexisting wüstite and periclase inclusions in diamond from the Juina area, Brazil (at.%; AEM data; from Kaminsky et al. 2009b)

Grain #	Mg	Mn	Fe	Total
Wüstite				
1	3.4	0	96.6	100.0
2	15.3	0.4	84.3	100.0
3	13.2	0	86.8	100.0
4	13.0	0.6	86.4	100.0
5	1.9	0	98.1	100.0
Periclase				
6	85.4	0.4	14.2	100.0
7	91.2	1.2	7.6	100.0
8	85.5	0.4	14.1	100.0
9	92.1	1.0	6.9	100.0
10	87.7	1.2	11.1	100.0
11	85.5	0.5	14.0	100.0
12	84.9	0.2	14.9	100.0
13	86.7	0.9	12.4	100.0
14	89.7	0.6	9.7	100.0
15	86.4	0.4	13.2	100.0
16	87.8	1.2	11.0	100.0
17	91.3	0.6	8.1	100.0
18	91.5	1.1	7.4	100.0
19	90.3	0.4	9.3	100.0
20	86.8	0.6	12.6	100.0

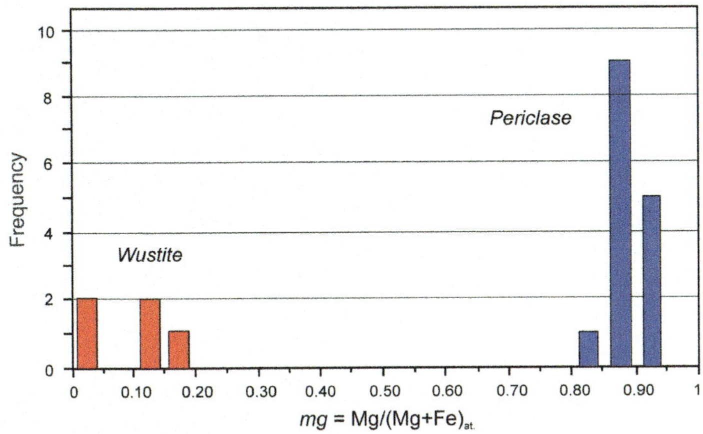

Fig. 4.16 Magnesium index *mg* in coexisting wüstite and periclase inclusions from diamond #8/103

polycrystalline periclase ($(Mg_{0.93}Fe_{0.07})O$) was observed in a lower-mantle inclusion in diamond (Kaminsky et al. 2015a) (Fig. 4.17). These findings in natural samples indicate that ferropericlase can decompose into Mg- and Fe-rich phases at certain *P–T* conditions, such that natural ferropericlase samples with intermediate compositions were formed, most likely, under pressure conditions of less than 83–86 GPa, i.e. at depths shallower than 1900–2000 km. A possible explanation for the dissociation is the metallization of the FeO end-member component that promotes this dissociation process. In this scenario, the highly conductive iron-rich (Mg,Fe)O magnesiowüstite and metallic FeO rocks located at the base of the mantle may play an important role for the dynamics and evolution of Earth's mantle and core (Ohta et al. 2012, 2014).

Metallization of B1-FeO. Theoretical studies predicted the metallization of B1-FeO at high pressures (Ringwood 1977; Sherman 1989; McWilliams et al. 2012); however, the location of this transition has not been predicted. The electronic transition, interpreted as metallization, was initially thought to coincide with the B1 → B8 transition (Fei and Mao 1994; Murakami et al. 2004b); however, with new data on the location and slope of the B1/B8 phase boundary (Fig. 4.15), it has become apparent that this is not the case. Other studies found that B1-FeO is a semiconductor up to ~70 GPa and 1500–1760 K, while B8-FeO is metallic (Ohta et al. 2010; Fischer et al. 2011a). More recent experiments in a LHDAC demonstrated that B1-structured FeO metallizes at around 70 GPa and 1900 K without any structural phase transition (Ohta et al. 2012). Despite a high electric conductivity above 85 GPa and 1300 K, more Mg-rich samples (mg = 0.05–0.20) remain insulators (Ohta et al. 2014).

The melting curve of $Fe_{1-x}O$ has been determined up to pressures and temperatures of 77 GPa and 3100 K (Fischer and Campbell 2010).

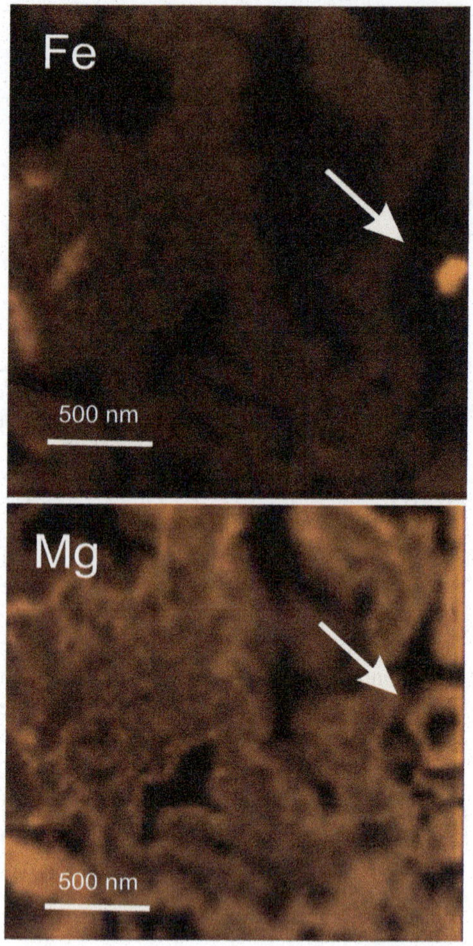

Fig. 4.17 Spherical inclusion of wüstite (core) + periclase (rim) in a dolomite matrix. *Irregular dark areas* are pores. Elemental map. Scale bar is 500 nm. Sample #8/108 from the Juina area, Brazil. From Kaminsky et al. (2016). Used with a permission of Springer

4.3.4 Physical Properties of Ferropericlase-Magnesiowüstite

The role of the chemical composition of ferropericlase. The substitution of FeO in (Mg,Fe)O inherently affects physical properties of the ferropericlase–magnesiowüstite system in the Earth's interior. Iron is the most abundant transition metal on the planet and can exhibit various spin and valence states based on thermodynamic environments. Ferrous iron is more than twice as heavy than the magnesium cation in the ferropericlase, but its ionic radius is only slightly larger than that of Mg^{2+}, such that the addition of FeO is expected to slightly expand the lattice parameters while the density is more significantly enhanced (Jacobsen et al. 2002).

Density. Based on static compression data at 70 GPa and 900 K (Fei and Mao 1994), the density of B8-type hexagonal FeO as a result of the B1 → B8 structure transition increases by 4%, which is consistent with the original evaluation of the shock compression data (Jeanloz and Ahrens 1980). The 4% density increase can be explained by geometric rearrangement from the rock salt to NiAs structure and the change in the *c*/*a* ratio of the B8 phase (Jackson et al. 1990).

Elastic properties. The bulk modulus (the incompressibility) of (Mg,Fe)O was studied as a function of pressure and composition using pressure–volume relations in XRD measurements and the V_P and V_S profiles using ultrasonic interferometry (Richet et al. 1989; Jacobsen et al. 2002; Lin et al. 2005, 2013; Zhang and Zhao 2005). Studies of non-stoichiometric, iron-rich (Mg,Fe)O showed that both adiabatic bulk modulus (K_S) and isothermal bulk modulus (K_T) decrease as a function of Fe concentration (Richet et al. 1989; Jacobsen et al. 2002). The trend is opposite for stoichiometric samples, where ultrasonic interferometry studies for iron-poor samples display a positive trend of K_S with increasing iron content (Jacobsen et al. 2002). In the wüstite end-member, K_T does depend on stoichiometry in the B1 phase, with $Fe_{0.99}O$ being much less compressible than $Fe_{<0.98}O$ (Fei 1996; Zhang and Zhao 2005). The initial studies of the elasticity of (Mg,Fe)O have shown that the shear modulus, *G*, is much more sensitive to iron content than the bulk modulus (e.g., Jackson et al. 2006).

Thermal and electrical conductivity. Ohta et al. (2017) examined the lattice thermal conductivity of (Mg,Fe)O containing 19 mol.% Fe up to 111 GPa and 300 K by means of the pulsed light heating thermoreflectance technique in a DAC. They demonstrated a strong reduction in the lattice thermal conductivity of ferropericlase due to iron substitution. Ohta et al. (2017) also measured the electrical conductivity of an identical ferropericlase sample up to 140 GPa and 2730 K and found that ferropericlase remains an insulator throughout the experimental conditions, indicating the electronic thermal conduction in ferropericlase is negligible. This data indicates that bridgmanite (and post-perovskite) is the best heat conductor in the Earth's lower mantle, and distribution of iron and its valence state among the lower mantle minerals are key factors to control the lower mantle thermal conductivity.

Electronic iron spin transition. Physical properties of ferropericlase can also be affected by the electronic spin transition of iron at high pressures, including changes in elasticity, iron partitioning and electrical and thermal conductivities (see Lin et al. 2013 for a review) that may contribute to seismic heterogeneities of the region (for a detailed discussion about the spin transition see Chap. 8). There have been a number of early experimental and theoretical studies on the incompressibility, equation of state and elastic constants of single-crystal ferropericlase across the spin transition at high pressures using various techniques, including Impulsive Stimulated Light Scattering (ISS), Brillouin Light Scattering (BLS), Inelastic X-ray Scattering (IXS), and Density Function Theory (DFT), but some results were not in agreement with others (Antonangeli et al. 2011; Crowhurst et al. 2008; Marquardt et al. 2009a, b; Yang et al. 2015; Wu et al. 2013; Wu and Wentzcovitch 2014). The most recent experimental results and DFT calculations have shown that ferropericlase exhibits significant softening in V_P, as well as the C_{11} and C_{12} elastic constants across the spin crossover at lower-mantle *P–T* conditions, although V_S and C_{44} were not affected by the transition (Wu et al. 2013; Yang et al. 2015). The spin transition was associated with a significant reduction of the aggregate V_P/V_S via the aggregate V_P softening because V_s softening did not visibly occur within the transition. These results suggest that the middle to lowermost parts of the lower-mantle may exhibit enhanced seismic heterogeneities due to the occurrence of the mixed-spin and low-spin ferropericlase (Yang et al. 2015).

Iron spin crossover in ferropericlase also influences the conductivity of ferropericlase: it reduces its lattice thermal conductivity, as well as its radiative conduction (Ohta et al. 2017).

An electronic high-spin to low-spin transition of iron in B8-structured wüstite (*fe* = 0.96) was experimentally observed to occur at ~119 GPa with a volume reduction of ~2.5% in situ X-ray emission spectroscopy XES and XRD measurements (Ozawa et al. 2011). While the crystal structure of FeO remains as B8 type across such volume reductions, the atomic arrangements of Fe and O change from the inverse to the normal NiA form, with considerable decrease in *c*/*a* axial ratio. Together with recent electrical resistance measurements, these results suggest that the spin crossover, inverse-normal structural change, and insulator–metal transition occur concurrently in B8-FeO at approximately 120 GPa (Ozawa et al. 2011).

4.4 Iron Partitioning Between Bridgmanite and Ferropericlase

4.4.1 General

High *P–T* experiments along an expected mantle geotherm for a pyrolitic composition demonstrated that the iron index, $fe = \mathrm{Fe/(Fe + Mg)_{at.}}$, of the pyrolitic lower-mantle ferropericlase should be 0.12–0.27 (Wood 2000; Lee et al. 2004),or even lower, at c. 0.10 (Kesson and Fitz Gerald 1991). These results suggest that the iron chemistry of lower-mantle ferropericlase may not vary significantly. However, although a majority of the studied natural lower-mantle ferropericlase grains hosted within deep-diamond inclusions have *fe* = 0.10–0.20, more than 40% of them vary in a 'forbidden' range with up to *fe* = 0.64 falling into the field of magnesiowüstite (e.g., Hutchison 1997; Kaminsky 2012); one of the analyzed samples has an iron index as high as 0.9, which is almost close to the end-member wüstite stoichiometry. Some grains of bridgmanite, coexisting with iron-rich ferropericlase, also exhibit relatively high iron contents. The discovery of these iron-rich ferropericlase and magnesiowüstite grains deviates from the traditional view of the pyrolitic lower mantle composition, which argues for a homogenous iron chemistry and mineralogy, and points to a much more complex chemistry of the lower mantle (Kaminsky and Lin 2017).

4.4.2 Experimental Works on Partitioning of Iron Between Bridgmanite and Ferropericlase

A number of experimental works on iron partitioning between bridgmanite and ferropericlase in a laser-heated DAC, subjected to relevant lower-mantle conditions of up to approximately 120 GPa have been performed over the last two decades. Almost all experiments modelled the suggested pyrolitic model of the lower mantle with iron index, fe = Fe/(Fe + Mg)$_{at.}$ = 0.05–0.10. The only exception was the work performed by Tange et al. (2009) in a multi-anvil apparatus with sintered diamond anvils, in which the starting material was very different from pyrolitic and had fe = 0.50; it resulted in a highly iron-rich association, bridgmanite with fe = 0.16–0.32 and ferropericlase with fe = 0.49–0.92. The other studies have shown disparate results on the distribution of Fe in the various minerals (Fig. 4.18a). For example, while some demonstrated a positive correlation of the iron index between bridgmanite and ferropericlase (0.070–0.234) (Katsura and Ito 1996; Kesson et al. 2002; Kobayashi et al. 2005; Murakami et al. 2005; Sakai et al. 2009), others found the opposite: a negative correlation (Irifune 1994; Kesson et al. 1998; Wood 2000; Irifune et al. 2010) (Fig. 4.18a). At fe_{fPer} = 0.05–0.20 for both trends, the positive correlation was obtained mainly in association with low-Fe bridgmanite (fe = 0.010–0.103), while the negative correlation was in association with high-Fe bridgmanite (fe = 0.081–0.167). Additionally, there were experimental results showing variations in fe_{Brd} = 0.011–0.077 that were virtually independent of fe_{fPer} (c. 0.10) (Auzende et al. 2008). The general trends (I and II) can be identified in a consolidated graph (Fig. 4.18a). The positive trend divides into two branches (trends Ia and Ib) with the same correlation but with different values of fe_{Brd}, which could be due to a different pressure range being examined. These experimental discrepancies remain unexplained.

Experimental evaluations of the partition coefficient of iron between ferropericlase and bridgmanite defined as $K_{\text{D}}^{\text{Brd-fPer}} = (\text{Fe/Mg})^{\text{Brd}}/(\text{Fe/Mg})^{\text{fPer}}{}_{\text{at.}}$ produce even more inconsistent results. The values of $K_{\text{D}}^{\text{Brd-fPer}}$ vary from 0.04 to 0.9

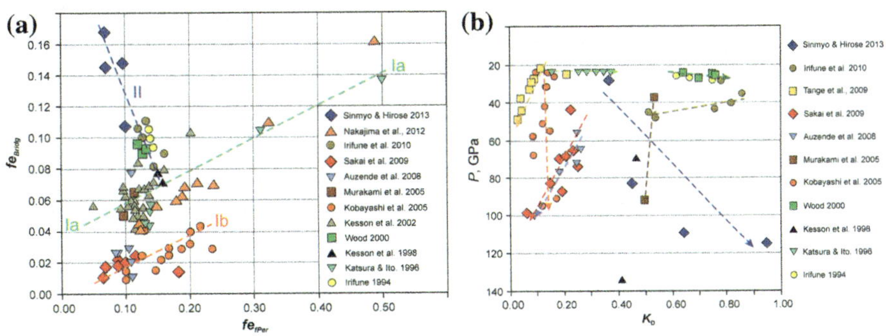

Fig. 4.18 Summary of literature experimental results of iron partitioning between bridgmanite and ferropericlase at high pressures and temperatures. **a** fe_{Brd} versus fe_{fPer}. **b** Variability of $K_{\text{D}}^{\text{Brd-fPer}}$ under different pressure conditions. From Kaminsky and Lin (2017). Used with a permission of the Mineralogical Society of America

(Fig. 4.18b). Some of the experiments have shown an increase of $K_D^{Brd-fPer}$ with pressure (Irifune 1994; Kobayashi et al. 2005; Sinmyo and Hirose 2013), whereas others have demonstrated the reverse (Andrault 2001; Kesson et al. 2002; Murakami et al. 2005; Tange et al. 2009; Irifune et al. 2010), and a third have not found any correlation between $K_D^{Brd-fPer}$ and pressure (Guyot et al. 1988; Katsura and Ito 1996; Auzende et al. 2008). Specifically, the concentration of Al^{3+} in bridgmanite has been shown to significantly affect its iron partitioning and Fe^{3+} content; it has been shown that the dissolution of Al^{3+} in bridgmanite at the top lower-mantle conditions can significantly enhance the Fe^{3+} occupancy in the A site of the lattice resulting in higher $K_D^{Bridg-fPer}$ (Irifune et al. 2010). The cause of these discrepancies may be due to the differing conditions in the experiments (great thermal gradients, pressure duration, chemical homogeneity, non-thermodynamic equilibrium, etc.) and in the compositions of the starting materials, including different concentrations of Fe and Al used in the experiments.

The discovery of iron spin crossover in ferropericlase led to the suggestion that $K_D^{Brd-fPer}$ would behave differently in relevant high *P–T* conditions found in the deep lower mantle, decreasing from the mid-lower mantle to bottom lower mantle conditions, such that most of the iron partitions into ferropericlase, leaving bridgmanite essentially iron-free (Badro et al. 2003; Speziale et al. 2005). More recent theoretical and experimental studies on a pyrolite composition have found that $K_D^{Brd-fPer}$ increases from approximately 0.5 at top-lower mantle conditions to 0.7–0.9 at about 800 km in depth due to the suggested coupled substitution of Al^{3+} and Fe^{3+} in bridgmanite, but then decreases to 0.4 or even lower due to the spin crossover of iron in ferropericlase at mid-lower mantle conditions (Irifune et al. 2010; Vilella et al. 2015; Xu et al. 2017). However, a high *P–T* DAC study, using a pyrolite composition, has shown that $K_D^{Brd-fPer}$ increases to 0.9 at bottom-lower mantle conditions; this is suggested to be due to the spin transition of Fe^{3+} in bridgmanite (Sinmyo and Hirose 2013). Recently, Fe-bearing bridgmanite has been reported to dissociate into Fe-rich phase, called H-Phase, and $MgSiO_3$-rich bridgmanite at deep lower mantle *P–T* conditions (Zhang et al. 2014), which can challenge the aforementioned conventional view of the iron index in the lower-mantle mineral assemblage. However, this report remains to be verified experimentally and theoretically. Thus far, a consensus on the iron partition coefficient across the spin transition in the lower-mantle assemblage has not yet been reached, especially under natural compositional, oxygen fugacity and *P–T* environments.

To elucidate the lower-mantle chemistry in these minerals, natural samples of bridgmanite manite and ferropericlase should be considered.

4.4.3 Natural Samples

To date, a number of bridgmanite, ferropericlase and other lower-mantle mineral grains have been identified as natural inclusions in diamonds, collected from Brazil, Canada, Guinea, Australia and some other countries, and subsequently analyzed (Kaminsky 2012 and references therein). Among them, there are 19 coexisting bridgmanite + ferropericlase pairs, which allow for studies into the distribution and partitioning of iron and other elements within the media from which those minerals originated within the lower mantle.

Ferropericlase. As demonstrated above, ferropericlase does not always have a suggested and experimentally modelled composition with *fe* = 0.08–0.11 or even closer, at c. 0.10 (Kesson and Fitz Gerald 1991). In the natural environment, however, ferropericlase may be much more iron-rich, up to *fe* = 0.64, and sometimes even up to 0.90; as such, this does not correspond to the pyrolitic composition of the formation media. Iron in ferropericlase is predominantly in a divalent form; the share of ferric iron is $Fe^{3+}/\Sigma Fe$ = 8–12 at.% (Kaminsky et al. 2015b). TEM studies with the use of EDX and EEL spectroscopy demonstrated that ferric iron in ferropericlase grains is located in exsolved non-stoichiometric Fe^{3+}-enriched clusters $Mg_{1+x}Fe^{3+}{}_{2-x}O_{4-x/2}$, varying in size from 1–2 nm to 10–15 nm and comprising ~3.6 vol.% of ferropericlase, while all remaining iron in ferropericlase is in the ferrous Fe^{2+} state (Kaminsky et al. 2015b) (Fig. 4.11a). These clusters, during the process of lowering pressure within the uppermost lower mantle, were released from the ferropericlase matrix and formed magnesioferrite $MgFe^{3+}{}_2O_4$ crystals, developing along dislocations in ferropericlase with a precise orientation relationship between both phases: $(022)_{mFer}//(022)_{fPer}$; $(11\text{-}1)_{mFer}//(11\text{-}1)_{fPer}$ and the zone axis for both phases is [2-11] (Kaminsky et al. 2015a) (Fig. 4.19).

Bridgmanite. Bridgmanite $^{VIII/XII}A^{2+}\ ^{VI}B^{4+}O^{2-}{}_3$, (where VI, VIII and XII are cation fold positions), in which A larger (mainly divalent), pseudo-dodecahedral (8/12-fold) site cations mostly include Mg, Fe^{2+}, Mn, Ni, Ca and some other elements; and B smaller, six-fold site cations include Si, Al and partly Fe^{3+}. The cation compositions for the analyzed samples arepresented in Table 4.8.

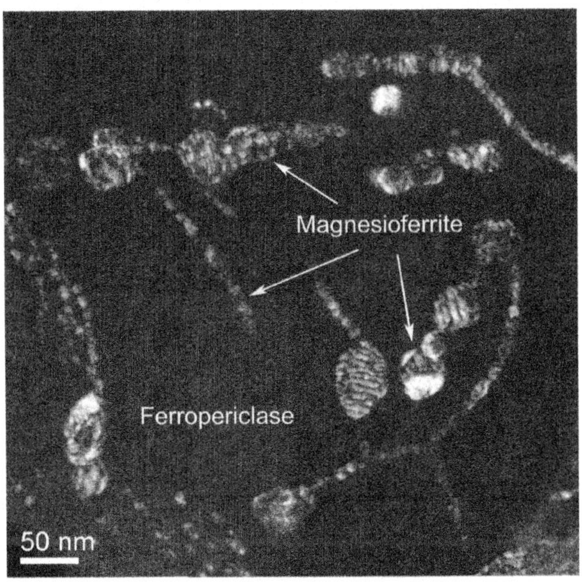

Fig. 4.19 Octahedral and cubic grains of magnesioferrite $MgFe^{3+}{}_2O_4$, developed along dislocations in ferropericlase. Dark field TEM image by R. Wirth (Kaminsky et al. 2015a). Used with a permission of the Canadian Mineralogist

Table 4.8 Cation fractions of natural lower-mantle bridgmanite coexisting with ferropericlase (in atomic numbers) (Kaminsky and Lin 2017)

Sample No.	Mg	Fe	Ni	Mn	Ca	Na	K	A_{total} (VIII/XII)	Si	Al	Ti	Cr	B_{total} (VI)	Refs[a]
Juina area, Brazil														
BZ120	0.878	0.064	0.000	0.002	0.001	0.002	0.000	0.948	0.943	0.024	0.003	0.006	0.976	1
BZ207	0.795	0.125	0.000	0.004	0.001	0.002	0.000	0.928	0.923	0.052	0.003	0.002	0.980	1
BZ210	0.750	0.071	0.000	0.013	0.012	0.034	0.006	0.885	0.856	0.197	0.000	0.016	1.069	1
BZ241	0.674	0.075	0.000	0.018	0.012	0.027	0.000	0.805	0.903	0.201	0.000	0.015	1.120	1
BZ242	0.406	0.055	0.000	0.018	0.078	0.149	0.000	0.707	0.915	0.247	0.000	0.023	1.185	1
BZ251	0.900	0.053	0.000	0.002	0.001	0.001	0.000	0.957	0.950	0.026	0.002	0.005	0.983	1
14-2	0.833	0.068	0.000	0.002	0.021	0.004	0.000	0.928	0.946	0.027	0.001	0.006	0.980	2
1-5	0.863	0.087	0.000	0.002	0.001	0.000	0.000	0.953	0.934	0.041	0.001	0.003	0.979	3
3-2	0.830	0.059	0.000	0.002	0.001	0.000	0.000	0.891	0.969	0.031	0.003	0.003	1.006	3
3-5	0.824	0.092	0.000	0.004	0.000	0.000	0.000	0.920	0.927	0.061	0.003	0.003	0.993	3
4-3	0.877	0.059	0.000	0.001	0.000	0.002	0.000	0.941	0.933	0.048	0.003	0.004	0.987	3
Northwest Territories, Canada														
DO2700100	0.940	0.040	0.000	0.001	0.007	0.000	0.000	0.988	0.956	0.014	0.000	0.004	0.973	4
DO2700300	0.938	0.041	0.000	0.001	0.001	0.002	0.000	0.984	0.981	0.014	0.000	0.004	0.998	4
DO27-97 14A	0.896	0.059	0.000	0.001	0.002	0.001	0.000	0.959	0.953	0.040	0.000	0.005	0.998	4
Kankan area, Guinea														
KK-103	0.891	0.047	0.000	0.001	0.001	0.001	0.000	0.942	0.955	0.012	0.000	0.002	0.970	5
KK-108	0.866	0.068	0.000	0.002	0.002	0.001	0.000	0.939	0.946	0.033	0.001	0.003	0.983	5
KK-16	0.892	0.046	0.000	0.002	0.001	0.002	0.000	0.943	0.957	0.011	0.000	0.004	0.972	5
KK-44	0.911	0.049	0.000	0.002	0.001	0.002	0.000	0.965	0.962	0.023	0.001	0.005	0.991	5
Eurelia area, South Australia														
FBS5-11	0.908	0.041	0.001	0.002	0.001	0.000	0.000	0.953	0.999	0.005	0.000	0.005	1.010	6

Note [a]References: (1) Hutchison (1997), (2) Zedgenizov et al. (2014a), (3) Hayman et al. (2005), (4) Davies et al. (2004), (5) Stachel et al. (2000), (6) Tappert et al. (2009a, b)

Most of the grains have compositions close to stoichiometric bridgmanite. Some deficit in both cation groups is probably caused by the presence of other, non-analyzed cations, such as P and REE, which are characteristic for this group of minerals. Judging by crystal-chemical calculations (Table 4.8), all iron substitutes for Mg^{2+} in the A site and, most likely, is in the divalent form Fe^{2+}, although studies based on Mössbauer spectroscopy by McCammon et al. (1997) suggested a significant portion of iron to be Fe^{3+}. There is no evidence for the presence of Fe^{3+} in the B site in natural bridgmanite. The three most Al-rich specimens (##BZ210; BZ241 and BZ242, all from Brazil) show an excess in alumina of 0.069–0.185. These amounts balance the deficit in cation site A of these specimens, suggesting that the location of this portion of Al most likely is in the site B, the 8/12-fold pseudo-octahedral coordination position.

4.4.4 Partition Coefficient of Iron in Natural Bridgmanite–Ferropericlase Assemblages

The analysis of the iron index *fe* in co-existing ferropericlase and bridgmanite grains and the partition coefficient of iron $K_D^{Brd-fPer}$ in all bridgmanite + ferropericlase assemblages shows that two groups of assemblages can be distinguished (Table 4.9; Fig. 4.20a). The first group (Group A; 84% of all pairs) forms an elongated cloud illustrating a general positive correlation between fe_{Brd} = 0.032–0.138 and fe_{fPer} = 0.116–0.363. The correlation trend is close to trend Ia from the experimental data, confirming the regularity in Fe exchange between the two major lower-mantle minerals. The second group (Group B; 16% of all grains) is composed of three specimens with almost constant fe_{fPer} = 0.173–0.193 and relatively high fe_{Brd}, varying from 0.088–0.120; all specimens are high-Al varieties. This group lies outside of any experimental trends. The increase in *fe* of both natural ferropericlase and bridgmanite from Group A is independent of the $K_D^{Brd-fPer}$ values (0.169–0.479), while Group B, which has *fe* values in the same range as Group A, has elevated values of $K_D^{Brd-fPer} = 0.433-0.657$ (Fig. 4.20b).

Such variations in the iron indices and the bulk iron contents in the co-existing major lower-mantle minerals differ drastically from what was expected for a pyrolitic lower mantle with a homogeneous distribution of iron in ferropericlase and bridgmanite (Kesson and Fitz Gerald 1991; Wood 2000; Lee et al. 2004). According to the experimental and theoretical data for a bulk pyrolitic composition, where the total iron content in the system was fixed at *fe* = 0.09–0.15 (Irifune et al. 2010; Vilella et al. 2015; Xu et al. 2017), we should expect a decrease of the iron content in bridgmanite with an increase of the iron content in ferropericlase. In reality, this is not the case, however, since both ferropericlase and bridgmanite demonstrate a simultaneous, well-correlated increase in *fe* values (Fig. 4.20a–d).

Both groups of assemblages are characterised by similar concentrations of Ni in ferropericlase, mainly corresponding to the bulk Ni contents in the mantle (6052–11,212 ppm from Group A and 9982–11,711 ppm from Group B) (Fig. 4.20c).

Table 4.9 Compositions of coexisting bridgmanite and ferropericlase included in lower-mantle diamonds (Kaminsky and Lin 2017)

Sample No.	Mineral association	Bridgmanite				Ferropericlase				$K_D^{Br-fPer}$	Refs[a]
		FeO, wt%	MgO, wt%	Al_2O_3, wt%	*fe*	FeO, wt%	MgO, wt%	Ni, ppm	*fe*		
Juina area, Brazil											
BZ120	Brd + fPer	4.63	35.39	1.23	0.069	42.81	55.25	6052	0.310	0.169	1
BZ207	Brd + fPer + Jeffbenite	9.02	32.03	2.66	0.138	43.14	54.84	7074	0.313	0.358	1
BZ210	Brd + fPer	5.14	30.21	10.04	0.088	27.20	69.22	9982	0.183	0.433	1
BZ241	Brd + fPer + Ruby	5.36	27.15	10.24	0.101	28.63	68.21	11,711	0.193	0.470	1
BZ242	Brd + fPer	3.95	16.35	12.58	0.120	25.50	69.30	11,083	0.173	0.657	1
BZ251	Brd + fPer	3.80	36.25	1.33	0.056	23.52	74.77	9825	0.152	0.333	1
14-2	Brd + fPer + Di?	4.87	33.59	1.37	0.086	31.50	65.60	9746	0.216	0.302	2
1-5	Brd + fPer	6.25	34.80	2.07	0.093	36.70	60.98	9137	0.258	0.298	3
3-2	Brd + fPer + CaSiPv + ‘Ol’	4.21	33.43	1.60	0.066	25.27	72.21	10,336	0.166	0.360	3
3-5	Brd + fPer + ‘Ol’	6.60	33.20	3.10	0.101	48.96	50.34	2437	0.363	0.204	3
4-3	Brd + fPer + ‘Ol’	4.27	35.36	2.45	0.064	25.67	71.11	10,631	0.168	0.381	3
Northwest Territories, Canada											
DO2700100	Brd + fPer + CaSiPv + SiMg	2.85	37.90	0.69	0.041	16.80	81.60	8882	0.104	0.365	4
DO2700300	Brd + fPer + ‘Ol’	2.97	37.80	0.72	0.042	23.60	73.70	11,083	0.154	0.245	4
DO27-97 14A	Brd + fPer + Ni	4.22	36.11	2.03	0.062	21.29	77.45	10,742	0.135	0.425	4
Kankan area, Guinea											
KK-103	Brd + fPer	3.39	35.92	0.61	0.050	20.11	76.98	11,004	0.129	0.388	5
KK-108	Brd + fPer Ga	4.90	34.90	1.67	0.037	22.15	75.61	10,049	0.143	0.479	5
KK-16	Brd + fPer + Siderite	3.34	35.94	0.55	0.050	22.68	75.34	11,212	0.146	0.343	5
KK-44	Brd + fPer + CaSiPv + ‘Ol’	3.52	36.73	1.19	0.051	20.15	76.45	10,532	0.130	0.363	5
Eurelia area, South Australia											
FBS5-11	Brd + fPer	2.93	36.58	0.25	0.043	23.42	74.04	9550	0.152	0.253	6

Note [a]Reference numbers are the same as in Table 4.8

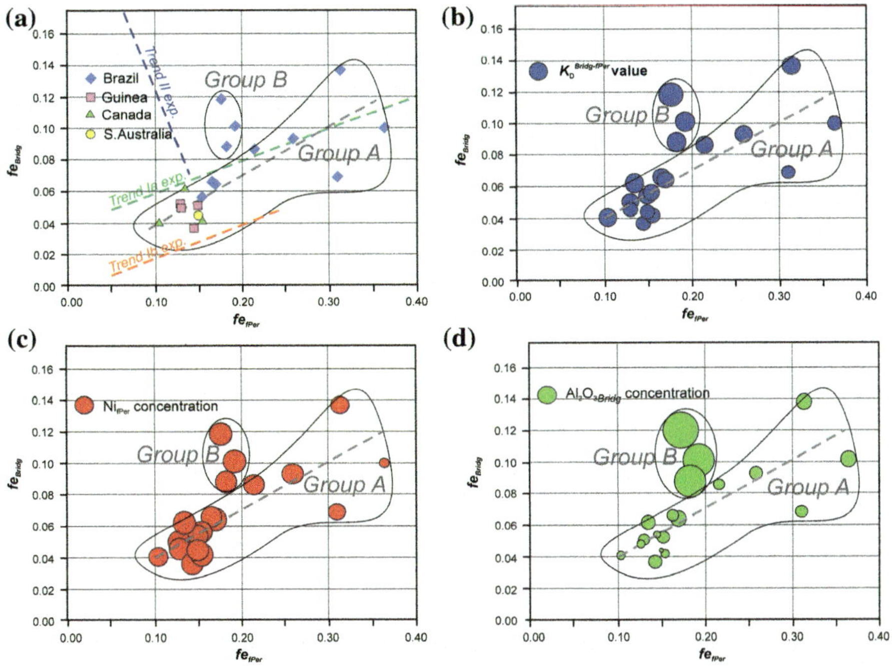

Fig. 4.20 Iron indices fe = Fe/(Fe + Mg)$_{at.}$ in coexisting ferropericlase and bridgmanite in lower-mantle diamond inclusions. **a** General plot. **b** With $K_D^{Brd-fPer}$ values, shown as *circles* proportional to the $K_D^{Brd-fPer}$ values. **c** With Ni concentrations in ferropericlase, shown as *circles* proportional to the Ni concentrations. **d** With Al_2O_3 concentrations in bridgmanite, shown as *circles* proportional to the Al_2O_3 concentrations. Four groups of samples from Brazil, Canada, Guinea, and Australia with the grains of ferropericlase and bridgmanite that associate with each other are used. From Kaminsky and Lin (2017). Used with a permission of the Mineralogical Society of America

However, one of the analyzed samples (#3-5 from Brazil) has a low concentration of Ni = 2437 ppm, and there are a number of low-Ni ferropericlase inclusions in diamond (Fig. 4.21). Analysis of the Ni content in ferropericlase as a function of the iron index shows that the ferropericlase grains with *fe* of approximately 0.2 in deep-diamond inclusions have Ni content close to the estimates of 8000–12,000 ppm for the bulk pyrolitic composition; however, the Ni content drastically decreases with increasing iron index in ferropericlase, especially for samples from Brazil (Fig. 4.10). The samples with *fe* of more than 0.6 from Brazil have Ni contents of less than 2000 ppm. Therefore, the analyses of natural ferropericlase grains show that the higher the iron index in the grain, the lower the Ni content.

There is a general (although not very strong) positive correlation between Ni and the partition coefficient of iron (Fig. 4.21). The lowest $K_D^{Brd-fPer}$ values (0.169 in sample # BZ120 and 0.204 in sample # 3-5) are characterized by the lowest Ni concentrations in ferropericlase, whereas the samples with the highest $K_D^{Brd-fPer}$ values (## BZ210, BZ241 and BZ242) are characterized by the highest Ni contents in ferropericlase.

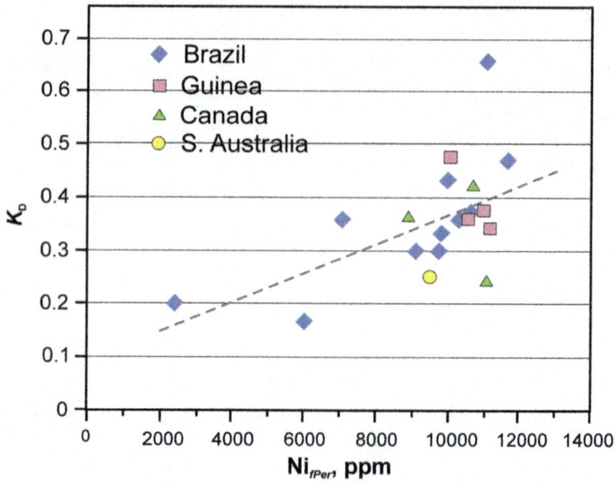

Fig. 4.21 Variations of Ni concentrations in ferropericlase. $K_D^{Brd-fPer}$ versus Ni content in ferropericlase. From Kaminsky and Lin (2017). Used with a permission of the Mineralogical Society of America

4.4.5 Role of Ni and Al in the Iron Partition Coefficient Distribution

The Ni concentration in ferropericlase indicates the presence of metallic iron phase (s) in the magmatic system of the deep mantle, as established by Frost et al. (2004). It was shown that the increasing weight fraction of the metallic phase (suggesting the conditions present within the lowermost lower mantle) leads to a decrease of Ni concentration in the lower-mantle material and also leads to a decrease in concentration of this element in coexisting ferropericlase and bridgmanite (Ryabchikov and Kaminsky 2013a, b). In this process, iron appears here as a dilutant of Ni dissolved in metal. According to experimental data (Frost et al. 2004; Frost and McCammon 2008), metal content in the lower mantle is estimated at 1 wt% (10,000 ppm) before the release of the metallic alloy. This implies that the ferropericlase grains with high Ni concentrations were formed in media that did not contain a metallic alloy (or its quantity was extremely small), suggesting that these originated within the uppermost lower mantle, while the low Ni ferropericlase grains were formed in the presence of a metallic alloy within lower parts of the lower mantle. More than 40% of the studied lower-mantle ferropericlase grains are low-Ni and high-Fe inclusions. These variations in Fe and Ni concentrations in ferropericlase point to a radial compositional gradient in the lower mantle at the time of the formation of these minerals, and the anti-correlation of Ni and Fe in ferropericlase may be applied as a qualitative criterion of the depth of its origin.

The correlation between the $K_D^{Brd-fPer}$ values and Ni concentrations in ferropericlase (Fig. 4.21) is not strong because of the presence of other elements in minerals, such as Al, Co, Mn, Na, among others. The major factor here is the Al impurity in ferropericlase and bridgmanite, which is the real cause of the existence of the two groups of associations (Fig. 4.20d). Al_2O_3 in ferropericlase and bridgmanite comprises 0.69–3.10 wt% in Group A and 10.04–12.58 wt% in Group B. In general,

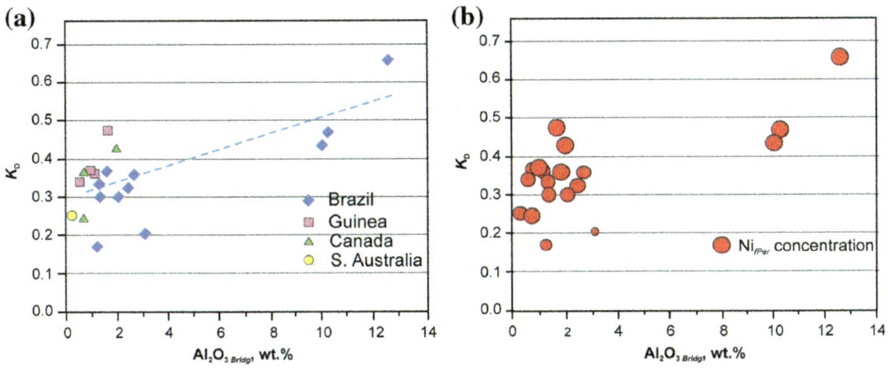

Fig. 4.22 Iron partitioning coefficient $K_D^{Brd-fPer}$ versus Al_2O_3 content in bridgmanite. **a** General plot. **b** With Ni concentrations in ferropericlase, shown as *circles*, the radii of which are proportional to the Ni contents. From Kaminsky and Lin (2017). Used with a permission of the Mineralogical Society of America

$K_D^{Brd-fPer}$ correlates with the Al_2O_3 content in ferropericlase and bridgmanite, which is particularly noticeable in the Brazilian samples (Fig. 4.22a). However, this correlation is quite weak (Fig. 4.22a, b) and is not get fully understood.

The enrichment of ferropericlase and bridgmanite with Al is strongly pressure dependent (Andrault et al. 2007). In the uppermost part of the lower mantle, Al-rich phases still exist, such as majorite and jeffbenite. In experiments at pressure conditions of 24–28 GPa, these minerals dissociate, and the released Al is incorporated into ferropericlase and bridgmanite (Irifune and Tsuchiya 2007). In the natural environment, the reverse scenario occurs: juvenile high-Al ferropericlase and bridgmanite at 24–28 GPa releases Al with the formation of Al-bearing phases, such as majorite and jeffbenite.

As a result, the following scenario can be outlined for the lower mantle. Bridgmanite has a high, 10–12 wt% concentration of Al_2O_3, which is consistent with its bulk concentration in the primitive mantle of 4–5 mol% (McDonough and Sun 1995). Al occupies its position in both cation sites, A and B. In each case, Al substitutes Si and divalent cations with oxygen-vacancy formation:

$$2^{VIII/XII}Al^{3+} + O^{2-} \rightarrow 2^{VIII/XII}(Mg, Fe)^{2+} + V^0; \text{ and}$$
$$2^{VI}Al^{3+} + V^0 \rightarrow 2^{VI}Si^{4+} + O^{2-};$$

In the case of Al positioning in both A and B sites, a charge-coupled mechanism (CCS) takes place, for which no oxygen vacancies are required for charge balance (Richmond and Brodholt 1998):

$$^{VI}Al^{3+} + ^{VIII/XII}Al^{3+} \rightarrow ^{VI}Si^{4+} + ^{VIII/XII}(Mg, Fe)^{2+};$$

which, according to theoretical calculations, dominates at lower-mantle pressures and temperatures (Akber-Knutson and Bukowinski 2004). The concentration of Al in ferropericlase and bridgmanite, particularly in the A site may indicate the depth of the minerals' origin because previous studies have shown that Al concentration in bridgmanite increases with increasing pressure along a representative lower-mantle geotherm (Irifune et al. 2010). Recently, an experimental study of the system $MgSiO_3$–Al_2O_3 under pressures up to 52 GPa and 2000 K, with the use of sintered diamond anvils combined with in situ synchrotron X-ray diffraction observations in a multi-anvil apparatus, demonstrated that the Al_2O_3 content in bridgmanite increases from 12 mol% at 27 GPa to 29 mol% at 52 GPa, suggesting that the Al_2O_3 content in bridgmanite can be used as a pressure indicator at pressures of above 30 GPa (Liu et al. 2016). As shown in Sect. 4.2.2, iron in ferropericlase and bridgmanite is in the A site, most likely in a ferrous divalent form. Ferropericlase with depth (and correspondingly increasing pressure) has a decreasing concentration of Ni and increasing *fe* values, reflecting the increase in the iron content in the lower mantle. The decrease of Ni content in ferropericlase may be used as another qualitative geobarometer in the lower mantle. In contrast to experimental conditions, most of which were performed with pyrolitic compositions (*fe* = 0.05–0.15), the results on natural samples show that iron indices in ferropericlase and bridgmanite have a positive correlation (Fig. 4.20), demonstrating the total enrichments of the lower-mantle media in iron. This enrichment is correlated with the Ni (negatively, Fig. 4.20c) and Al_2O_3 (positively, Fig. 4.20d) concentrations, each of which is an independent criterion of the increasing pressure. One may conclude that the Fe content of the lower mantle increases with depth, in contrasr to the pyrolitic composition.

This scenario plays only a small part of a more complicated picture for existing phases within the lower mantle, which incorporates other mineral phases and is characterized by changeable conditions within an open, natural system. It explains the controversies in the aforementioned high P–T experimental results carried out for standard closed systems.

4.4.6 Regional Differences in the Distribution of Iron in Ferropericlase and Bridgmanite

There are observed differences in the iron index of ferropericlase values and related characteristics from different regions. For example, ferropericlase from Canada, South Australia and South Africa have *fe* = 0.10–0.20, which is close to the pyrolite values. Ferropericlase from Guinea, along with similar *fe* values, also has higher *fe* values of 0.25. Moreover, iron-rich ferropericlase from Brazil (reaching the composition of magnesiowüstite) comprises almost half of all grains in this region (c. 46.5%). It has been suggested that such differences represent the different depths of the host diamonds sampled in the lower mantle. In a pyrolitic lower mantle, the average amount of iron in the lower-mantle minerals are commonly believed to be close to 10%, chiefly distributed between ferropericlase and bridgmanite and ferropericlase phases (Irifune et al. 2010). The analysis of the iron indices in these phases have shown that the total iron content in some of these minerals can be much higher than what is expected for a 'normal' lower mantle. As shown in Fig. 4.9, the occurrence of these 'anomalous', high-Fe ferropericlase and high-Fe ferropericlase and bridgmanite pairs with *fe* = 0.17–0.36 and 0.07–0.14, respectively, indicates that the lower mantle is chemically inhomogeneous with respect to the iron content at least in certain regions where these assemblages originated.

4.4.7 Major Differences in Iron Partitioning Between Natural Samples and Mineral Physics Results

Recent theoretical calculations considering a wide range of oxygen fugacity conditions and different iron spin and valence states in the lower-mantle pyrolite assemblage have further indicated that the Fe^{3+} content in ferropericlase and bridgmanite decreases with increasing depth in the lower mantle; at lowermost lower mantle conditions, Fe^{3+} content in ferropericlase and bridgmanite becomes almost negligible (Xu et al. 2017). However, our analyses of the natural ferropericlase and bridgmanite and ferropericlase assemblages show that most of the iron in the A site of the ferropericlase and bridgmanite lattice is in the Fe^{2+} state, which contradicts these mineral physics studies.

High *P–T* experiments have shown that the partition coefficient $K_D^{Brd-fPer}$ is observed to increase from approximately 0.5 at 23 GPa up to 0.85 at approximately 28 GPa ($\sim$800 km in depth. This can be associated with the Al^{3+}–Fe^{3+} coupled substitution in ferropericlase and bridgmanite in a pyrolitic composition, where iron ions partition almost equally between the ferropericlase, and the ferropericlase and bridgmanite lattices (Irifune et al. 2010). Results from such high P–T experiments showed that the $K_D^{Brd-fPer}$ value stays almost constant at $\sim$0.85 at pressures of 28–40 GPa ($\sim$800–1000 km in depth) and then notably decreases to 0.4–0.5 with increasing pressures of 40–50 GPa (1000–1200 km in depth) (Irifune et al. 2010). This dramatic decrease is most likely associated with the spin transition of Fe^{2+} in ferropericlase, which results in a volume collapse of about 2% (Lin et al. 2013) promoting iron ions in ferropericlase and bridgmanite to favorably partition into the low-spin ferropericlase, which is energetically more stable than the high-spin ferropericlase and bridgmanite and ferropericlase. Without the alumina substitution in ferropericlase and bridgmanite at such conditions, the $K_D^{Brd-fPer}$ values decrease to approximately 0.2 (Sakai et al. 2009), i.e., lower than the value suggested from high *P–T* experiments using a pyrolitic composition (Irifune et al. 2010). That is, in an alumina-poor environment in the lower mantle, such as the olivine-rich peridotite environment, the $K_D^{Brd-fPer}$ value is expected to remain low at approximately 0.2

with or without the spin transition effect. If the lower mantle is chemically pyrolitic, these changes in the iron partitioning indicate that the deeper parts of the lower mantle (especially toward the lowermost parts) would likely contain iron-enriched, low-spin ferropericlase and iron-poor, alumina-rich ferropericlase and bridgmanite. By contrast, in the upper parts of the lower mantle, especially at approximately 800 km in depth, iron would almost equally partition between bridgmanite and ferropericlase (Irifune et al. 2010). However, a comparison of these experimental results with our analyses on natural samples clearly shows contradicting $K_D^{Brd-fPer}$ values; the natural samples show very diverse $K_D^{Brd-fPer}$ values that cannot be simply explained by the alumina dissolution in ferropericlase and bridgmanite and the spin transition of Fe^{2+} in ferropericlase. These major differences in the iron indices and $K_D^{Brd-fPer}$ values demonstrate the complex, inhomogeneous iron speciation and chemistry in the lower mantle, which are well beyond the knowledge of current mineral physics studies.

4.4.8 Conclusions

One may conclude, based upon determined iron contents in natural ferropericlase and bridgmanite, and on the distribution of the partition coefficient of iron in natural bridgmanite + ferropericlase assemblages that some areas in the deep lower mantle are iron-rich and differ markedly from a pyrolitic composition. In the lowermost lower mantle, ferropericlase is also Ni-poor, demonstrating the presence in the mineral-forming environment of a free metallic phase; bridgmanite is Al-rich.

The juvenile iron partitioning in the bridgmanite + ferropericlase association $K_D^{Brd-fPer}$ is as low as 0.1–0.2. During the crystallization of diamonds at a relatively more shallow depth in the lower mantle, in which $K_D^{Bridg-fPer}$ increases to 0.4–0.5, and even as high as 0.7, they trap and deliver to the surface such associations. This regularity, established in geological samples, was supported by recently published theoretical calculations by Muir and Brodholt (2016), showing a decrease of $K_D^{Brd-fPer}$ with depths from 0.32 to 0.06.

The details of the element partitioning between natural ferropericlase and ferropericlase, in the lower mantle (at least in some areas), are as follows:

- According to crystal-chemical calculations, iron in ferropericlase and bridgmanite is ferrous Fe^{2+} in the A site, substituting for Mg^{2+};
- Almost all iron in ferropericlase is ferrous Fe^{2+}; the share of ferric Fe^{3+} iron in ferropericlase is $Fe^{3+}/\Sigma Fe$ = 8–12 at.%; Fe^{3+} is concentrated in exsolved clusters of $Mg_{1+x}Fe^{3+}{}_{2-x}O_{4-x/2}$, while all remaining iron in ferropericlase is in the ferrous, Fe^{2+} state;
- Iron contents in both ferropericlase and bridgmanite increase with depth (pressure), reflecting the increasing Fe content in the lower part of the lower mantle;

- Al content in ferropericlase and bridgmanite from the lower part of the lower mantle is at ~10–12 wt% Al_2O_3;
- Al in bridgmanite is mainly in the cation B site and partly in the cation A site, in both cases substituting Si, Mg and Fe with vacancy formation; and in the case of Al positioning in both, B and A sites, a charge-balanced reaction takes place.

These observations in natural samples cannot be simply explained by the dissolution of Al in ferropericlase and bridgmanite and the spin transition of Fe^{2+} in ferropericlase. The differences between observations in natural samples and experimental results should be taken into account in future calculations and experimental works.

4.5 $CaSiO_3$ Perovskite

4.5.1 General

CaSi-perovskite $CaSiO_3$ is considered to be the third most abundant mineral in the lower mantle with a 5–10 wt% abundance in the pyrolite model (Akaogi 2007; Irifune et al. 2010; Tsuchiya and Kawai 2013). The percentage volume of CaSi-perovskite in the subducted mid-oceanic ridge basalt could be as high as 21–29%, such that its presence will lower the Mg/Si ratio of the lower mantle and thus influence the mineralogy of the region (Funamori et al. 2000; Hirose et al. 1999; Perrillat et al. 2006). Ca cation in CaSi-perovskite is much larger in the ionic radius than two other common abundance cations, Mg and Fe, and is not soluble in bridgmanite, in which its A and B sites cannot accommodate the larger Ca cation. In the relevant lower-mantle *P–T* conditions, CaSi-perovskite is stable in the cubic crystal structure from the transition zone to the lowermost part of the lower mantle (Shim et al. 2000a). In deep-diamond inclusions, CaSi-perovskite forms small, colorless or milky-white grains with an anhedral or characteristic 'negative' euhedral shape (Fig. 4.23). It associates with bridgmanite, ferropericlase and coesite (formerly stishovite), as well as with CaTi-perovskite, jeffbenite, merwinite, majorite, 'olivine', and native Ni (Kaminsky 2012; Zedgenizov et al. 2016). Within diamonds, delivered to the surface, CaSi-perovskite acquires a retrograde structure, like that of larnite or walstromite (Gasparik et al. 1994; Joswig et al. 1999).

Some samples, where $CaSiO_3$ associates not with bridgmanite and ferropericlase, but with Ca_2SiO_4 and $CaSi_2O_5$, were identified within several Guinean and Brazilian diamonds (Stachel et al. 2000; Hayman et al. 2005; Anzolini et al. 2016a; Burnham et al. 2016; Zedgenizov et al. 2016). This association of walstromite $CaSiO_3$ + larnite β-Ca_2SiO_4 + titanite $CaSi_2O_5$, on the basis of the in situ XRD and micro-Raman spectroscopy, combined with thermodynamic calculations, is believed to be formed at ~9.5–11.5 GPa and 1500 K. In the natural environment, this may be a retrograde product of perovskite-structured $CaSiO_3$ at a pressure of 10–12 GPa.

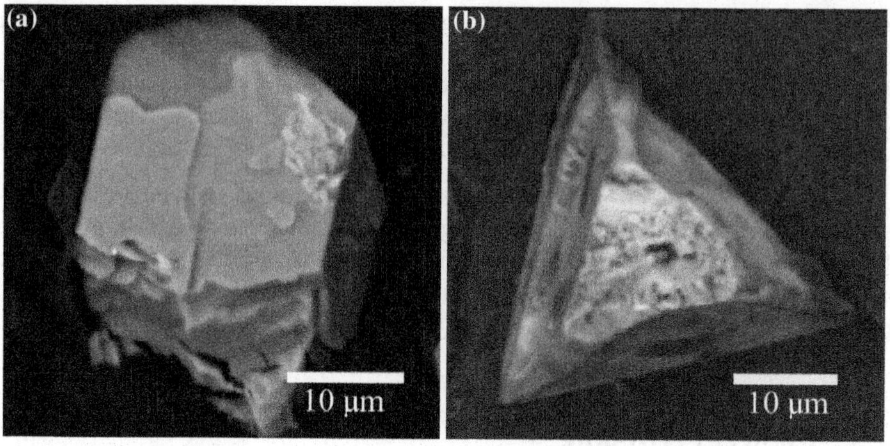

Fig. 4.23 Grains of CaSi-perovskite extracted from diamonds of the Rio Soriso area. SEM image. From Hayman et al. (2005)

4.5.2 Chemical Composition of Natural $CaSiO_3$ Perovskite

During the course of retrograde phase transitions, the composition of CaSi-perovskite remains unchanged and remarkably clean, with only minor admixtures of Mg, Ti, Al, Fe, Sr and K (Tables 4.10 and 4.11).

The major admixtures are Ti and Al. Most samples have low concentrations of these elements, less than 0.7 wt% Al_2O_3 and less than 0.6 wt% TiO_2, forming Group 1 (Fig. 4.24). Some samples have elevated concentrations of TiO_2, from 2.13 to 6.65 wt%, with a similar Al_2O_3 content (0.20–0.56 wt%), forming Group 2. Experimental data indicate that $CaSiO_3$ and $CaTiO_3$ are immiscible phases at low (below 9 GPa)-pressure conditions (Kubo et al. 1997). It is possible that at the lower-mantle conditions, the solubility of Ti increases with pressure. After the release of inclusions from host diamonds, dissociation occurs: Ti-phase forms either isolated $CaTiO_3$ grains in the $CaSiO_3$ matrix or symplectitic intergrowths of both phases (Zedgenizov et al. 2016).

A separate group is formed by $Ca(Si,Ti)O_3$ compositions with TiO_2 = 23–53 wt % reaching an almost stoichiometric $CaTiO_3$ composition. In individual diamonds, Ca–Ti-perovskite was found either as isolated inclusions or in association with majoritic garnet, SiO_2, ilmenite and a Si–Mg phase (Kaminsky et al. 2001; Hayman et al. 2005; Zedgenizov et al. 2014a, 2016). Inclusions of Ca–Ti-perovskite contain minor SiO_2 (1.1–3.9 wt%) and Al_2O_3 (0.6–2.3 wt%). The negative Al–Ti trend in this group indicates the replacement of Si with Ti in cation Group B. In this case, an oxygen vacancy should be expected from the result of the substitution: $2^BSi^{4+} \rightarrow 2^BAl^{3+} + V^0$. Some of these compositions may be titanite.

In high-pressure experiments performed on high-Al CaSi-perovskite ($CaSi_{0.5-0.6}Al_{0.4-0.5}O_{2.75-2.8}$) at pressure conditions below 18 GPa (i.e., corresponding to the

Table 4.10 Representative compositions of CaSi-perovskite inclusions in diamond from different areas (wt%)

Country	Brazil					Canada	Guinea	
Area	Juina				Machado	Northwest Territories	Kankan	
Location	Sao Luiz placer			Corrigo Chicoria placer	Placer	Panda pipe	Placer	
Sample No.	BZ115	13	31-1	5-104a	P28-1b	PA-50c	KK-44g	KK-87a
Mineral association	CaSiPrv	CaSiTiPrv + Brd + Jeffbenite	CaTiSiPrv + Maj + SiO2 + Ky	CaSiPrv + Ni	CaSiPrv + Di	CaSiPrv + fPer	CaSiPrv + Brd + fPer + ‘Ol’	CaSiPrv + fPer
SiO_2	50.59	47.30	3.94	51.23	49.98	51.13	51.33	51.11
TiO_2	0.01	6.65	52.70	0.04	0.03	0.01	0.01	0.07
Al_2O_3	0.08	0.32	2.34	0.06	0.04	0.09	0.06	0.66
Cr_2O_3	0.01	0.03	0.00	n.a.	<0.04	0.01	0.00	0.03
FeO	0.16	0.09	0.00	0.06	0.16	0.04	0.04	0.10
NiO	0.02	n.a.	n.a.	0.03	0.02	0.00	0.00	0.00
MnO	0.02	0.00	0.00	0.05	<0.04	0.04	0.02	0.05
MgO	0.14	0.00	0.00	n.a.	0.49	0.10	0.03	0.09
CaO	48.49	43.00	40.10	48.37	47.22	47.90	47.70	45.80
SrO	n.a.	n.a.	n.a.	n.a.	n.a.	0.10	0.21	0.85
Na_2O	0.02	0.02	0.00	n.a.	0.04	0.02	0.07	0.05
K_2O	n.a.	0.01	0.00	n.a.	0.08	0.10	0.03	0.73
P_2O_5	n.a.	n.a.	n.a.	n.a.	n.a.	0.01	0.00	0.02
Total	99.54	97.47	99.14	99.84	98.06	99.54	99.50	99.57
fe	0.325	–	–	0.428	0.155	0.197	0.379	0.386
mg	0.675	–	–	0.572	0.845	0/803	0.621	0.614
References	Hutchison (1997)	Zedgenizov et al. (2014a)		Kaminsky et al. (2001)	Burnham et al. (2016)	Tappert et al. (2009a)	Stachel et al. (2000)	

Note: *n.a.* not analysed

Table 4.11 Minor elements in CaSi-perovskite (wt%)

Oxides	Brazil				Guinea	Canada	
	Rio São Luiz	Rio Vermelho and Chicoria	Rio Soriso	Juina-5 pipe	Kankan	DO-27 and A21 pipes	Panda pipe
MgO	0–0.46	n.a.	0.05–0.35	0–0.30	0.03–0.09	0	0.10
TiO_2	0–6.65	0.04–0.26	0.10–2.13	0.02–0.64	0.01–0.07	0	0.01
Al_2O_3	0–1.52	0.03–0.11	0.21–0.26	0.14–6.27	0.03–0.66	0–0.02	0.09
SrO	n.a.	0.01–0.06	n.a.	n.a.	0.06–0.85	n.a.	0.10
K_2O	0–0.02	0–0.07	0–0.14	0.01–0.11	0.01–0.73	0–0.07	0.10
References	Wilding et al. 1991, Harte et al. (1999)	Kaminsky et al. (2001)	Hayman et al. (2005)	Thomson et al. (2014)	Stachel et al. (2000)	Davies et al. (2004)	Tappert et al. (2009a)

Note: *n.a.* not analysed

Fig. 4.24 Al_2O_3 versus TiO_2 in natural CaSi-perovskite

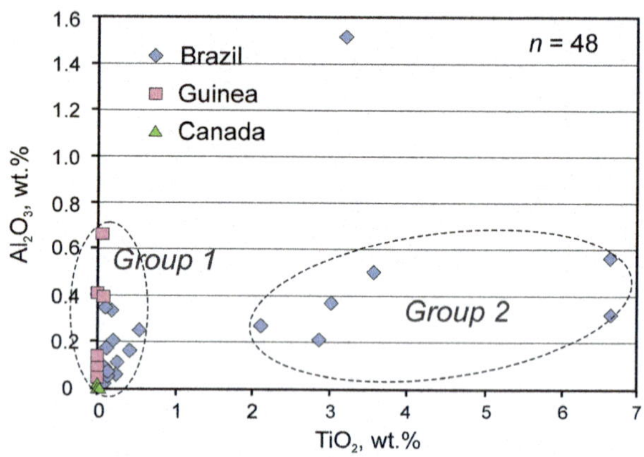

transition zone), oxygen defects occur in pseudo-cubic (111) layers in well-ordered ten- or eight-fold superstructures (Bläß et al. 2007).

Similar to bridgmanite, a charge-coupled substitution (CCM) $2Al^{3+} \rightarrow Si^{4+} + Ca^{2+}$ may also be suggested; however, for the CaSi-perovskite, the occupation of a larger calcium site by a relatively small trivalent aluminium cation is less favorable and limits the maximum amount of Al incorporated by a charge-coupled substitution to, at most, 6 mol% (Takafuji et al. 2002). The amount of Al incorporated into natural CaSi-perovskite may be limited to this amount. For CaSi-perovskite samples analyzed to date, a compositional gap between CaSi- and CaTi-perovskites occurs (Fig. 4.24), but it remains unclear whether this gap exists in nature. Furthermore, two low-Ti samples of CaSi-perovskite show anomalously high Al concentrations, 1.52 and 6.27 wt%

Al_2O_3 (Fig. 4.24). The nature of this anomaly remains unclear. The explanation that these samples belong to the mafic association has not been excluded.

The Mg concentration in CaSi-perovskite does not exceed 0.35% MgO. In some experiments, up to 10% Mg can be substituted into CaSi-perovskite at 55 GPa (Armstrong et al. 2012). This effect was enhanced with pressure, temperature, and the substitution of Ti^{4+} for Si^{4+}, with Mg–Ca forming a complete solution at 97 GPa for modest amounts of Ti substitution (Ti/(Ti + Si) = 0.05). However, in analyzed natural $Ca(Si,Ti)O_3$ compositions MgO was not detected at all.

CaSi-perovskite/melt partition coefficients for REE are much higher than those for bridgmanite (some by a factor of 1000) (Righter et al. 2014), causing high concentrations of REE (up to 2–3 magnitude orders) in CaSi-perovskite (Fig. 4.25). In contrast to bridgmanite and ferropericlase-magnesiowüstite, the REE patterns in CaSi-perovskite show significant enrichment in LREE (e.g., ratios of La/Yb = 3.6–37.7; Fig. 4.25). Some CaSi-perovskite grains show positive Eu-anomalies, which are possibly attributable to the composition of subducted lithospheric slabs. Such grains most likely belong to the mafic association. This type of pattern may be explained by a significant increase in the crystal-liquid partition coefficient for LREE, compared with HREE, with increasing pressure (Wang et al. 2000). In this case, LREE enrichment may be a potential qualitative geobarometer. The general enrichment of CaSi-perovskite in incompatible elements is caused by much larger (roughly by one order of magnitude) crystal-liquid partition coefficients of trivalent and tetravalent cations in the CaSi-perovskite in comparison with bridgmanite

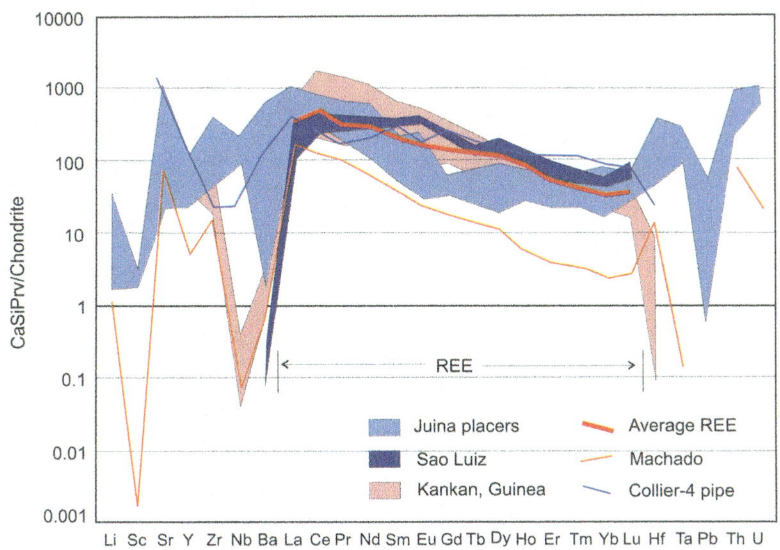

Fig. 4.25 Chondrite-normalised trace element patterns in natural CaSi-perovskite. Data from Wang et al. (2000), Kaminsky et al. (2001), Stachel et al. (2000), Bulanova et al. (2010), Harte et al. (1999), Burnham et al. (2016). Chondrite composition was derived from McDonough and Sun (1995). Modified after Kaminsky (2012), with additions

(Wood and Corgne 2009). The patterns of partitioning of trace elements into $CaTiO_3$ and $CaSiO_3$ perovskites are virtually identical (Corgne and Wood 2002). The major implication of the partitioning data is that, despite being a volumetrically minor phase, CaSi-perovskire is the major host for the heat-producing elements Th and U in the deep Earth, as well as being the principal reservoir for the REE (Corgne et al. 2005). Perry et al. (2017), on the basis of DFT calculations, suggested that CaSi-perovskite may be a concentrator of U and Th in the lower mantle.

4.5.3 Crystal Structures and Phase Transitions

Cubic structure. In early theoretical predictions and high-pressure XRD experiments including those performed in DAC, CaSi-perovskite was thought to have a cubic structure (Fig. 4.26a) with a space group *Pm*3*m* under pressures of up to 134 GPa at room temperature (Liu and Ringwood 1975; Mao et al. 1989; Tarrida and Richet 1989; Wang and Weidner 1994; Wentzcovitch et al. 1995; Warren et al. 1998; Shim et al. 2000a, b). No structural transformations were observed in these studies, and CaSi-perovskite was considered as 'an invisible' component of the lower mantle in view of the similarity of its density with that given by the standard Earth model, PREM (Tarrida and Richet 1989).

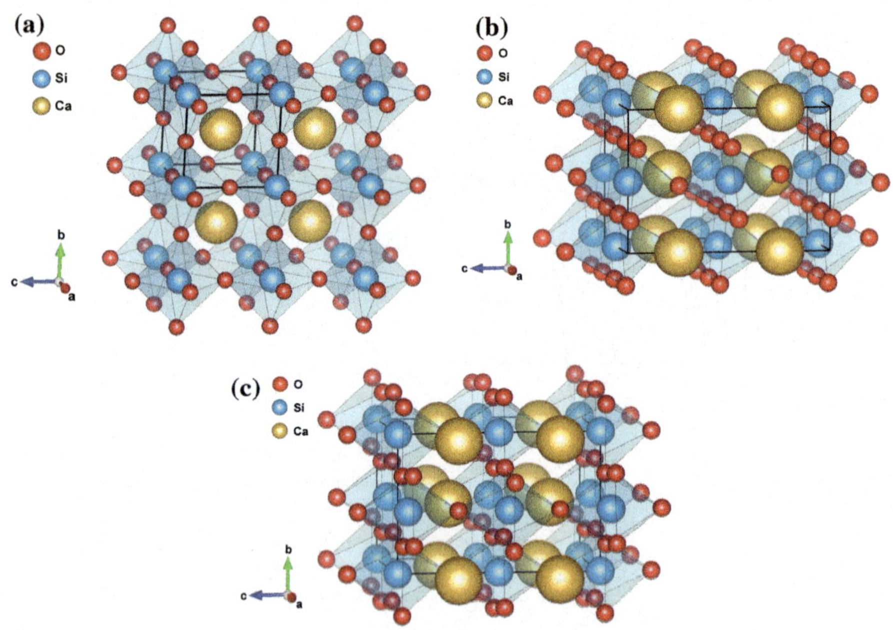

Fig. 4.26 Crystal structures of $CaSiO_3$ polymorphs: **a** Cubic B1 (space group *Pm*3*m*); **b** Tetragonal (space group *I*4/*mcm*); **c** Orthorhombic (space group *Pbnm*). Lattice parameters from Jung and Oganov (2005)

Tetragonal structure. Stixrude et al. (1996) were the first to propose, on the basis of theoretical first-principle linear-response calculations, that, as a result of slight rotations of the SiO_6 octahedra, a slightly distorted tetragonal structure with a space group *I4/mcm* should be the stable phase of $CaSiO_3$ at high pressures and 0 K, instead of the cubic phase predicted previously (Fig. 4.26b). Stixrude et al. (1996) suggested that the cubic perovskite structure undergoes a phase transition to tetragonal $CaSiO_3$ at above 2200 K at 80 GPa. However, this transition temperature estimate was apparently too high, according to further experiments and calculations (Shim et al. 2002; Kurashina et al. 2004). Later, Shim et al. (2002), using a LHDAC coupled with synchrotron angle-dispersive XRD, shows that the symmetry of $CaSiO_3$ at 20–46 GPa and at room temperature was indeed tetragonal, but the space group *I4/mcm* was not consistent with their experimental results. Subsequent theoretical first-principle calculations show again that the *I4/mcm* tetragonal structure is stable up to at least 150 GPa at 0 K (Caracas et al. 2005; Jung and Oganov 2005; Adams and Oganov 2006); this was confirmed by XRD studies that show that the cubic phase is also stable under high pressure-temperature conditions relevant to the lower mantle (Kurashina et al. 2004; Ono et al. 2004, 2005; Murakami et al. 2005; Komabayashi et al. 2007). This structural transition in CaSi-perovskite was also observed by Ono et al. (2004) and Murakami et al. (2005) in a multi-component peridotite system using XRD in a laser-heated DAC. They demonstrated that CaSi-perovskite takes a cubic structure on heating above 2000 K while it had a tetragonal structure at room temperature. In XRD experiments, the cubic phases at high temperatures exhibits a diffraction peak with a small full width at half maximum (FWHM); across the tetragonal transition, this peak splits into the 004 and 220 peaks in the tetragonal structure (Sun et al. 2016). This transition is shown to be a second-order structural phase transition. However, these studies did not put tight constraints on the *P–T* conditions of the phase transition boundary.

The summary of experimental data on $CaSiO_3$ is presented in Table 4.12; the general diagram of the established phase transactions is presented in Fig. 4.27.

According to experiments performed by Kurashina et al. (2004) and Komabayashi et al. (2007) in pure CaSi-perovskite, at a fairly low temperature of 490–580 K at 52–60 GPa, the cubic CaSi-perovskite transforms to the tetragonal $CaSiO_3$ phase. Temperature uncertainties in Kurashina et al. (2004) were significantly large because of the use of the LHDAC with a large temperature gradient and uncertainties in temperature measurements. Komabayashi et al. (2007) determined the phase transition in pure CaSi-perovskite by a combination of externally heated DAC (EHDAC) and in situ synchrotron XRD measurements. Contrary to the laser-heated DAC techniques, the uncertainty in temperature is very small (approximately 5 K) in the present EHDAC experiments.

Theoretical computations based on density functional theory suggested much higher transition temperatures at high pressures ranging from 1150 K near 0 GPa to 2450 K at 136 GPa and $\sim$1600 K at 60 GPa (Stixrude et al. 2007). The DFT calculations do not agree as well with those based on molecular dynamics

Table 4.12 Comparison of theoretical and experimental studies of $CaSiO_3$

Authors	Structure	Method[a]
Wentzcovitch et al. (1995)	Cubic	LDA pseudopotential calculation
Chizmeshya et al. (1996, 1998)	Cubic is unstable at lower-mantle conditions	DFT within LDA
Stixrude et al. (1996)	Tetragonal Cubic is unstable at all pressures	Linear augmented plane wave with LDA
Warren et al. (1998)	Cubic	LDA pseudopotential calculations
Shim et al. (2000a)	Cubic	LHDAC with double-side heating technique and energy-dispersive XRD
Shim et al. (2000a, b)	Tetragonal	LHDAC with high-resolution synchrotron energy dispersive XRD
Akber-Knutson et al. (2002)	Orthorhombic at all mantle pressures	DFT—VIB method calculations
Magyari-Köpe et al. (2002a, b)	Orthorhombic	Total-energy calculations
Kurashina et al. (2004)	Low *T*: tetragonal High *T*: cubic With Al admixture *T* increases and orthorhombic $CaSiO_3$ appears	LHDAC with In situ XRD
Ono et al. (2004, 2005)	Low *T*: tetragonal High *T*: cubic	LHDAC with in situ XRD
Caracas et al. (2005)	Tetragonal	DFT calculations
Hirose et al. (2005)	Low *T*: tetragonal High *T*: cubic With Al admixture *T* increases	LHDAC with In situ XRD
Murakami et al. (2005)	Cubic under high $T = 2450$ K up to 114 GPa	LHDAC with double-side heating technique and angle-dispersive XRD
Jung and Oganov (2005)	Orthorhombic up to 14.2 GPa, then tetragonal	DFT with GGA-PAW calculations
Adams and Oganov (2006)	Low *T*: orthorhombic Medium *T*: tetragonal High *T*: cubic	First-principles molecular dynamics calculations
Li et al. (2006)	Low *T*: orthorhombic, Medium *T*: tetragonal High *T*: cubic	First-principles molecular dynamics calculations
Komabayashi et al. (2007)	Low *T*: tetragonal High *T*: cubic	EHDAC with in situ XRD
Stixrude et al. (2007)	Low *T*: tetragonal High *T*: cubic Low P: orthorhombic	DFT calculations
Sun et al. (2016)	Low *T*: tetragonal High *T*: cubic	LHDAC with angle dispersive XRD

Notes: [a]*DFT* density functional theory; *EHDAC* externally-heated diamond anvil cell; *GGA* generalised gradient approximation; *LDA* local density approximation; *LHDAC* laser-heated diamond-anvil cell; *PAW* projector augmented wave; *VIB* variationally induced breathing; *XRD* X-ray diffraction

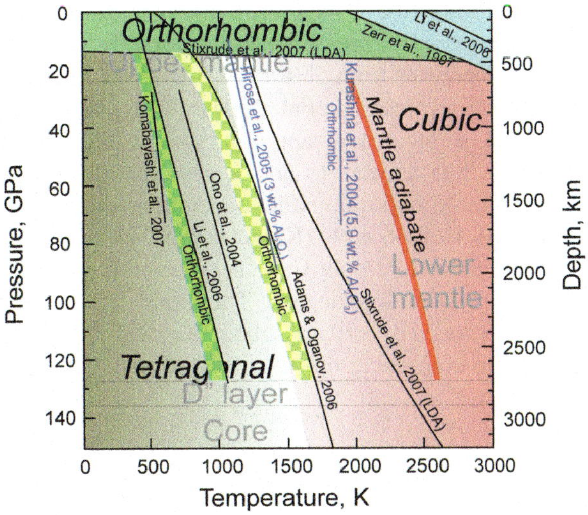

Fig. 4.27 Phase diagram for pure (*black lines*) and Al-containing (*blue lines*) $CaMgSiO_3$ under high pressure and temperature. Data from Zerr et al. (1997), Kurashina et al. (2004), Ono et al. (2004), Hirose et al. (2005), Adams and Oganov (2006); Komabayashi et al. (2007), Stixrude et al. (2007). Mantle adiabate after Katsura et al. (2010)

calculations (Adams and Oganov 2006; Li et al. 2006), and, in any case, these results show that the phase boundary is lower than the expected normal mantle geotherm. In the natural environment, near the normal geotherm conditions within an interval of 30–106 GPa and up to 2600 K, pure CaSi-perovskite has a cubic structure, and no phase transition occurs (Fig. 4.27). This means that the tetragonal to cubic phase transition in $CaSiO_3$ is not pressure-dependent but temperature-dependent with a slightly positive dT/dP slope of 1.1 (±1.3) K/GPa (Komabayashi et al. 2007), confirming earlier calculations with the use of DFT calculations based on the Variationally Induced Breathing (VIB) method, showing that structural and energy differences between the orthorhombic and cubic lattices under static conditions are small, and increase only slightly with pressure (Akber-Knutson et al. 2002).

Orthorhombic structure. At the same time, Magyari-Köpe et al. (2002b) calculated, using the total-energy method, that the *Pbnm* orthorhombic phase of $CaSiO_3$ (Fig. 4.26c) at low temperatures is energetically more favourable than the cubic polymorph, and that the transition between these phases shows weak pressure dependence. Based on calculated diffraction patterns, Li et al. (2006) concluded that the orthorhombic structure (the space group was not identified) is stable at 0 K but transforms to a tetragonal structure at room temperature with a Clapeyron slope of approximately 5 K/GPa from the orthorhombic phase to the tetragonal phase (Fig. 4.27). At high pressure, the orthorhombic phase is the stable room temperature phase, while at low pressure the tetragonal field covers room temperature. To date, there has been no experimental confirmation of the orthorhombic-cubic phase transformation, and DFT calculations suggest that the observation of an orthorhombic phase in the molecular dynamics simulations is an artifact of finite reciprocal space sampling (Stixrude et al. 2007).

The admixture of Al_2O_3 enhances the distortion of the perovskite structure, and increases the transition temperature remarkably. The cubic transition boundary in $CaSiO_3$ increases from 550–1900 K at 60 GPa with increasing Al_2O_3 content from 0–5.9 wt% (Kurashina et al. 2004). For $CaSiO_3$ perovskite with 3 wt% Al_2O_3, the transition is located at approximately 1200 K at 50 GPa (Hirose et al. 2005). However, this appears to not be applicable in natural ultramafic systems because CaSi-perovskite in ultramafic association contains less than 1 wt% Al_2O_3; it varies from 0.01 to 0.66 wt% (Kaminsky 2012) such that the expected transition temperature in such Al-poor $CaSiO_3$ perovskite in the lower mantle is too low to be of any geophysical significance. The cubic–tetragonal structure transition of $CaSiO_3$ may be expected only in subducted cold lithospheric slabs for Al-rich $CaSiO_3$ perovskite containing up to 4.49 wt% Al_2O_3 (see Sect. 5.2.2).

4.5.4 Physical Properties

This phase transition in CaSi-perovskite from tetragonal to cubic is a second-order structural phase transition without any volume change since the structure of low-temperature phase (tetragonal) merges continuously with that of the high-temperature cubic phase with increasing temperature (Komabayashi et al. 2007). On the other hand, this phase transition may cause drastic changes in physical properties; in particular, it may cause a significant reduction in the shear modulus if it is accompanied by an elastic softening (Stixrude et al. 1996; Caracas et al. 2005). It is known that many distorted perovskites exhibit ferroelastic behaviour, in which acoustic velocity, thermal expansion, and heat capacity show the large anomalies near the phase transitions to higher-symmetric structures (Carpenter and Salje 1998). If the phase transition in CaSi-perovskite is also ferroelastic-type, a large decrease in *S*-wave velocity is expected to occur at the phase boundary.

Such a transition is possibly accompanied by significant anomalies in elastic properties (Stixrude et al. 1996) and may, therefore, have relevance for the interpretation of seismic profiles. The tetragonal to cubic transition in CaSi-perovskite, like in SiO_2 transitions (see Sect. 4.6), may be a ferroelastic-type that is associated with elastic anomalies, and may cause large seismic anomalies in a wide depth range (Hirose et al. 2005; Carpenter 2006). Acoustic velocity in such perovskites drops remarkably (up to 30%), across the structural transition from low to high symmetry phases, which are well pronounced for shear waves V_s, rather than for compressional waves V_p, resulting in unusually large ratios d(ln V_s)/d(ln V_p) which may explain the lateral heterogeneities of seismic velocities in the lower mantle (Yeganeh-Haeri et al. 1989; Carpenter and Salje 1998). Phase transition in CaSi-perovskite possibly causes significant seismic anomalies. Since the ρ and V_Φ profiles of CaSi-perovskite are very close to PREM seismic profiles (Dziewonski and Anderson 1981), the variation in the fraction of $CaSiO_3$-perovskite with respect to the relative bridgmanite to ferropericlase ratio in the lower mantle has been suggested to not be seismically detectable (Sun et al. 2016).

4.6 Silica SiO_2

4.6.1 General

Free silica SiO_2 is not characteristic of the upper mantle that typically comprises ultramafic to mafic rock assemblages. It was not expected in the lower-mantle ultramafic assemblage model (Fig. 3.3). However, it appeared to be a common member of the natural ultramafic association where it coexists as an inclusion in diamond with ferropericlase and CaSi-perovskite and comprises 1.1–20% (average: 8.4%) (Kaminsky 2012). Silica inclusions were identified in all sets of the lower-mantle minerals observed in diamond from all regions and areas: 2.6–7.3% from Rio São Luiz (Harte et al. 1999; Zedgenizov et al. 2014a), 1.1% from Rio Soriso (Hayman et al. 2005), 4.8% from other placer deposits in the Juina area, Brazil (Kaminsky et al. 2001), 8.2% from the Kankan area, Guinea (Stachel et al. 2000), and 10–20% from Canadian kimberlites (Davies et al. 2004; Tappert et al. 2005a, b).

The composition of natural silica inclusions is found to be close to pure SiO_2, but some grains contain small (less than 0.09 wt%) amounts of Al, Ti, Cr, Fe, Mn, Ni, Mg, and Ca. Due to the lack of in situ structural determinations of silica from the juvenile ultramafic association, the crystal forms of these silica inclusions in association with characteristic lower-mantle minerals, such as ferropericlase and CaSi-perovskite, remain unkown. Since CaSi-perovskite is stable only at pressures greater than 14–16 GPa, i.e. in the stability field of stishovite, the SiO_2 should be stishovite. However, because of the instability of stishovite in these inclusions of silica after extraction from host diamonds, it typically shows up as a retrograde quartz structure.

4.6.2 Phase Transitions in Silica at High Pressures and Temperatures

Silica, a framework silicate consisting of SiO_4 tetrahedra, is one of the most common building blocks of crystal Earth materials. Quartz, the stable form of SiO_2 at ambient conditions, is expected to undergo phase transitions at high pressures with an increased coordination number from four to six, where silicon is in the octahedral coordination environment similar to that in perovskite. Theoretical studies indicate that a large number of structurally related and thermodynamically stable phases of SiO_2 are possible at high pressures, and that many of these phases possess similar total energies (Karki et al. 1997a; Teter et al. 1998; Klug et al. 2001; Tsuchiya et al. 2004a; Oganov et al. 2005; Driver et al. 2010). However, only three of the high-pressure SiO_2 polymorphs were experimentally proven and/or found in natural lower-mantle conditions: tetragonal stishovite, orthorhombic $CaCl_2$-polymorph and α-PbO_2 polymorph (seifertite).

Stishovite, a post-coesite SiO_2 phase, stable at pressures above ~7 GPa, was named after the Russian scientist Sergei Stishov who first discovered it in quenched samples from shocked silica in gas gun shockwave experiments (Stishov and Popova 1961). Later, it was discovered in nature in Meteor Crater, Arizona (Chao et al. 1962). Stishovite has a tetragonal rutile TiO_2 structure with a space group *P*4₂/*mnm* (Stishov and Belov 1962). This structure consists of slightly distorted SiO_6 octahedra that share edges to form chains running parallel to the *c* axis (Fig. 4.28a). Each octahedron is corner linked to four neighbouring chains. The structure can also be described as a distorted hexagonal close-packed array of O^{2-} anions with half of the octahedral sites occupied (Wicks and Duffy 2016). Static high-pressure experiments established that stishovite is stable at pressures from 7–9 GPa to 43–70 GPa (Kingma et al. 1995, 1996; Zhang et al. 1996; Andrault et al. 1998b; Nomura et al. 2010).

Orthorhombic $CaCl_2$-structured silica. The possibility of a post-stishovite phase was extensively debated in the 1980–1990s (Park et al. 1988; Hemley et al.

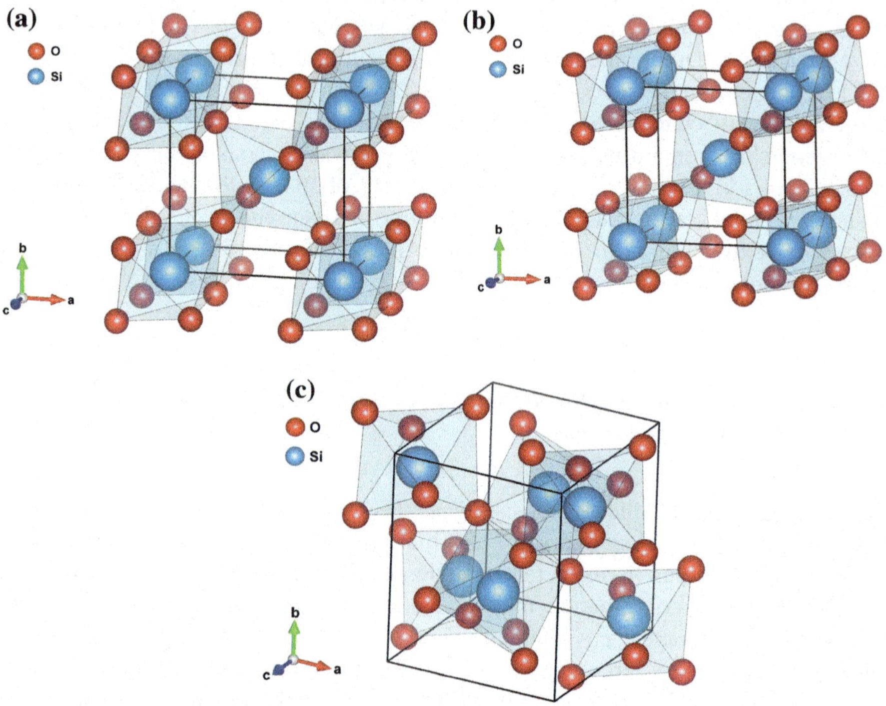

Fig. 4.28 Crystal structures of high-pressure SiO_2 polymorphs. **a** Stishovite with a tetragonal rutile TiO_2 structure (space group $P4_2/mnm$); **b** $CaCl_2$-type orthorhombic structure (space group *Pnnm*), oxygen escapes from the diagonal of the (*a*, *b*) plane, thus changing the symmetrical arrangement from tetragonal to orthorhombic; **c** Seifertite orthothombic structure (space group *Pbcn*). Lattice parameters from Yamanaka et al. (2002) for stishovite, from Andrault et al. (2003) for the $CaCl_2$-structured polymorph, and from Dera et al. (2002) for seifertite

1994 and references therein). Theoretical calculations predicted the transformation of the tetragonal structure into the orthorhombic $CaCl_2$ structure initially at a megabar pressure (Cohen 1987). Transition of stishovite into a $CaCl_2$ structure with the transformation of the tetragonal structure into orthorhombic (*Pnnm* space group; Fig. 4.28b) was first proven experimentally by Tsuchida and Yagi (1989). They observed it at a pressure of 108–124 GPa using a laser-heated DAC and powder XRD. The subsequent experiments using XRD and Raman spectroscopy (Kingma et al. 1995, 1996; Andrault et al. 1998b; Hemley et al. 2000) established the stishovite → $CaCl_2$-structured phase transformation at a much lower pressure of 43–54 GPa at room temperature; and theoretical calculations confirmed the same pressure range, at 47–55 GPa (Karki et al. 1997a; Teter et al. 1998). The orthorhombic $CaCl_2$ structure differs from the stishovitic one only by a small rotation of the octahedral chains (Fig. 4.28b).

The *P–T* phase boundary for the post-stishovite transition has been examined a number of times. Tsuchiya et al. (2004a) performed first-principles calculations and determined the stishovite-$CaCl_2$-structured phase transition boundary with a positive Clapeyron slope to be P (GPa) = 56.1 + 0.00585 × T (K), suggesting that the transition occurs at 1670 km depth along the normal mantle geotherm, where it may be responsible for the seismic heterogeneity. High *P–T* experiments in LHDAC equipment and in situ XRD observed that the phase transition occurred at 80 GPa at high temperature, corresponding to a depth of ~1750 km along the typical mantle geotherm; the boundary was represented by a linear equation: P (GPa) = (51 ± 2) + (0.012 ± 0.005) × T (K) (Ono et al. 2002). Ono et al. concluded that the post-stishovite phase transition is not the cause of such heterogeneities in the mid-lower mantle. However, the boundary was not tightly constrained in their experiments. In other experiments, the transition occurs at 45–60 GPa under 500–1000 K and increases up to 75–90 GPa under 1500–3000 K with the boundary equation 11.1 MPa/K (Nomura et al. 2010). According to the Nomura et al. (2010) calculations, the transition and the accompanying seismic inhomogeneity should occur within the 1530–1830 km boundary, depending on the position of geotherm. The summary on the stishovite → $CaCl_2$-structured SiO_2 transition obtained from experimental data and theoretical calculations is presented in Table 4.13 and Fig. 4.29.

Concentration of aluminium in stishovite. According to experiments, stishovite may contain up to 5 wt% Al_2O_3 (Litasov et al. 2007). The concentration of aluminium in SiO_2 influences the position of the phase transition. The phase transition boundary of Al-bearing stishovite is lower than that of pure SiO_2 by ~25 GPa: it is located at ~62 GPa under 2000 K, which corresponds to ~1500-km depth (Hirose et al. 2005) (Fig. 4.29).

Concentration of H_2O in stishovite is related to the Al admixture in silica. While Al-free stishovite contains only 16–30 wt ppm H_2O, synthetic stishovite with 4.4 wt% Al_2O_3 contains ~3010 wt ppm H_2O (Litasov et al. 2007). Most hydrogen in stishovite is associated with Al^{3+} substitutional defects on the octahedral (Si^{4+}) site. Hydrogen can occupy 40% of vacancies created by incorporation of Al^{3+} at 20 GPa (Litasov et al. 2007). It is also possible that Al forms the AlOOH

Table 4.13 Summary on experimental and theoretical studies of stishovite-to-$CaCl_2$-structured SiO_2 transition

Authors	Experimental method[a]	Theoretical method[a]	Diffraction method[a]	Starting material; Al_2O_3, in SiO_2	P–T conditions of transfer
Tsuchida and Yagi (1989)	LHDAC		Powder in situ XRD	Quartz or amorphous SiO_2	108–124 GPa
Kingma et al. (1995)	DAC with neon medium		In situ Raman spectroscopy	Natural quartz	50 ± 3 GPa at room temperature
Kingma et al. (1996)	DAC without a medium		Synchrotron XRD	Quartz	43 GPa
Karki et al. (1997a)		Pseudopotential method within the LDA			47 GPa
Andrault et al. (1998b)			Angle-dispersive synchrotron XRD	Synthetic quartz	54 GPa at 300 K
Teter et al. (1998)		Fundamental crystal chemistry and first-principles total-energy calculations			55 GPa
Hemley et al. (2000)	DAC with hydrogen medium		Single-crystal synchrotron XRD		~50 GPa
Ono et al. (2002)	LHDAC		In situ XRD		80 GPa at 1900 km (mantle geotherm)
Tsuchiya et al. (2004a, b)		Density-functional linear response theory			58 GPa at 0 K and 72 Gpa at 2700 K
Hirose et al. (2005)	LHDAC		In situ angle-dispersive XRD	MORB glass or gel	62 Gpa and 2000 K
Lakshtanov et al. (2007)	DAC with neon medium		Synchrotron angle-dispersive XRD	5 wt% Al_2O_3	30 GPa
Nomura et al. (2010)	LHDAC		Angle-dispersive XRD	Powdered synthetic quartz	71 GPa at 2150 K (mantle geotherm)
Yamazaki et al. (2014)	Multi-anvil with SDA		Synchrotron XRD	Sintered at 12 GPa stishovite	62.3 GPa at 300 K
Umemoto et al. (2016)		DFT using the PBE-type GGA and molecular dynamics simulations		6.25 mol% AlOOH	30–50 GPa at 1000–1500 K

Notes: [a]*DAC* diamond-anvil cell; *DFT* density functional theory; *GGA* generalized gradient approximation; *LDA* local density approximation; *LHDAC* laser-heated diamond-anvil cell; *SDA* sintered diamond anvils; *XRD* X-ray diffraction

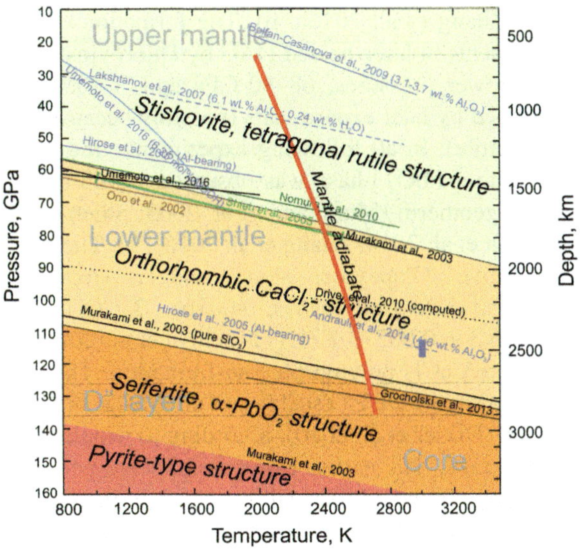

Fig. 4.29 Phase diagram for SiO_2 under high pressure and temperature conditions. Data from Ono et al. (2002), Murakami et al. (2003), Hirose et al. (2005), Shieh et al. (2005), Lakshtanov et al. (2007), Bolfan-Casanova et al. (2009), Driver et al. (2010), Nomura et al. (2010), Grocholski et al. (2013), Umemoto et al. (2016). Mantle adiabate after Katsura et al. (2010)

phase, which is highly soluble in stishovite (e.g., Chung and Kagi 2002; Panero and Stixrude 2004) (see Sect. 5.5.3). Lakshtanov et al. (2007) showed that the incorporation of water into Al-rich SiO_2 significantly reduces the post-stishovite transition pressure in SiO_2: in hydrous stishovite containing 5 wt% Al_2O_3 transformation occurs at 23–24 GPa (at 300 K) (Lakshtanov et al. 2007; Bolfan-Casanova et al. 2009), which is far lower than in pure silica. Umemoto et al. (2016), using first-principles static calculations and molecular dynamics simulations concluded that hydrogen bonds and hydrogen mobility play a crucial role in lowering the post-stishovite transition pressure. A cooperative redistribution of hydrogen atoms is the main mechanism responsible for the transition pressure reduction in hydrous Al-stishovite. Stishovite, containing 6.25 mol% AlOOH, transforms into $CaCl_2$-structured phase at pressures from 30–32 GPa (1000 K) to 50 GPa (1300 K) (Umemoto et al. 2016).

The $CaCl_2$-structured phase of SiO_2 is not observed in the natural environment to date.

Seifertite. In the lowermost mantle conditions, the pure $CaCl_2$-structured SiO_2 transforms into another orthorhombic α-PbO_2 ('skrutinyite')-structured polymorph named seifertite (El Goresy et al. 2008). In this phase, distorted SiO_6 octahedra are arranged into kinked chains extending along the *c* axis (Fig. 4.28c), resulting in a small density increase compared with stishovite and the $CaCl_2$-type phase (Wicks and Duffy 2016). This transition of $CaCl_2$-structured SiO_2 into an orthorhombic structure with space group *Pnc*2 was predicted to occur at ~100–120 GPa using theoretical calculations (Belonoshko et al. 1996; Karki et al. 1997a). Later theoretical calculations by Teter et al. (1998), based on fundamental crystal chemistry and first-principles total-energy calculations, demonstrated that the space group of this phase should be *Pbcn*, which was confirmed in all further analyses (Sharp et al.

1999; Zhang et al. 2016a, b). The formation of α-PbO_2 phase was confirmed by experiments in laser-heated DAC by Dubrovinsky et al. (1997, 2001b, 2003), but at much lower pressures, 68–80 GPa and ~1000–2000 K. These results were not confirmed by later experiments (possibly because of the wide range of metastability of seifertite). In all following experiments, the transformation of $CaCl_2$-structured SiO_2 into α-PbO_2 phase was observed within the range of 90–140 GPa along the mantle geotherm (Murakami et al. 2003; Shieh et al. 2005; Grocholski et al. 2013; Andrault et al. 2014; Zhang et al. 2016a, b). The boundary of the transformation has a positive Clapeyron slope and is represented by a linear equation P (GPa) = 98 + (0.0095 ± 0.0016) T (K) (Murakami et al. 2003) or $P = 106.3 + 0.00579\,T$ (K) (Tsuchiya et al. 2004a), which is almost parallel to the stishovite/$CaCl_2$ polymorph transformation. These data correspond to calculations with the use of DFT (Tsuchiya et al. 2004a, b; Oganov et al. 2005) and the QMC method (Driver et al. 2010), according to which the phase transition should occur at 90–120 GPa. The earlier experiments were performed without the pressure transmitting medium and thermal insulation in the diamond-anvil cell, which can result in severe deviatoric stresses and thermal gradients. In most recent experimental works, using insulating noble gas media in DAC and in situ XRD measurements, the formation of the *Pbcn*-structured α-PbO_2 phase was observed at the lowermost mantle conditions (113–140 GPa), most likely within the D″ layer (Grocholski et al. 2013; Andrault et al. 2014; Zhang et al. 2016a, b). The summary on theoretical calculations and experimental data on the formation of seifertite is presented in Table 4.14 and Fig. 4.29.

The space group of the seifertite orthorhombic structure has been under debate due to its similarity to several other space groups in the orthorhombic system. When the phase was first experimentally reported in DAC experiments at pressures of between 68 and 85 GPa and temperatures above ~2000 K (Dubrovinsky et al. 1997, 2003), the few peaks observed in the powder XRD pattern did not allow for an unambiguous structure determination, such that the theoretically predicted space group *Pnc*2 using the molecular dynamics theory was used to explain the diffraction patterns (Belonoshko et al. 1996). To date, however, an unambiguous structure determination of seifertite with the space group *Pbcn* within its stability field is established with a single-crystal XRD (Zhang et al. 2016a, b), which is in agreement with predicted first-principle calculations (Karki et al. 1997a; Teter et al. 1998) and found in meteorites (Dera et al. 2002; El Goresy et al. 2008) (Fig. 4.29).

Seifertite is remarkable in that it can be metastably preserved to 1 bar from 140 GPa after high temperature quenching, instead of converting back to its lower-pressure forms, such as quartz, stishovite or densified glass. The greater molar volume of seifertite found in early experiments supports a metastable synthesis of the phase outside of its stability field (Grocholski et al. 2013). However, to date no seifertite grains were identified in lower-mantle diamonds, even though several grains of silica polymorphs have been studied using synchrotron XRD. This implies that the transformation of $CaCl_2$-structured SiO_2 into α-PbO_2-structured seifertite takes place deeper than the lower mantle boundary, or that seifertite inclusions are still yet to be found in diamond.

Table 4.14 Summary on theoretical and experimental studies of seifertite

Authors	Theoretical method[a]	Experimental method[a]	Diffraction method[a]	Natural sample	Starting material; Al_2O_3, in SiO_2	P–T conditions of transfer	Space group of seifertite
Belonoshko et al. (1996)	Molecular dynamics					Above ~120 GPa in a wide range of *T*	*Pnc*2 (intermediate between *Pbcn* and $P2_1/c$)
Karki et al. (1997a)	Pseudopotential method within the LDA					98 GPa	*Pnc*2
Dubrovinsky et al. (1997)		LHDAC	XRD		Silica gel and quartz	68 ± 5 GPa at 2000 ± 50 K	*Pnc*2 (based on theoretical considerations)
Dubrovinsky et al. (2001b)		Electrically and laser heated DAC	In situ XRD		α-crystobalite	45–53 GPa and >80 GPa at high *T*	*Pnc*2
Dubrovinsky et al. (2003)		Electrically heated DAC			Silica gel 99.99% purity	75–80 GPa at 950 K	
Teter et al. (1998)	Fundamental crystal chemistry and first-principles total-energy calculations						*Pbcn*
Sharp et al. (1999)			Single-crystal selected-area electron diffraction	Martian meteorite Shergotty	Al_2O_3 = 1.12 wt %; Na_2O = 0.4 wt %		*Pbcn*
El Goresy et al. (2000)			XRD	Martian meteorite Shergotty			Mixture of phases *Pbcn* and $P2_1/c$
Dera et al. (2002)			Powder XRD	Martian meteorite Shergotty	Al_2O_3 = 0.8–1.6 wt%; Na_2O = 0.2–0.5 wt%		*Pbcn*

(continued)

Table 4.14 (continued)

Authors	Theoretical method[a]	Experimental method[a]	Diffraction method[a]	Natural sample	Starting material; Al_2O_3, in SiO_2	P–T conditions of transfer	Space group of seifertite
Murakami et al. (2003)		LHDAC without a medium	In situ synchrotron XRD		Synthetic amorphous pure SiO_2	121 GPa at 2400 K	
Tsuchiya et al. (2004a, b)	DFT					~120 GPa	
Oganov et al. (2005)	Density-functional perturbation theory					~90 GPa	
Shieh et al. (2005)		LHDAC	In situ angle-dispersive XRD		Pure silica glass and stishovite	113–117 GPa at 2160–2320 K and 124–128 GPa at 2250–2500 K	
El Goresy et al. (2008)			Selected-area electron diffraction	Martian meteorites Shergotty and Zagami	Al_2O_3 = 0.8–1.6 wt%; Na_2O = 0.2–0.5 wt%	Shock event with $P > 35$ GPa	*Pbcn* or *Pb2n*
Driver et al. (2010)	Quantum Monte Carlo method				Pure SiO_2	86–122 GPa at 2300–2600 K	
Miyahara et al. (2013)			XRD	Lunar meteorite NWA4734	Na_2O = 0.59–0.89 wt%; Na_2O = 0.21–0.26 wt%	Dynamic event	*Pbcn*
Grocholski et al. (2013)		LHDAC with insulating noble gas medium	In situ XRD		Pure (99.8%) or 10-mol% Al_2O_3 amorphous SiO_2	130–140 GPa at 2500 K	
Andrault et al. (2014)		LHDAC	Synchrotron in situ XRD		Na_2O = 4–6 wt%	113–119 GPa at 2500 K	
Zhang et al. (2016a, b)		LHDAC with neon medium	Single-crystal XRD		$(Mg_{0.6}Fe_{0.4})SiO_3$ + 5% SiO_2	129 GPa at 2500 K	*Pbcn*

Notes: [a]*DAC* diamond-anvil cell; *DFT* density functional theory; *LDA* local density approximation; *LHDAC* laser-heated diamond-anvil cell; *XRD* X-ray diffraction

On the surface of the Earth, seifertite was identified in Martian shergottites Sherotty, Zagami and NWA 4734 (Sharp et al. 1999) and the Lunar Asuka 881757 meteorite (Miyahara et al. 2013). Martian seifertite has the same (as in experiments) orthorhombic *Pbcn* structure (Dera et al. 2002; El Goresy et al. 2008) and minor admixtures of Na_2O (0.2–0.5 wt%) and Al_2O_3 (0.8–1.6 wt%). The formation of seifertite in meteorites is explained by shock events under dynamic pressures of 35–80 GPa (El Goresey et al. 2008; Miyahara et al. 2013). A disagreement between these estimates and much higher pressures of origin of seifertite under static conditions may be due to the extreme metastability of seifertite (Grocholski et al. 2013).

High alumina content in $CaCl_2$-structured SiO_2 has the same effect as in stishovite: according to experimental data, Al lowers the transition of $CaCl_2$-structured SiO_2 into seifertite by 5–10 GPa compared with pure silica: to 113–119 GPa at 2500 K (Hirose et al. 2005; Andrault et al. 2014).

However, the naturally occurring stishovite and seifertite contain only a small amount of Al admixture; for example, there is well below 1% in stishovite in lower-mantle diamond inclusions (Wirth et al. 2007), and 0.8–1.6 wt% Al_2O_3 in seifertite in Martian meteorite (El Goresy et al. 2000, 2008). Therefore, these silica polymorphs in nature are expected to have a chemical composition very close to pure SiO_2; and the Al_2O_3 dissolution is not expected to significantly affect the phase transition boundary.

Seifertite is stable up to 150–260 GPa; at such pressure, seifertite should obtain a **pyrite-type structure** (Karki et al. 1997a; Murakami et al. 2003; Kuwayama et al. 2005).

4.6.3 Physical Properties

The presence of free silica in the deep mantle was suggested to cause seismic velocity anomalies that are faster, by a few percent, relative to surrounding ultramafic peridotite assemblages (Bina 2010 and references therein).

The stishovite → $CaCl_2$ polymorph transition is a second-order ferroelastic transition, in the terminology of Wadhawan (1982), like in CaSi-perovskite (see Sect. 4.5.3). It consists of an orthorhombic distortion of tetragonal stishovite and displays large shear modulus softening (Cohen 1991; Karki et al. 1997a; Carpenter et al. 2000; Tsuchiya et al. 2004a, b; Jiang et al. 2009; Asahara et al. 2013). As a result, even a small fraction of free silica may produce a detectable seismic signal in the mid-lower mantle (Karki et al. 1997a).

Theoretical calculations of the volume change during the stishovite → $CaCl_2$ polymorph transition are controversial. A single Birch-Murnaghan equation of state suggested that the phase transition occurs without a substantial change in volume (Andrault et al. 1998b), while the pseudopotential method within the local-density approximation showed that the phase transition is associated with a volume decrease (Karki et al. 1997a), and the shear wave velocity V_s decreases rapidly (by

60%) between 40 and 47 GPa prior to the transition from stishovite to the $CaCl_2$ structure at 47 GPa (Karki et al. 1997b). Further experiments proved that the volume of the $CaCl_2$ phase is smaller than that in stishovite across the phase transition, and $CaCl_2$ polymorph is a significantly denser than stishovite in the high-pressure range (Andrault et al. 2003). Carpenter et al. (2000) established that seismic wave velocities, especially *S*-wave speed, are significantly reduced near this phase transition: *S*-wave velocity drops by ~25% at the phase transition and by 12–15% 100 km above and below the phase transition, i.e., the transition occupies the interval of ~200 km (Carpenter et al. 2000; Komabayashi et al. 2007). This seismic effect may be expected at the 1600–1800 km depth (see Fig. 4.29).

The $CaCl_2$ phase → seifertite (α-PbO_2) structural transition in SiO_2 (*Pnnm* ↔ *Pbcn*), like the stishovite → $CaCl_2$ phase transition ($P4_2/mnm$ ↔ *Pnnm*), is also of the ferroelastic type. Seifertite within its stability field is 1.5% denser than $CaCl_2$-structured SiO_2 (Grocholski et al. 2013) and approximately 10% lower density and 10% larger bulk sound velocity than the surrounding lowermost mantle (Tsuchiya et al. 2004a). The transformation of the $CaCl_2$ phase to seifertite is accompanied by a discontinuous change of 1–2% in elastic wave velocity and a decrease by a factor of two in anisotropy. Even a small amount of silica in the lower mantle may contribute significantly to observed seismic anisotropy (Karki et al. 1997b).

4.6.4 Origin of Silica Polymorphs in the Lower-Mantle Ultramafic Association

Stishovite SiO_2 and periclase–wüstite solid solutions $(MgO\text{–}FeO)_{ss}$, associate paradoxically in primary inclusions of lower-mantle diamonds. Under the conditions of the Earth's crust and upper mantle, such oxide assemblages are chemically impossible (forbidden) because the oxides MgO and FeO and SiO_2 react to produce intermediate silicate compounds, enstatite and ferrosilite. However, in the lower mantle silica is permanently present in association with bridgmanite and ferropericlase. Such occurrence in the lower-mantle associations was refered to as a 'stishovite paradox' (Litvin 2014; Litvin et al. 2016) and is not yet fully understood.

Initially, a model of the decomposition of bridgmanite with the resultant formation of ferropericlase and stishovite has been proposed:

$$(\mathrm{Mg,Fe})\mathrm{SiO_3} \rightarrow (\mathrm{Fe,Mg})\mathrm{O} + \mathrm{SiO_2}.$$

$$\text{Fe}-\text{bridgmanite} \rightarrow \text{Ferropericlase} + \text{Stishovite}$$

Such a transformation has been experimentally simulated, but only for Fe-rich systems, at 17–23 GPa (Ito and Takahashi 1989; Fei et al. 1996; Akaogi 2007). Ito and Takahashi (1989) estimated the limiting composition for bridgmanite in equilibrium with SiO_2 and ferropericlase as having mg = 0.89 and the composition of the co-existing ferropericlase at *mg* = 0.42. However, the compositions of known ferropericlase associated with SiO_2 in a natural environment are more magnesium-rich: *mg* = 0.70 in Sao Luiz, Brazil (sample #BZ-103a; Harte et al. 1999) and *mg* = 0.86 in Kankan, Guinea (sample KK-31a; Stachel et al. 2000).

An explanation of formation of free silica as the result of decomposition of post-perovskite within the D″ layer was offered by Knittle and Jeanloz (1991). According to this model, post-perovskite (Mg,Fe)SiO_3 reacts with molten iron at the core–mantle boundary (within the D″ layer). As a result of this reaction, metallic alloys, FeO, FeSi and free silica may occur:

$$(\mathrm{Mg,Fe})\mathrm{SiO_3} + \mathrm{Fe} \rightarrow \mathrm{MgSiO_3} + \mathrm{SiO_2} + \mathrm{FeO} + \mathrm{FeSi}.$$

Another model of the formation of SiO_2 in the ultramafic association (based on the fractional crystallization of the lower-mantle material) was proposed by Litvin (Litvin 2014; Litvin et al. 2016). According to this model, high iron content in the lower mantle may occur as a result of the crystallization of ferropericlase-magnesiowüstite accompanied by the increase of its *fe* index. During the course of fractional crystallization in ultramafic magma within the lower mantle, permanent fractionation with the removal of newly formed minerals results in the progressive evolution of the total composition of the system. The total composition is successively consistent with the changing compositions of residual melts. Fractional crystallization of ferropericlase-magnesiowüstite with decreasing temperature should be accompanied by an increase in the concentration of the relatively low-temperature component of the solid solution (in our case, FeO) in residual melts. The FeO concentration in residual melts is critical for initialization of the reaction Bridg + Liq = Sti + MgWus in peritectic Liq + MgWus + Bridg + CaSiPrv + Sti. This may explain the 'stishovite paradox', since stishovite appears in the peritectic composition. After the disappearance of bridgmanite, as an individual phase, the figurative point of the melt migrates along the monovariant curve Liq + MgWus + CaSiPrv + Sti, which is accompanied by an increase of *fe* in ferropericlase-magnesiowüstite associated with stishovite. This model is in fact another variety of the bridgmanite decomposition mechanism.

All these models are based on the assumptions from a pyrolitic composition of the lower mantle, analogous to the upper mantle composition. However, the enrichment of the lower mantle, particularly its lowermost part in Si may be caused by its chemical stratification, layering and very limited mass transport between the lower and the upper mantle. In geological samples from all areas free silica is always present, and this may be considered to be evidence of the enrichment of the juvenile lower mantle in SiO_2. The initial enrichment of the lower mantle with Si was suggested by Kaminsky (2012) during a study of natural parageneses in the lower mantle. This model was confirmed by the distribution of sound velocities in the lower mantle, which fit its enrichment in Si compared with the upper mantle (Murakami et al. 2012) and may be caused by the presence of free silica in the lower mantle (Bina 2010). Such interpretation is consistent with earlier proposals from both geochemical and isotopic studies and seismic data (e.g., Anderson 2002; Cammarano and Romanowicz 2007; Matas et al. 2007; Ricolleau et al. 2008). It has been suggested that the source of silica in the lower mantle may be the Earth's core, which originally included ~6 wt% SiO_2 but then became depleated in silicon by crystallizing SiO_2 and $MgSiO_3$ (Ozawa et al. 2016).

This model is also confirmed by planetary geochemical evidence. If the Earth was accreted from material such as the CI chondrites, which have the Mg/Si ratio of approximately 1.074, the lower mantle should be chemically distinct from the upper mantle (Javoy 1995; Williams and Knittle 2005), which has the Mg/Si ratio close to 1.27 (Ringwood 1975; Anderson 1983), i.e. lower mantle should be enriched in Si. In addition, the CI-chondrite model of the Earth's composition has been challenged, and the enstatite-chondrite model has been proposed (Javoy et al. 2010; Kaminski and Javoy 2013, 2015). Enstatite chondrite is significantly richer in silica than CI chondrites (Mg/Si = 0.69) and contains free silica in its composition (Javoy et al. 2010). The normative mineralogy of the lower mantle, calculated on the enstatite chondrite compositions, contains free silica ranging from 5% for low-iron (EL) enstatite chondrite to 12% for high-iron (EH) enstatite chondrite compositions (Frost and Myhill 2016).

Thermodynamic modeling with the use of HeFESTo code, based on experimental data and performed for the $MgSiO_3$–$FeSiO_3$ (enstatite–ferrosilite) compositions at lower-mantle pressures and ~2000 K by Dorfman et al. 2013 demonstrated that within a wide compositional range (*fe* = 0.1–1), SiO_2 is stable in association with bridgmanite: initially in the form of stishovite (from 24 to ~50 GPa), then as the $CaCl_2$-structured phase. This confirms that SiO_2 may occur in the ultramafic lower mantle without fractional crystallization and may be in equilibria with other lower-mantle phases.

In such a case, no special scenario of SiO_2 origin in the lower mantle is needed; *free silica should be considered as one of the major initial mineral components of the lower mantle.*

4.7 Mg–Cr–Fe, Ca–Cr and Other Orthorhombic Oxides

4.7.1 General

In one of the diamond grains from the Juina area, Brazil, a microxenolith of a supposed lower-mantle rock was found (Kaminsky et al. 2015a). It was composed of iron carbide, ferropericlase and two orthorhombic oxides $Mg(Cr,Fe)_2O_4$ and $CaCr_2O_4$, which are new mineral phases. This association was formed during several stages and it is considered as a rock microfragment of the lower mantle. In addition to the carbide-oxide fragment, some other phases were identified as inclusions in that diamond, such as orthorhombic MgO and an assemblage of periclase plus wüstite, among others.

Another series of lower-mantle orthorhombic oxides were recently identified as an inclusion in diamond, in association with nitride Fe_3N, and are suggested to originate within the lowermost mantle (Kaminsky and Wirth 2017).

4.7.2 Chemical Compositions

An elemental map of the carbide-orthorhombic oxides assemblage, using X-ray fluorescence intensities of the C–K_α, O–K_α, Mg–K_α, Ca–K_α, Cr–K_α and Fe–K_α, illustrates different element distributions in these mineral phases within the lower-mantle microxenolith (Fig. 4.30). The chemical compositions of the minerals, calculated from the EDX spectra, are presented in Table 4.15. In addition to major elements, EDX spectra demonstrate the presence of minor admixtures: Mg, Al, Ti, V, and Fe in Ca–Cr oxide; Al and V in Mg–Cr–Fe oxide; unusually high Mn (5.52 at.%) and Cr in ferropericlase.

Based on this analysis, an empirical formula for the new Mg-bearing oxide can be derived as $(Mg_{0.90}Mn_{0.18})_{1.08}(Cr_{1.37}Fe^{3+}{}_{0.39}V_{0.11}Al_{0.05})_{1.92}O_4$ or simplified as $Mg(Cr,Fe)_2O_4$. This formula is isomorphic to magnesium chrome spinel from kimberlites (e.g., Barnes and Roeder 2001) and is similar to xieite from the Suizhou meteorite (Chen et al. 2003b, 2008) with Mg instead of divalent Fe in its composition. The derived formula for Ca-bearing oxide is $(Ca_{1.07}Mg_{0.02}Mn_{0.02})_{1.11}(Cr_{1.71}Fe^{3+}{}_{0.06}V_{0.06}Ti_{0.03}Al_{0.03})_{1.89}O_4$ or simplified as $CaCr_2O_4$ calcium chrome spinel. In addition to the Cr, Al, Mg and Fe substitution in natural chromite, the orthorhombic phases are likely to incorporate Ca, Ti, Fe, Na, Si (as in $CaFe_2O_4$, $CaTi_2O_4$ and $NaAlSiO_4$), and possibly other divalent and trivalent transition elements and rare earth elements, thus having a major effect on element partitioning of the deep-Earth geochemistry (Chen et al. 2003a, b).

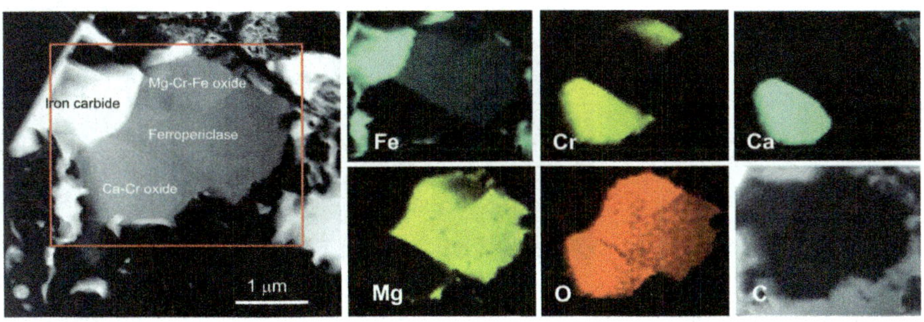

Fig. 4.30 Polyphase mineral inclusion in diamond from the Juina area, Brazil. General view (*left image*): an assemblage of major mineral phases forming an allotriomorphic texture and surrounded by graphite in the inclusion; *Right panels* elemental maps for Fe, Cr, Ca, Mg, O and C of the polyphase inclusion. From Kaminsky et al. (2015a). Used with a permission of the Canadian Mineralogist

Table 4.15 Chemical compositions of oxides (at.%)

Elements	From lower-mantle microxenolith			From association with iron nitride Fe_3N	
	Ferropericlase	Orthorhombic Mg–Cr–Fe oxide	Orthorhombic Ca–Cr oxide	Orthorhombic Mn–Fe–Cr oxides	
Mg	63.08	30.09	0.79	–	–
Ca	–	–	35.72	–	–
Cr	0.30	45.78	57.02	23.07	56.17
Fe	31.10	12.92	1.93	28.09	13.31
Mn	5.52	6.08	0.69	46.05	25.29
Al	–	1.58	0.80	–	–
Ti	–	–	1.09	–	–
V	–	3.55	1.96	1.31	3.38
Si	–	–	–	1.48	1.85
Total	100.00	100.00	100.00	100.00	100.00
mg = Mg/(Fe + Mg)$_{at.}$	0.670	0.700	0.290	0	0

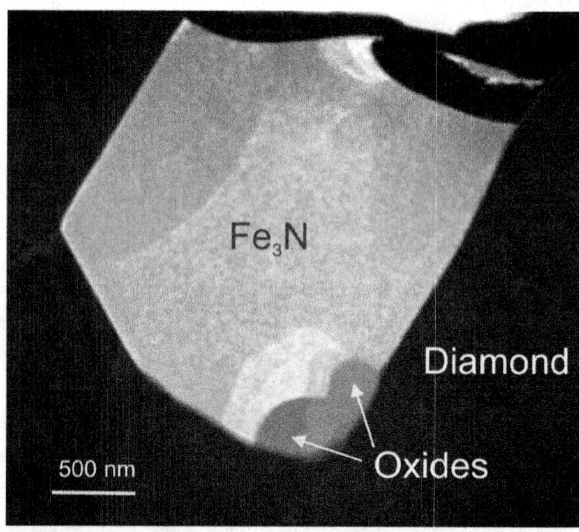

Fig. 4.31 A grain of iron nitride, Fe_3N, intergrown with orthorhombic oxides from an inclusion in a diamond from Rio Soriso, Juina area, Brazil. TEM dark field image. Photo made by R. Wirth. After Kaminsky and Wirth (2017)

Two orthorhombic oxides identified in association with iron nitride Fe_3N (Fig. 4.31) are characterized by the absence of Mg and high concentrations of MnO = 25.29–46.05 at.% (Table 4.15). Their occurrences indicate great variability of post-spinel oxides in the lower mantle.

4.7.3 Crystal Structures of the Orthorhombic Oxides and Their Stability Range

Reid and Ringwood (1969, 1970) proposed two orthorhombic structures of $CaFe_2O_4$-type (CF) and $CaTi_2O_4$-type (CT) structures, which are denser than spinel, as being top candidates for post-spinel phases in the Earth's lower mantle. Both CF and CT structures contain dodecahedral and octahedral sites. The difference between these structures lies only in slight modifications of the linkage to the SiO_6 polyhedra (Fig. 4.32). The new $Mg(Cr,Fe)_2O_4$ oxide has an orthorhombic CT structure with lattice parameters (in Å): $a = 10.654$; $b = 9.401$; $c = 2.905$.

$CaCr_2O_4$ oxide also has an orthorhombic structure, with lattice parameters (in Å): $a = 10.170$; $b = 9.017$; $c = 2.874$ (Kaminsky et al. 2015a). Its *d*-spacings and angles between lattice planes do not fit exactly to either a CT or a CF structure. It is most likely to be the high-temperature CF (β) variety (Lee and Nassaralla 1997) (Fig. 4.33a, b); however, the CT structure should not be excluded (Fig. 4.33c). Although $CaCr_2O_4$ is one of the important components of materials for many high-temperature applications in metallurgy, ceramics and solid oxide fuel cells (Róg et al. 2007; Zhai et al. 2016), there are no natural analogues for this oxide to date.

Experimental investigations have demonstrated spinel-structured $CaAl_2O_4$ and $MgAl_2O_4$ transformations into a CF structure at pressures above 26.5 GPa at the top of the lower mantle conditions (Akaogi et al. 1999) and the $MgAl_2O_4$ structure transformations into a CT structure at pressures in excess of 40 GPa in the mid-lower mantle conditions (Funamori et al. 1998; Chen et al. 2003a and references therein). The only natural orthorhombic post-spinel polymorph of chromite with a CF structure, with an ideal formula of $FeCr_2O_4$, was discovered in shock veins of the

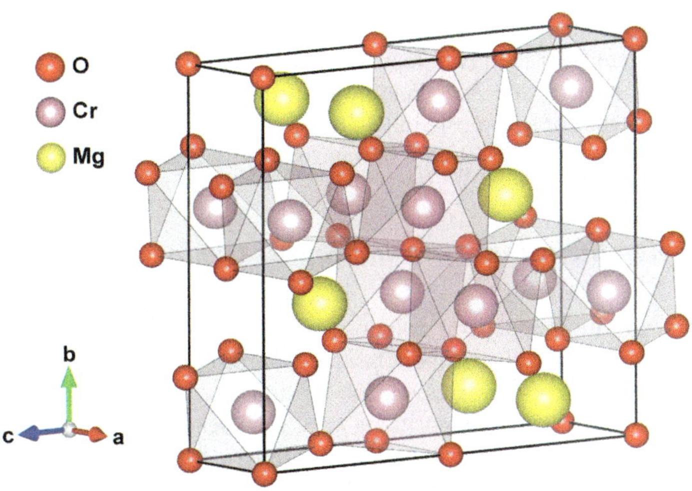

Fig. 4.32 Orthorhombic CT crystal structure (space group *Cmcm*) of the $Mg(Cr,Fe)_2O_4$ oxide. Lattice parameters from Yamanaka et al. (2008)

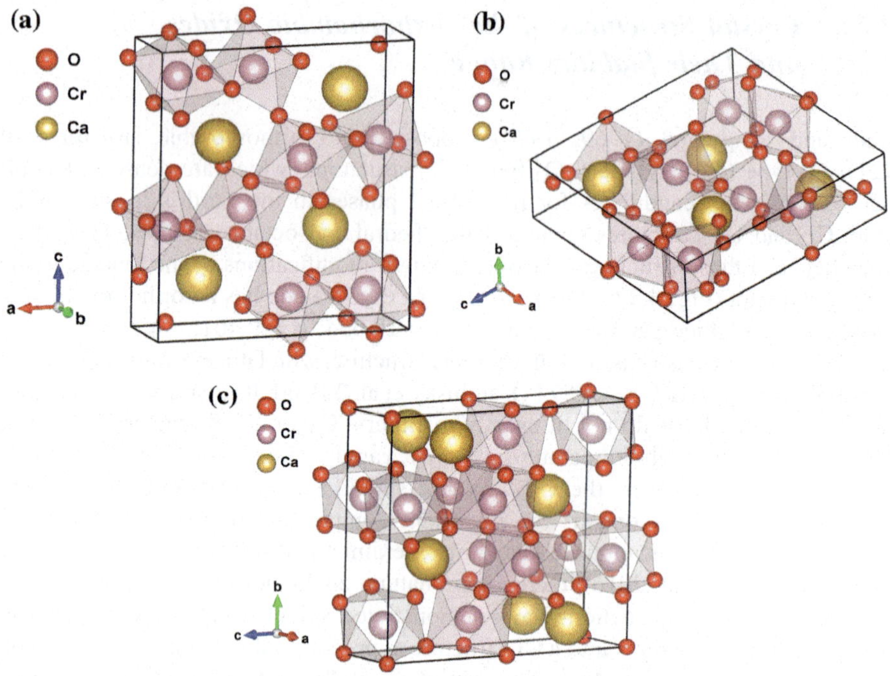

Fig. 4.33 Possible structures of the $CaCr_2O_4$ oxide. *Left* **a** and *center* **b** Orthorhombic CF structure (space group *Pnma*) in differently oriented axes; *Right* **c** Orthorhombic CT structure (space group *Cmcm*). Lattice parameters from Yamanaka et al. (2008)

Suizhou meteorite and termed xieite (Chen et al. 2003b, 2008). Ishii et al. (2014) suggested that xieite does not belong to the CF-type but exhibits a modified CF-type structure. On the other hand, the CT structure is considered to be the most probable post-spinel structure for the lower mantle. Minerals with such a structure have been shown to coexist under the *P–T* conditions of the lower mantle (Yamanaka et al. 2008). The existence of such post-spinel phases suggests that the pressure conditions for their formation are at ≥ 40 GPa or approximately at a depth of ≥ 1000 km.

4.7.4 *Other Inclusions with the Spinel Composition*

Mineral grains with the general composition of chromian spinel (Mg,Fe)(Cr,Al, $Fe)_2O_4$, in association with ferropericlase and MgSi-perovskite, were identified in diamonds from the Pandrea pipes and placer deposits in Juina, Brazil (Kaminsky et al. 2001, 2009a). Their compositions vary significantly and are quite different from the chrome spinel inclusions in lithospheric diamonds (Table 4.16 and Fig. 4.34). It is possible that these inclusions with various compositions may have an orthorhombic structure (CF or CT) stable within the mid-lower mantle.

MREE and HREE abundances in these grains are close to chondritic, and the REE patterns are characterized by LREE-depletion (Fig. 4.35).

Table 4.16 Representative analyses of inclusions in diamond with chemical compositions of spinels (wt%)

Mineral	'Chromite'			'Al-spinel'	'Magnetite'	
Country	Brazil			Canada	Brazil	
Area	Juina			NW Territories	Juina	
Location	Pandrea-2 pipe		Corrigo Chicoria placer	Panda (Ekati)	Rio Soriso placer	Juina-5 pipe
Sample No.	P2-2	P2-17	Juina 5-17b	PA-54a	1-4K	Ju5-102
Assemblage	2Chr + MgCaSi-prv	2Chr + fPer + nativeFe	2Chr + 2fPer	Spl + fPer + 'Ol'	Mag	Mag + Brd + Jeffbenite + Ilm + Ttn + FeMgSi
SiO_2	0.14	0.04	0.57	0.40	0.16	0.30
TiO_2	0.58	0.12	10.73	0.10	2.36	15.79
Al_2O_3	19.75	15.1	5.37	57.42	4.75	9.96
Cr_2O_3	46.73	55.08	36.34	8.47	n.a.	0.11
V_2O_3	0.22	0.18	n.a.	n.a.	58.21	n.a.
FeO	22.29	16.73	33.53	9.10	30.80	64.81
MnO	0.77	0.49	0.79	0.07	0.58	0.40
NiO	0.03	0.01	n.a.	0.23	n.a.	n.a.
MgO	9.2	12.46	10.72	23.77	1.46	2.63
CaO	0.01	0.00	n.a.	0.00	0.15	0.02
ZnO	0.46	0.07	n.a.	n.a.	n.a.	n.a.
Na_2O	0.04	0.01	n.a.	0.00	n.a.	0.03
K_2O	0.01	0.oo	n.a.	0.00	n.a.	0.00
Total	100.23	100.29	98.05	99.56	98.47	94.04
fe	0.576	0.429	0.637	0.177	0.922	0.932
mg	0.424	0.571	0.363	0.823	0.078	0.068
cr#	0.613	0.710	0.819	0.090	0.000	0.007
References	Kaminsky et al. (2009b)		Kaminsky et al. (2001)	Tappert et al. (2009a)	Hayman et al. (2005)	Thomson et al. (2014)

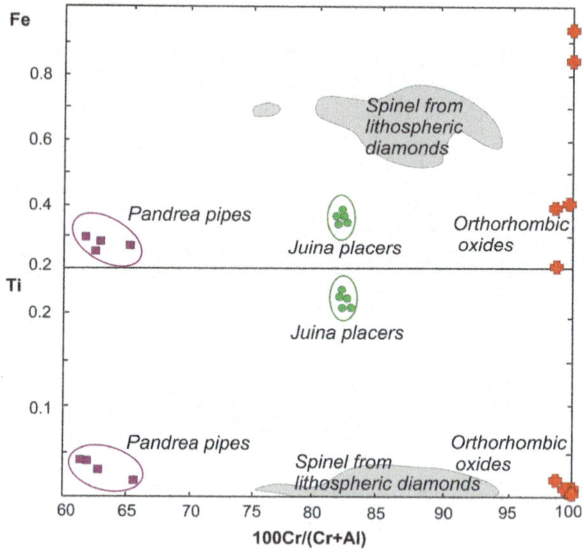

Fig. 4.34 Fe and Ti (cations) versus $Cr/(Cr + Al)_{at.}$ in inclusions with spinel compositions. Modified after Kaminsky (2012) with additions

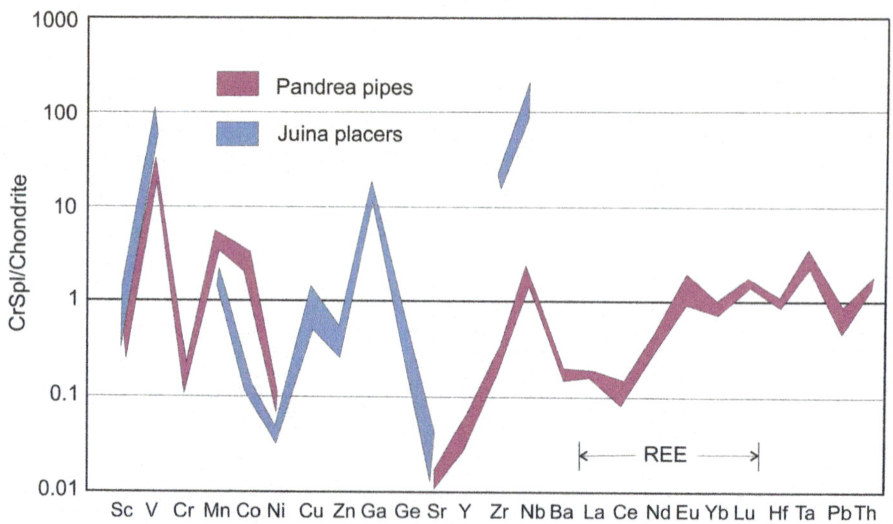

Fig. 4.35 Trace element patterns in inclusions with chromian spinel compositions. Data from Kaminsky et al. (2001, 2009a). Chondrite composition was derived from McDonough and Sun (1995). After Kaminsky (2012)

Other than the grains with the general chrome spinel composition, grains with alumospinel with *mg* = 0.76–0.82 have been identified in diamond from the Panda, Canada and Collier-4, Brazil kimberlite pipes (Tappert et al. 2005a; Bulanova et al. 2010). In addition, grains with the composition of magnetite were also identified in diamonds from the Rio Soriso placer and the Juina-5 kimberlite pipe in Brazil, accompanied (in the latter place) by bridgmanite, jeffbenite, ilmenite and titanite (Thomson et al. 2014). Other than Fe cation substitution, major cations in these minerals are Si (0.16–0.28 wt% SiO_2), Ti (1.50–3.61 wt% TiO_2), Al (3.44–6.92 wt % Al_2O_3), Cr (2.10–3.03 wt% Cr_2O_3), Mg (1.46–3.58 wt% MgO), and Na (0–0.11 wt% Na_2O). Some of them with TiO_2 up to ~16 wt% (Thomson et al. 2014) may be attributed as titanomagnetite.

4.8 Jeffbenite

4.8.1 General

Jeffbenite $(Mg,Fe^{2+},Ca)_3(Al,Fe^{3+},Cr)_2$ $(Si,Ti)_3O_{12}$ (IMA No. 2014-097; Nestola et al. 2015, 2016b), a tetragonal almandine-pyrope phase formerly known as TAPP, was not expected to exist in the lower-mantle conditions with peridotite, pyrolite or MORB compositions (e.g., Irifune et al. 2010). It was discovered as small inclusions of 30–100 μm in diameter in diamonds from the São Luiz placer deposit in the Juina area, Brazil (Harte and Harris 1994; Harris et al. 1997; Harte et al. 1999; Zedgenizov et al. 2014a). Subsequently, it was identified in diamonds from other placers and kimberlite pipes of the same area and in Guinea placer, Kankan, and was suggested as a primary mineral phase in the upper-mantle and the lower parts of the transition zone (Kaminsky et al. 2001, 2009a, b; Hayman et al. 2005; Bulanova et al. 2010). Jeffbenite occurs as cubo-octahedral or elongate-tabular, apple-green grains with a stoichiometric composition similar to pyrope-almandine garnet. Some grains exhibit inclusions of ilmenite and ilmenite + spinel, which are possibly retrograde (Zedgenizov et al. 2014a).

As of yet, jeffbenite has not been synthesized experimentally in ultramafic systems at high *P*–*T* conditions since it appeared unstable in LHDAC experiments at pressures higher than 10–13 GPa and temperatures of 1300–1700 K (Armstrong and Walter 2012). Some researchers have suggested that the origin of jeffbenite is related to a retrograde process during the ascent of magmatic materials to the Earth's surface (Finger and Conrad 2000; Brenker et al. 2002; Armstrong and Walter 2012). An alternative hypothesis for its origin is that jeffbenite is a reaction product of majorite in the lower mantle (Harte 2010; Harte and Hudson 2013). Prior to the observation of the occurrence of an aluminous phase in the lower mantle, all Al was thought to reside in bridgmanite (Irifune 1994; Kesson et al. 1995), such that jeffbenite was thought to be the only aluminous phase to occur in a juvenile lower-mantle ultramafic association. Based on the association of jeffbenite with typical lower-mantle mineral phases, such as bridgmanite, ferropericlase and others (Harte and Harris 1994; Harris et al. 1997; Harte et al. 1999; Hutchison et al. 2001; Hayman et al. 2005; Kaminsky et al. 2001; Zedgenizov et al. 2014a), here we have grouped jeffbenite as a lower-mantle phase.

4.8.2 Chemical Composition

Chemical compositions of jeffbenite are presented in Table 4.17. Based on the compositional analysis, jeffbenite lacks evidence for majoritic solid solution and is very low in Ca (usually 0.03–0.54 wt% CaO; one sample from Guinea has 11.72 wt% CaO, partly replacing MgO (see Table 4.17), sample KK-108a), as compared with upper-mantle peridotitic or eclogitic garnets. The increasing admixture of Ti (up to 7.55 wt% TiO_2) is accompanied by decreasing the Si content, demonstrating the isomorphism between these elements. Ti-free jeffbenite is believed to form single-phase inclusions under relatively higher pressure conditions (Anzolini et al. 2016b), but it is not confirmed by geological data, according to which Ti-free jeffbenite occurs in association with bridgmanite and NaAl-silicate (Hutchison 1997; Table 4.17, sample BZ259A2).

Jeffbenite has very variable iron, magnesium and aluminium contents (FeO = 4.43–23.39 wt%; MgO = 12.86–37.01 wt%; Al_2O_3 = 7.79–24.17 wt%) and iron and magnesium indices (*fe* = 0.084–0.453; *mg* = 0.547–0.916). Ferric iron is the abundant form of iron ions: Fe^{3+}/Fe_{tot} = 66–74% (McCammon et al. 1997). The distribution of *fe* in jeffbenite is bimodal; two groups of samples can be distinguished by this parameter: low-Fe (*fe* = 0.084–0.217) and high-Fe (*fe* = 0.365–0.453) (Fig. 4.36a). These two groups of jeffbenite have different geochemical characteristics. The low-Fe group of jeffbenite (*fe* = 0.084–0.217) has high Al (15.60–24.17 wt% Al_2O_3) and elevated concentrations of Cr (1.33–3.00 wt % Cr_2O_3), while the second high-Fe group (*fe* = 0.365–0.453) usually has low Al (7.79 vs. 19.92 wt% Al_2O_3) and low Cr content (0.02–0.06 wt% Cr_2O_3) (Fig. 4.36b). The first group with the low Fe content is associated with typical lower-mantle minerals such as bridgmanite and ferropericlase, suggestive of jeffbenite being a primary lower-mantle mineral phase. The second group with the high Fe content is associated with garnet, spinel, magnesite and perhaps orthopyroxene, most likely originating in the transition zone or of the retrograde process.

Jeffbenite also contains minor amounts of Ni (0.01–0.07 wt% NiO), Mn (0.14–0.96 wt% MnO), Ca (0.01–0.13 wt%; in one sample 11.72 wt% CaO); Na (0.02–1.13 wt% Na_2O) and K (0–0.02 wt% K_2O). Concentrations of REE trace elements in jeffbenite, apart from Nb, Zr, and Hf, are less than those in chondrites; the trends of the REE patterns are, however, similar to chondritic (Fig. 4.37).

4.8.3 Crystal Structure

Jeffbenite was first examined to display a tetragonal crystal structure with a space group *I*-42*d* (*a* = 6.526 Å and *c* = 18.182 Å) and a chemical formula $^{VIII\text{-}IV}(Mg,Fe^{3+})^{II}(Al,Cr,Mn)_2{}^{II}(Mg,Fe^{2+})_2Si_3O_{12}$, where the first cation group occupies a tetrahedral coordination with a four-fold symmetry and the second and the third groups have an octahedral coordination with a two-fold symmetry (Harris et al. 1997). Strictly speaking, the tetrahedral site is a distorted eight-fold 'capped'

Table 4.17 Representative compositions of jeffbenite (wt%)

Country	Brazil						Guinea
Area	Juina					Machado	Kankan
Location	Collier-4 pipe	Sao Luiz placer		Rio Vermelho placer	Rio Soriso placer	Placer	Placer
Sample No.	J4-1	BZ207A	BZ259A2	Juina 4-104	1-5A2	P33-1b	KK-108a
Assemblage	Jeffbenite + Magnesite	Jeffbenite + Brd + fPer	2Jeffbenite + Brd + NaAl-silicate	Jeffbenite + Brd	Jeffbenite + Brd + fPer + 'Ol'	Jeffbenite + Brd + 'Ol'	Jeffbenite + 2Brd + 2fPer + Ga
SiO_2	36.05	39.93	41.83	37.75	39.99	30.49	47.00
TiO_2	3.56	4.23	0.02	7.55	4.71	2.76	1.01
Al_2O_3	17.82	20.01	23.15	16.96	19.03	13.08	7.79
Cr_2O_3	0.01	1.34	2.40	1.74	2.74	0.07	0.02
FeO	20.12	9.51	4.43	10.06	6.87	14.80	14.53
NiO	n.a.	0.02	0.01	n.a.	n.a.	0.04	0.03
MnO	0.37	0.27	0.65	0.16	0.14	<0.02	0.41
MgO	17.99	25.16	26.91	24.39	25.75	37.01	12.86
CaO	0.04	0.03	0.11	0	0.04	<0.01	11.72
Na_2O	0.10	0.02	0.15	n.a.	n.a.	0.08	1.13
K_2O	0.04	0.01	0.01	n.a.	n.a.	0.03	0.80
Total	96.10	100.53	99.67	98.61	99.27	98.36	97.28
fe	0.385	0.175	0.084	0.188	0.130	0.183	0.388
mg	0.615	0.825	0.916	0.812	0.870	0.817	0.612
References	Nestola et al. (2016b)	Hutchison (1997)		Kaminsky et al. (2001)	Hayman et al. (2005)	Burnham et al. (2016)	Stachel et al. (2000)

Note: *n.a.* not analysed

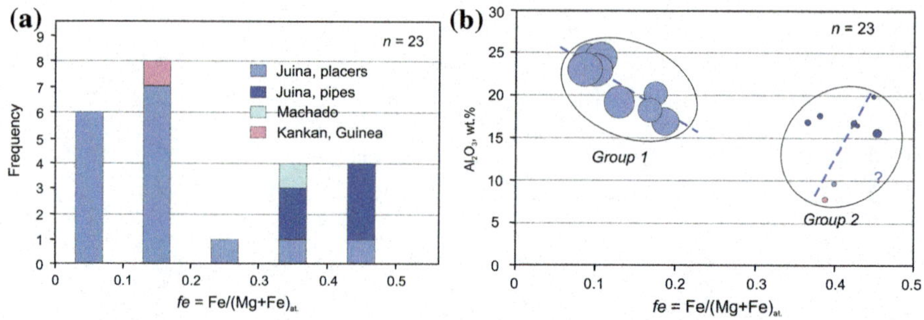

Fig. 4.36 Chemical characteristics of jeffbenite. **a** Distribution of iron index *fe* in jeffbenite. **b** Al_2O_3 versus *fe* in jeffbenite. Cr concentrations in jeffbenite are shown as *circles*, the radii of which are proportional to the Cr contents. Data from Hutchison (1997), Stachel et al. (2000), Kaminsky et al. (2001), Hayman et al. (2005), Bulanova et al. (2010), Thomson et al. (2014), Zedgenizov et al. (2014a), Burnham et al. (2016), Nestola et al. (2016b)

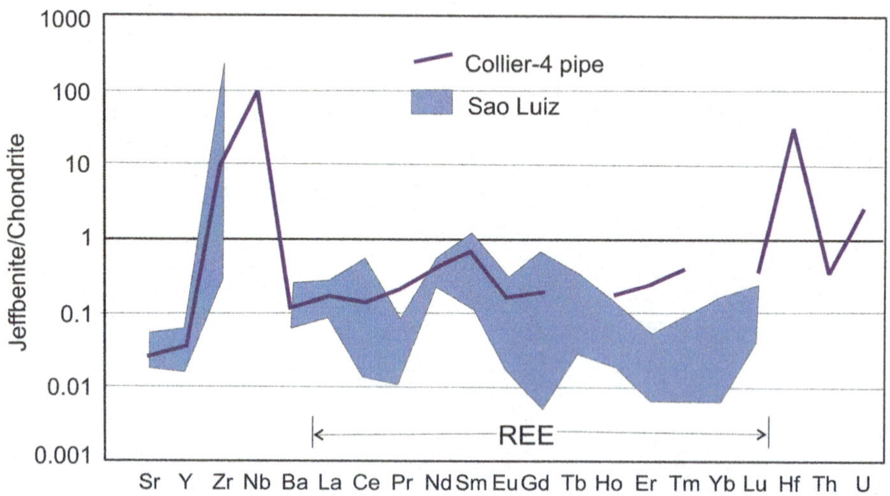

Fig. 4.37 Chondrite-normalised trace element patterns in jeffbenite from Brazil. Data from Hutchison et al. (2001), Bulanova et al. (2010), Harte et al. (1999). Chondrite composition was derived from McDonough and Sun (1995). After Kaminsky (2012). © Elsevier

coordination. These cation occupations in the lattice lead to small charge deficiencies at all cation sites, making a small amount of cation transfer in the lattice possible.

Finger and Conrad (2000) further studied one of the earlier examined jeffbenite samples and, in general, confirmed the initial assignment of the tetrahedral crystal structure. However, they proposed a revised model for the cation occupancies and simplified the formula as $^{VIII\text{-}IV}(Mg,Fe)^{VIII}(Al,Mg,Fe)_4[^{IV}(SiAl)O_4]_3$. The latest structural study of jeffbenite provided a more general formula of $Mg_3Al_2Si_3O_{12}$ with unit cell parameters of a = 6.5231 Å and c = 18.1756 Å (Nestola et al. 2015, 2016b). Considering the fact that jeffbenite always contains some other cation substitutions in nature, the proposed general formula can be presented as (Mg,Fe^{2+},

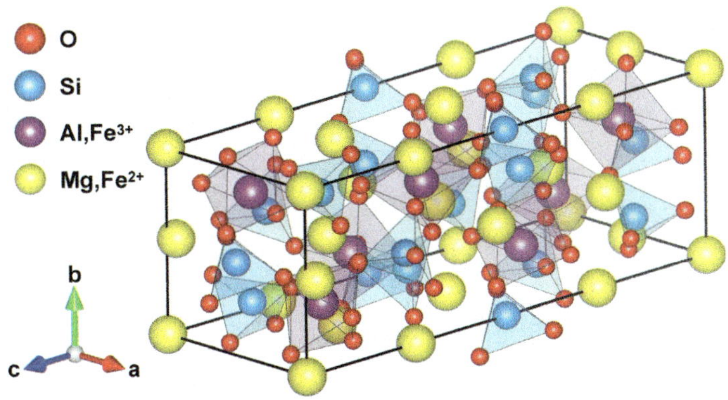

Fig. 4.38 Tetragonal crystal structure of jeffbenite (space group $I\bar{4}2d$). Lattice parameters from Nestola et al. (2016b)

$Ca)_3(Al,Fe^{3+},Cr)_2(Si,Ti)_3O_{12}$ (Fig. 4.38). It should be noted that all structural studies were performed for the low-Fe high-Cr 'juvenile' group of jeffbenite.

Jeffbenite was suggested to have a limited *P–T* stability field in moderately aluminous bulk compositions of the uppermost lower mantle (Harris et al. 1997). Recent experiments demonstrated that the stability field of jeffbenite extends to higher pressures than previously determined (Anzolini et al. 2016b).

4.9 Majorite Garnet

4.9.1 General

Majorite garnet, or majorite $Mg_3(Mg,Fe,Al,Si)_2Si_3O_{12}$ (in which there is an excess of Si in octahedral coordination) is the major mineral of the transition zone (e.g., Harte 2010). In majorite grains from the Monastery and Jagersfontein pipes, South Africa, the silica excess is as high as X_{Si} = 3.2–3.6 (Moore and Gurney 1985; Chinn et al. 1998; Tappert et al. 2005b). Some experimental studies have demonstrated that majorite is stable at pressures of up to 28 GPa (Irifune and Ringwood 1993). The association of majorite, in diamond from the Juina placers in Brazil, with perovskite and ilmenite (Wilding et al. 1991; Kaminsky et al. 2001) allows for its possible presence in the uppermost parts of the lower mantle. However, it is difficult to distinguish lower-mantle majorite; most of them are from the transition-zone.

4.9.2 Chemical Composition

The chemical compositions of the observed majorite varies within a wide range of *mg* = 0.402–0.895; mostly within the range of *mg* = 0.55–0.70 (Table 4.18;

Table 4.18 Representative compositions of majorite (wt%)

Country	Brazil							Canada
Area	Juina						Machado	Slave
Location	Sao Luiz				Corrigo Chicoria	Collier 4	Placer	Pipe DO27
Sample No.	BZ237C	28	34	C11	Juina 5-108b	J9-1	P23	97 28(1)
Assemblage	Maj + Brd	Maj + CaSiPrv + CaSiTiPrv + SiO_2	Maj + CaSiPv	Maj	Maj + CaSiprv + Ilm	3Maj	Maj	5Maj + Ni-sulph
SiO_2	43.37	44.36	41.94	39.41	41.46	42.31	49.13	42.73
TiO_2	0.01	0.34	1.66	1.39	1.95	1.06	0.42	0.79
Al_2O_3	21.47	20.36	18.76	20.40	18.82	21.45	12.41	19.31
Cr_2O_3	2.96	0.08	0.07	0.04	0.12	0.07	0.21	0.07
FeO	6.02	13.74	13.32	18.91	13.37	14.90	5.14	14.16
NiO	0.02	n.a.	n.a.	n.a.	n.a.	0.02	<0.07	n/a
MnO	0.89	0.28	0.29	0.32	0.3	0.3	0.31	0.29
MgO	22.20	12.69	10.77	7.13	12.46	10.05	24.62	9.87
CaO	1.72	7.29	12.26	9.94	10.7	9.48	8.14	13.05
Na_2O	0.27	1.50	0.71	0.93	0.67	0.85	0.30	0.11
K_2O	0.01	0.01	0	0.02	n.a.	0	<.04	n.a.
Total	98.94	100.64	99.78	98.48	99.85	100.49	100.68	100.38
fe	0.132	0.378	0.409	0.598	0.376	0.454	0.105	0.446
mg	0.868	0.622	0.591	0.402	0.624	0.546	0.895	0.554
References	Hutchison (1997)	Zedgenizov et al. (2014a)			Kaminsky et al. (2001)	Bulanova et al. (2010)	Burnham et al. (2016)	Davies et al. (2004)

Note: *n.a.* not analysed

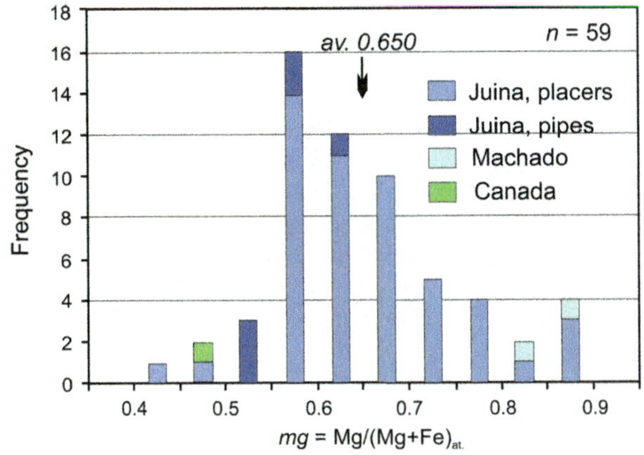

Fig. 4.39 Distribution of magnesium index *mg* in majorite inclusions in diamond. Data from Wilding (1990), Hutchison (1997), Davies et al. (2004), Kaminsky et al. (2001), Bulanova et al. (2010), Burnham et al. (2016)

Fig. 4.39). The major admixture is sodium (0.27–1.12 wt% Na_2O), which corresponds to the experimental data regarding the synthesis of majorite with up to 12 mol% of Na in a garnet solid solution (Bobrov et al. 2008a; Dymshits et al. 2010). Na-majorite may be an important concentrator of sodium in the upper part of the lower mantle and transition zone. Sodium majorite $Na_2MgSi_5O_{12}$ has tetragonal structure where excess Si occupies the octahedral site that is compensated for by the entry of Na in the X site of the garnet structure (Bobrov et al. 2008b; Bindi et al. 2011).

Concentrations of trace elements in majorite are usually one to two magnitudes higher than in chondrite, except Ni and Ba, which form negative anomalies (Fig. 4.40).

The patterns of REE in majorite from different regions are also distinct (Fig. 4.40). Majorite from Brazil and Guinea have almost linear, chondritic-like patterns (with a small predominance of LREE in majorite from Guinea), while majorite from South Africa (pipes Monastery and Jagersfontein) are significantly depleted with respect to LREE; the most contrasting distribution for the samples is present in those from the Jagersfontein pipe. Many majorite grains have a positive or a negative Eu anomaly, most likely indicating their 'eclogitic' genesis, with formation of these grains likely to have occurred within subducted lithosphere. This is confirmed by very light isotopic compositions of the hosting diamonds (from −17‰ to −24‰ $\delta^{13}C$ PDB). The other majorite grains, which lack Eu-anomalies, may belong to the ultramafic association.

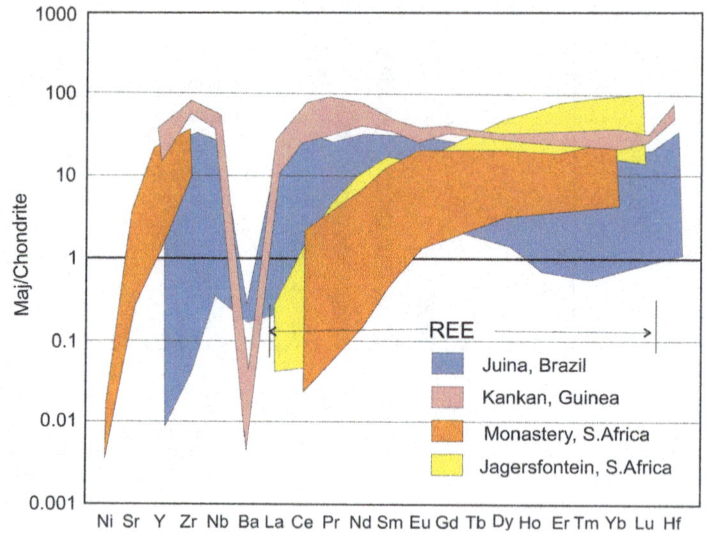

Fig. 4.40 Chondrite-normalised trace element patterns in majorite. Data from Moore et al. (1986), Hutchison et al. (2001), Kaminsky et al. (2001), Stachel et al. (2000), Tappert et al. (2005b), Burnham et al. (2016). Chondrite composition from McDonough and Sun (1995). Modified after Kaminsky (2012), with additions

4.10 Ilmenite

4.10.1 General

Ilmenite (Fe,Mg,Mn)TiO_3 is identified in inclusions from Brazilian diamonds together with bridgmanite, ferropericlase, CaSi-perovskite, jeffbenite and titanite, and is attributed to the lower mantle and transition zone origin (Kaminsky et al. 2001, 2009a; Thomson et al. 2014). Three compositional varieties, common ilmenite, Mg-ilmenite (picroilmenite) (Fe,Mg)TiO_3 and Mn-ilmenite (manganilmenite) (Fe,Mn)TiO_3, occur as inclusions in diamond. Picroilmenite is a common inclusion in upper-mantle diamonds, as well as in all kimberlite pipes worldwide. By contrast, common ilmenite and manganilmenite were not found in diamonds or kimberlites until recently. Of particular interest is manganilmenite, which was previously encountered only as a late- or post-magmatic, metasomatic phase in kimberlites, carbonatites, agpaitic and ultramafic pegmatites, and some other rocks.

Low-magnesium ilmenite (with only 0.11–0.14 wt% MgO) was first reported as an inclusion in three Brazilian diamonds by Meyer and Svisero (1975) who noted that its composition is close to stoichiometry and it does not have high Mg concentrations, characteristic for kimberlitic picroilmenite. The mineral has only one major impurity, MnO (0.64–0.75 wt%), which is markedly different from kimberlitic ilmenite. The exact locality of those samples was not mentioned in the literature, however, but it might be related to the Juina area in which extensive

Fig. 4.41 Manganoilmenite from the Juina area, Brazil. **a** General view of grains from the Pandrea-3 kimberlite. **b** Internal structure with minor inhomogeneities. Black-scattered electron image. From Kaminsky and Belousova (2009). © Elsevier

mining for diamonds started in the 1970s. This finding was confirmed in the 1990s, when a series of low-Mg manganoan ilmenite inclusions in Juina placer diamonds was found (Kaminsky et al. 2001). At the same time, inclusions of low-Mg, Mn-ilmenite were found in placer diamonds from Guaniamo, Venezuela (Kaminsky et al. 1997, 2000; Sobolev et al. 1998), and later manganoan ilmenite was found as an inclusion in a diamond from a primary deposit kimberlitic pipe Pandrea-7 in the Juina area (Kaminsky et al. 2009a). In Juina diamonds, manganilmenite was identified, for the first time, as a primary-magmatic phase with a homogeneous, internal structure. In one of the diamond crystals from the Pandrea-7 pipe, manganilmenite occurs with picroilmenite. A great number of Mn-ilmenite grains were identified in this and other kimberlitic rocks from the same area (Fig. 4.41a). They appeared to be low-Mg, manganoan ilmenite, analogous to the composition of the grains included in diamonds, and were considered to be a new kimberlite/diamond indicator mineral (Kaminsky and Belousova 2009). However, the internal structure of the Mn-ilmenite grains shows some inhomogeneous features (Fig. 4.41b).

4.10.2 Chemical Composition and Ilmenite Varieties

The chemical compositions of the ilmenite grains are presented in Table 4.19. All three compositional varieties show significant non-stoichiometry by as much as 10–13% in some samples, which is most likely caused by a high proportion of Fe^{3+} cation substitution in the lattice. Manganilmenite has a MnO content from 0.42 to 6.01 wt% and an elevated vanadium admixture (V_2O_3 = 0.21–0.43 wt%); in one grain the MnO content reaches 11.46 wt%. MgO content in all manganilmenite grains varies from less than 0.04 wt% (the analytical detection limit) to 0.81 wt%, while in picroilmenite it is approximately 9.33–11.94 wt%. Analysis of the MnO and MgO contents shows no correlation between each other and with Fe.

Table 4.19 Representative compositions of ilmenites from the Juina area, Brazil (wt%)

Location	Pandrea-7 pipe		Rio Sao Luis placer			Corrigo Chicoria placer	
Sample No.	P7-29		Juina 1-4	Juina 1-32	Juina 1-34	Juina 5-108	Juina 5-56
Mineral association	Pilm + Mn-Ilm		Ilm + fPer	Mn-ilm	Ilm	Mn-Ilm + CaSiPrv + Maj	Pilm
SiO_2	0.03	0.02	0.01	0.01	0.00	0.02	0.04
TiO_2	49.08	55.78	50.22	52.51	50.41	50.13	51.93
Al_2O_3	0.86	0.16	0.00	0.05	0.00	0.00	0.47
Cr_2O_3	0.29	0.00	0.00	0.00	0.00	0.06	0.66
V_2O_3	0.43	0.13	n.a.	n.a.	n.a.	n.a.	n.a.
FeO	38.92	38.08	46.21	39.62	48.24	42.43	32.72
MnO	0.27	2.81	0.73	5.37	0.44	6.01	0.25
NiO	0.05	0.00	0.00	0.00	0.00	0.00	0.11
MgO	9.33	0.12	0.12	0.81	0.40	0.00	11.94
CaO	0.02	0.01	0.03	0.00	0.00	0.03	0.00
ZnO	0.02	0.04	n.a.	n.a.	n.a.	n.a.	n.a.
Na_2O	0.02	0.01	n.a.	n.a.	n.a.	n.a.	n.a.
K_2O	0.01	0	n.a.	n.a.	n.a.	n.a.	n.a.
Total	99.33	97.16	97.32	98.37	99.49	98.68	98.12
fe	0.697	0.926	0.980	0.852	0.977	0.874	0.603
mg	0.298	0.005	0.005	0.031	0.014	0.000	0.392
mn	0.005	0.069	0.016	0.117	0.009	0.126	0.005
References	Kaminsky et al. (2009b)		Kaminsky et al. (2001)				

Note: *n.a.* not analysed

All manganilmenite grains found earlier in carbonatites, gabbro, pegmatites, and other rocks were proposed to be formed at the latest magmatic or postmagmatic, metasomatic stages (Simpson 1929; Vincent and Phillips 1954; Omori and Hasegawa 1955; Snetsinger 1969; Elsdon 1975; Mitchell 1978; Gaspar and Wyllie 1984). They either form reaction rims on grains of picroilmenite and other ore minerals, or compose laths in the groundmass. The manganilmenite from Juina diamonds as well as from the sources of kimberlites is a primary mineral phase, and can be distinguished from the secondary manganilmenite based on their compositional differences (Table 4.20).

Trace element patterns in manganilmenite and picroilmenite are plotted in Fig. 4.42. They show resemblance to each other mineral with very little variation. Both manganilmenite and picroilmenite have high concentrations of V, Zr, Nb, Ta, Th, and U, while Ni, Cr, Co, Ta, and Nb in manganilmenite are less abundant than that in picroilmenite.

Table 4.20 Chemical characteristics of manganilmenite from different localities (Kaminsky and Belousova 2009)

Characteristic	Inclusions in diamond	Kimberlite, Juina, Brazil	Jacupiranga carbonatite, Brazil	Other igneous rocks	Metamorphosed mafic/ultramafic rocks, Western Australia
MnO, wt%	0.42–11.46	0.63–2.49	2.29–7.87	1.44–15.15	1.15–7.38
MgO	0.0.81	0–0.24	15.69–23.46	0.0.46	0.01–0.49
TiO_2	50.13–56.15	55.49–57.79	55.83–59.24	46.23–52.02	42.80–52.66
V_2O_3	0.13	0.21–0.39	–	0.06	0.01–0.43
NiO	0–0.12	0–0.02	4.56–6.80	–	–
$MnTiO_3$, mol %	0.9–23.5	1.3–5.0	4.2–14.6	3.2–40.4	3.0–16.4
$MgTiO_3$, mol %	0–2.0	0–0.6	52.6–75.6	0–6.0	0.1–2.3
References	Kaminsky et al. (2001, 2009a)	Kaminsky and Belousova (2009)	Mitchell (1978)	Simpson (1929), Vincent and Phillips (1954), Omori and Hasegawa (1955), Snetsinger (1969), Elsdon (1975)	Cassidy et al. (1988)

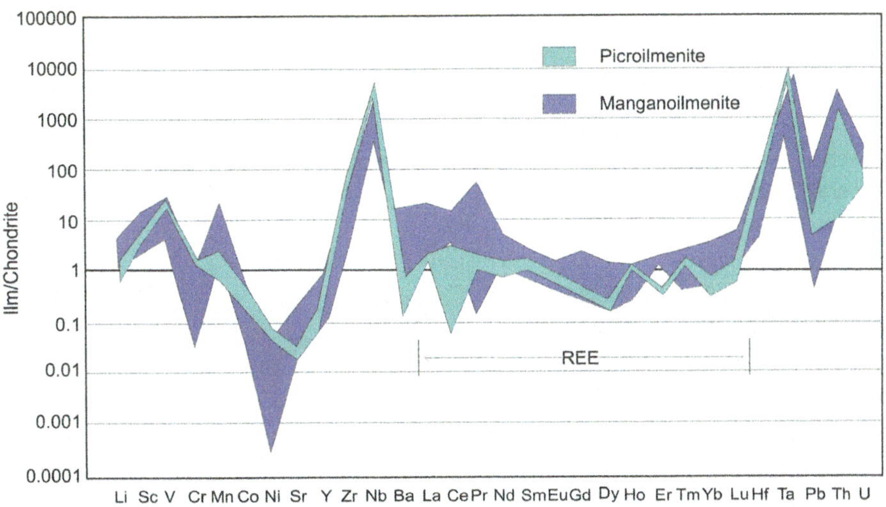

Fig. 4.42 Chondrite-normalised trace element patterns in manganilmenite and picroilmenite. Data from Kaminsky et al. (2001, 2009a). Chondrite composition from McDonough and Sun (1995). Modified after Kaminsky (2012)

4.10.3 *Crystal Structure and Phase Stability of Ilmenite*

The ilmenite phase with trigonal structure (space group $R\bar{3}$) in the $MgSiO_3$–$FeSiO_3$ system was found in meteorites and named 'akimotoite' after Prof. Syun-iti Akimoto of the University of Tokyo (Tomioka and Fujino 1997, 1999). Akimotoite is believed to be a significant mineral in the Earth's mantle at depths of approximately 600–800 km (at pressures between approximately 21 and 28 GPa). The ilmenite crystal structure can be described as a corundu structure with Mg^{2+} and Si^{4+} located in an ordered alternative fashion in the six-co-ordinated Al^{3+} sites (Liu 1977). At lower-mantle pressures, ilmenite transitions into the bridgmanite structure, which is a 3D framework of the corner-shared SiO_6 octahedra with Mg^{2+} in the pseudo-dodecahedral sites.

When in association with the majoritic garnet (Akaogi et al. 2002; Fei and Bertka 1999), the addition of Al_2O_3 widens the stability field of majorite and suppresses the stability field of ilmenite at high pressures and relatively high temperatures (Akaogi et al. 2002). Across the ilmenite–bridgmanite transition, the addition of Al_2O_3 extends the transition boundary to greater pressures (deeper parts of the lower-mantle conditions). In the ilmenite solid solution, the addition of Al_2O_3 slightly increases the *a*-axis but decreases the *c*-axis and molar volume substantially (Akaogi et al. 2002). According to calculations using generalised gradient approximations, ilmenite may be stable only in the uppermost levels of the lower mantle and in the transition zone under a pressure of 20–26 GPa (Yu et al. 2011).

4.11 Other Minerals

A series of other minerals were found in association with major lower-mantle minerals and may be considered to be accessory mineral phases in the deep-Earth mantle. Some of them may occur locally.

4.11.1 *Olivine Polymorphs*

A mineral phase with a composition identical to olivine $(Mg,Fe)_2SiO_4$ was identified in diamonds from Brazilian kimberlite pipes and placer deposits, Canadian kimberlite pipes, Guinea, and Australia, where it forms colorless grains in association with bridgmanite, ferropericlase, CaSi-perovskites, jeffbenite and some other mineral phases (Wilding et al. 1991; Stachel et al. 2000; Kaminsky et al. 2001; Hayman et al. 2005; Tappert 2005a, 2009a, b; Bulanova et al. 2010; Thomson et al. 2014). These grains have wide variations in Mg (mg = 0.848–0.965) (Table 4.21). The distribution of mg in 'olivine' grains is similar to the one in bridgmanite (Fig. 4.43a) with an almost identical value range and average mg = 0.908 (0.917 in bridgmanite) and bimodality. The latter is most likely caused by different compositions of this mineral in different areas: in Brazil, this phase has less magnesium

Table 4.21 Representative analyses of mineral phase with a composition identical to olivine (wt%)

Country	Brazil					Canada	Guinea	Australia
Area	Juina					Northwest Territories	Kankan	Orroroo
Location	Pipe Juina-5	Corrigo Chicoria	Sao Luiz placer	Rio Soriso placer	Rio Soriso	Pipe DO27	Placer	Dyke Eurelia K7
Sample No.	Ju5-73	Juina 5-5	80	3-2S	4-3C	DO270300	KK-44e	FBS5-12
Assemblage	'Ol' + fPer	'Ol' + fPer	'Ol' + Brd + CaSiPrv	'Ol' + Brd + fPer + CaSiPrv	'Ol' + Brd + fPer	'Ol' + 3Brd + fPer	'Ol' + 3Brd + fPer + CaSiPrv	'Ol'
SiO_2	41.55	40.31	40.5	39.42	42.74	41.70	41.02	40.78
TiO_2	0.00	n.a.	n.a.	n.a.	0.07	0.00	0.01	0.02
Al_2O_3	0.02	0.03	n.a.	n.a.	0.8	0.03	0.04	0.00
Cr_2O_3	0.14	0.15	n.a.	n.a.	n.a.	0.00	0.07	0.07
FeO	9.25	10.79	14.1	11.15	4.61	3.52	6.15	8.45
NiO	0.45	0.37	0.24	n.a.	n.a.	0.08	0.09	0.47
MnO	0.09	n.a.	n.a.	0.12	0.15	0.09	0.11	0.11
MgO	48.22	47.42	44.1	48.85	51.96	53.80	52.26	48.81
CaO	0.00	n.a.	n.a.	0.07	n.a.	0.05	0.01	0.06
Na_2O	0.05	0.04	n.a.	n.a.	n.a.	0.00	0.02	0.00
K_2O	0.01	n.a.	n.a.	n.a.	n.a.	n.a.	0.00	0.00
Total	99.79	99.11	99.02	99.61	100.33	99.27	99.83	98.77
fe	0.097	0.113	0.152	0.113	0.047	0.035	0.062	0.088
mg	0.903	0.887	0.848	0.887	0.953	0.965	0.938	0.912
References	Thomson et al. 2014	Kaminsky et al. (2001)	Zedgenizov et al. (2014a)	Hayman et al. (2005)		Davies et al. (2004a)	Stachel et al. (2000)	Tappert et al. (2005a)

Note: *n.a.* not analysed

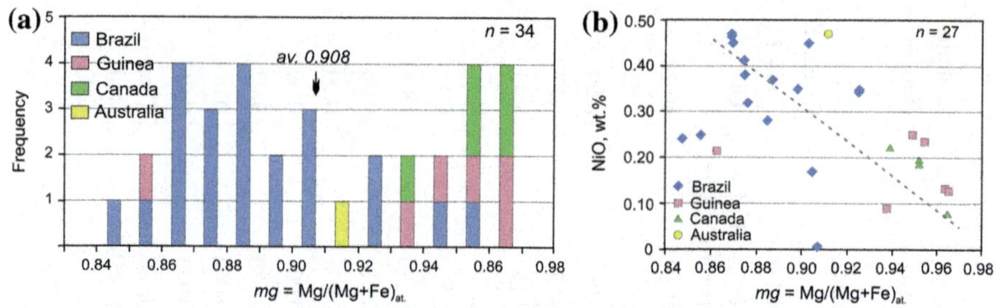

Fig. 4.43 Chemical characteristics of mineral phase with a composition of olivine. **a** Distribution of magnesium index *mg* in the mineral. **b** NiO versus *mg*. Data from Wilding (1990), Hutchison (1997), Stachel et al. (2000), Kaminsky et al. (2001), Davies et al. (2004), Hayman et al. (2005), Tappert et al. (2009a, b), Bulanova et al. (2010), Thomson et al. (2014), Zedgenizov et al. (2014a)

than in other areas. The Mg concentration shows a non-strong negative correlation with Ni (Fig. 4.43b).

This phase is most likely a product of a retrograde transformation from a lower-mantle mineral, or a retrograde reaction fPer + Brd → Ol. The latter is less possible because it associates (in diamond) not only with ferropericlase but also with bridgmanite, sometimes in a 'touching' association. None of the 'olivine' grains were studied structurally. It is not excluded that these grains are, in fact, ringwoodite from the uppermost layers from the lower mantle.

4.11.2 *Titanite*

Titanite ($CaTiSiO_5$) was identified in placers of Rio Chicoria (Kaminsky et al. 2001) and Rio Soriso (Hayman et al. 2005) in the Juina area, Brazil, where it associates with CaSi-perovskite. Its chemical composition (in wt%) is: 33.04 SiO_2; 29.70 TiO_2; 3.16 Al_2O_3; 0.06 FeO; 27.75 CaO. Low total for bulk composition (93.70%) suggests the existence of other, non-analyzed elements in the mineral. Structural analysis of this mineral in diamond inclusions has not been performed, although it is possible that it is a polymorph modification of $CaTiSiO_5$.

4.11.3 *Merwinite*

The inclusion of merwinite ($Ca_{2.85}Mg_{0.96}Fe_{0.11}Si_{2.04}O_8$) was first detected by Raman spectroscopy in the core zone of one Sao-Luiz diamond, Brazil in association with the inclusions of walsromite-structured $CaSiO_3$ and olivine of $(Mg_{0.86}Fe_{0.14})_2SiO_4$ composition (Zedgenizov et al. 2014b). Merwinite was found to contain microinclusions of calcite.

The existing experimental data (Sharygin et al. 2012) indicate that under mantle conditions merwinite can be only formed in specific, Ca-rich and Mg- and Si-depleted environments, which differ from any known mantle lithology. Such chemical conditions can occur during interaction of Ca-rich carbonatite melts with ultramafic mantle, which would eventually lead to crystallization of $CaSiO_3$ and merwinite together with late olivine after re-equilibration of Ca and Mg (Zedgenizov et al. 2014b, 2016).

4.11.4 Native Nickel

Native nickel (Ni^0) associated with CaSi-perovskite was identified in diamond from Rio São Luiz (Wilding et al. 1991) and Rio Chicoria (Kaminsky et al. 2001) in the Juina area, Brazil. Native nickel is also found in pipe DO27, Canada, occurring with ferropericlase and MgSi-perovskite (Davies et al. 2004). Its peculiar chemical feature is a noticeable admixture of potassium (in the sample from the Chicoria, 0.21 wt% K_2O). Other admixtures include (in wt%): Fe (1.01–1.29 FeO), Ti (0.02–0.07 TiO_2), Cr (0.01 Cr_2O_3), Mn (0.07 MnO), and Na (0.02 Na_2O).

4.11.5 Native Iron

Native iron (Fe^0) was identified in diamonds from the Rio Soriso placer in association with low-Ni pyrrhotite (Hayman et al. 2005) and from the Pandrea-1 pipe in association with ferropericlase and chrome spinel (Kaminsky et al. 2009a), both from the Juina area, Brazil; and from the placer Kankan in Guinea (Stachel et al. 2000). The mineral is notably rich in Cr (0.77–1.62 wt%), as well as in Ni (0.22–0.31 wt%), Mg (0.10–0.57 wt%), Mn (0.13–0.36 wt%), Ti (0.01–0.07 wt%) and Al (0.05–0.14 wt%).

4.11.6 Unidentified Si–Mg Phase

An unidentified mineral of Si–Mg with a Si/Mg atomic ratio of ~4/3 (1.37) was observed in a diamond from the Chicoria placer in Brazil, occurring in association with perovskite, manganilmenite, and majorite (i.e., a lower-mantle origin is not obvious) (Kaminsky et al. 2001). The composition of the mineral consists only of Si (61.95 wt% SiO_2) and Mg (30.33 wt% MgO), with admixtures of Al (0.10 wt% Al_2O_3) and Fe (0.45 wt% FeO). Considering the low analytical totals (92.71 wt%) for this material, the presence of volatiles in this phase may be suggested.

4.11.7 Moissanite

Moissanite SiC, another potential lower mantle phase, was identified in diamonds from the São Luiz placer deposit (Wilding et al. 1991), but it is found to be unassociated with other minerals.

4.12 Conclusions

The juvenile ultramafic association predominates in the lower mantle. It is composed of bridgmanite + ferropericlase + CaSi-perovskite + free silica. This association, in contrast to the upper mantle, consists not of silicates but predominantly of oxides, and the traditional term 'silicate mantle' is not applicable to the lower mantle. It should be referred to as the 'oxide mantle'. Of particular importance, the presence of iron-rich ferropericlase-magnesiowüstite and free silica in the ultramafic association disagrees with the proposed pyrolitic composition of the lower mantle.

Bridgmanite has a high-Mg composition (mg = 0.84–0.96) and variable admixture of Al (0.25–12.58 wt% Al_2O_3). Two groups of bridgmanite are distinguished: low-Al (0.25–3.10 wt% Al_2O_3) and high-Al (8.34–12.58 wt% Al_2O_3). Natural high-Mg bridgmanite does not experience any phase transitions in the lower mantle down to the D″ border. The addition of Fe in bridgmanite causes an increase in its density, bulk modulus and bulk sound velocity, independently of the Al concentration, at lower-mantle pressures. The incorporation of Al either has no effect on the thermodynamic properties of bridgmanite or slightly decreases its bulk modulus; in any case, it does not appear to be significant for seismologic interpretations. The orthorhombic structure of bridgmanite is stable within the entire lower mantle down to the D″ layer, where it transforms into the post-perovskite phase.

Ferropericlase occurs in geological samples more frequently than was expected from experimental and theoretic data. Its chemical composition is different to that predicted, with magnesium index mg varying widely from 0.90 to 0.36, most likely caused by magma fractional crystallization in the lower mantle. Ferropericlase remains stable within the entire lower mantle with the cubic rocksalt structure; however, high Fe-rich varieties at subsolidus temperatures may obtain rhombohedral and hexagonal structures.

Iron contents in both ferropericlase and bridgmanite increase with depth, reflecting the increasing Fe content in the bulk composition of the lowermost mantle.

The occurrence of ~8.4% free silica (stishovite and $CaCl_2$–structured polymorph) is most likely cause by more silica-rich composition of the lower mantle compared with the upper mantle. The presence of free silica may create faster seismic velocities by a few percent. At a depth of 1600–1800 km (70–80 GPa pressure range), stishovite transforms into the $CaCl_2$-structured polymorph, which should cause the reduction of seismic velocities (particularly V_s).

In addition to major minerals, a variety of other mineral phases occurs in the lower mantle: Mg–Cr–Fe, Ca–Cr and other orthorhombic oxides, jeffbenite, ilmenite (including picroilmenite and manganilmenite), titanite, native Ni and Fe, moissanite, and some others.

References

Adams, D. J., & Oganov, A. R. (2006). Ab initio molecular dynamics study of $CaSiO_3$ perovskite at P–T conditions of Earth's lower mantle. *Physical Review B, 73,* 184106.

Akaogi, M. (2007). Phase transitions of minerals in the transition zone and upper part of the lower mantle. In E. Ohtani (Ed.), *Advances in High Pressure Mineralogy* (Vol. 421, pp. 1–13). Geological Society of America Memoirs.

Akaogi, M., Hamada, Y., Suzuki, T., Kobayashi, M., & Okada, M. (1999). High pressure transitions in the system $MgAl_2O_4$–$CaAl_2O_4$: A new hexagonal aluminous phase with implication for the lower mantle. *Physics of the Earth and Planetary Interiors, 115,* 67–77.

Akaogi, M., Tanaka, A., & Ito, E. (2002). Garnet–ilmenite–perovskite transitions in the system $Mg_4Si_4O_{12}$–$Mg_3Al_2Si_3O_{12}$ at high pressures and high temperatures: Phase equilibria, calorimetry and implications for mantle structure. *Physics of the Earth and Planetary Interiors, 132,* 303–324.

Akber-Knutson, S., & Bukowinski, M. S. T. (2004). The energetics of aluminum solubility into $MgSiO_3$ perovskite at lower mantle conditions. *Earth and Planetary Science Letters, 220,* 317–330. doi:10.1016/S0012-821X(04)00065-2

Akber-Knutson, S., Bukowinski, M. S. T., & Matas, J. (2002). On the structure and compressibility of $CaSiO_3$ perovskte. *Geophysical Research Letters, 29*(3), 1034–1037.

Anderson, D. L. (1983). Chemical composition of the mantle. *Journal of Geophysical Research, 88,* B41–B52. doi:10.1029/Jb088is01p00b41

Anderson, D. L. (2002). The case for irreversible chemical stratification of the mantle. *International Geology Review, 44*(2), 97–116.

Andrault, D. (2001). Evaluation of (Mg, Fe) partitioning between silicate perovskite and magnesiowüstite up to 120 GPa and 2300 K. *Journal of Geophysical Research, 106*(B2), 2079–2087.

Andrault, D., Angel, R. J., Mosenfelder, J. I., & Le Bihan, T. (2003). Equation of state of stishovite to lower mantle pressures. *American Mineralogist, 88,* 301–307.

Andrault, D., Bolfan-Casanova, N., Bouhifd, M. A., Guignot, N., & Kawamoto, T. (2007). The role of Al-defects on the equation of state of Al–(Mg, Fe)SiO_3 perovskite. *Earth and Planetary Science Letters, 263,* 167–179. doi:10.1016/j.epsl.2007.08.012

Andrault, D., Bolfan-Casanova, N., & Guignot, N. (2001). Equation of state of lower mantle (Al, Fe)-$MgSiO_3$ perovskite. *Earth and Planetary Science Letters, 193*(3–4), 501–508.

Andrault, D., Fiquet, G., Guyot, F., & Hanfland, M. (1998a). Pressure-induced Landau-type transition in stishovite. *Science, 282*(5389), 720–724.

Andrault, D., Neuville, D. R., Flank, A. M., & Wang, Y. (1998b). Cation sites in Al-rich $MgSiO_3$ perovskites. *American Mineralogist, 83,* 1045–1053.

Andrault, D., Pesce, G., Bouhifd, M. A., Bolfan-Casanova, N., Hénot, J.-M., & Mezouar, M. (2014). Melting of subducted basalt at the core-mantle boundary. *Science, 344*(6186), 892–895.

Andrault, D., Petitgirard, S., Lo Nigro, G., Devidal, J.-L., Veronesi, G., Garbarino, G., et al. (2012). Solid-liquid iron partitioning in Earth's deep mantle. *Nature, 487,* 354–357.

Antonangeli, D., Siebert, J., Aracne, C. M., Farber, D. L., Bosak, A., Hoesch, M., et al. (2011). Spin crossover in ferropericlase at high pressure: A seismologically transparent transition? *Science, 331,* 64–67.

Anzolini, C., Angel, R. J., Merlini, M., Derzsi, M., Tokár, K., Milani, S., et al. (2016a). Depth of formation of $CaSiO_3$-walstromite included in super-deep diamonds. *Lithos, 265,* 138–147. doi:10.1016/j.lithos.2016.09.025

Anzolini, C., Drewitt, J., Lord, O. T., Walter, M. J., & Nestola, F. (2016b). New stability field of jeffbenite (ex-"TAPP"): Possibility of super-deep origin. AGU Fall Meeting Abstract MR33A-2675, 1 p.

Armstrong, L. S., & Walter, M. J. (2012). Tetragonal almandine pyrope phase (TAPP): Retrograde Mg-perovskite from subducted oceanic crust? *European Journal of Mineralogy, 24*(4), 587–597.

Armstrong, L. S., Walter, M. J., Tuff, J. R., Lord, O. T., Lennie, A. R., Kleppe, F. R., et al. (2012). Perovskite phase relations in the system CaO–MgO–TiO_2–SiO_2 and implications for deep mantle lithologies. *Journal of Petrology, 53*(3), 611–635. doi:10.1093/petrology/egr073

Asahara, Y., Hirose, K., Ohishi, Y., Hirao, N., Ozawa, H., & Murakami, M. (2013). Acoustic velocity measurements for stishovite across the post-stishovite phase transition under deviatoric stress: Implications for the seismic features of subducting slabs in the mid-mantle. *American Mineralogist, 98*(11–12), 2053–2062. doi:10.2138/am.2013.4145

Auzende, A.-L., Badro, J., Ryerson, F. J., Weber, P. K., Fallon, S. J., Addad, A., et al. (2008). Element partitioning between magnesium silicate perovskite and ferropericlase: New insights into bulk lower-mantle geochemistry. *Earth and Planetary Science Letters, 269,* 164–174. doi:10.1016/j.epsl.2008.02.001

Badro, J., Fiquet, G., Guyot, F., Rueff, J.-P., Struzhkin, V. V., Vankó, G., et al. (2003). Iron partitioning in Earth's mantle: Toward a deep lower mantle discontinuity. *Science, 300*(5620), 789–791. doi:10.1126/science.1081311

Barnes, S. J., & Roeder, P. L. (2001). The range of spinel compositions in terrestrial mafic and ultramafic rocks. *Journal of Petrology, 42,* 2279–2302.

Bass, J. D. (2008). Recent progress in studies of the elastic properties of earth materials. *Physics of the Earth and Planetary Interiors, 170,* 207–209.

Bassett, W. A. (2001). The birth and development of laser heating in DACs. *Review of Scientific Instruments, 72,* 1270–1272.

Becerro, A. I., McCammon, C. A., Langenhorst, F., Angel, R. J., & Seifert, F. (1999). Oxygen vacancy ordering in $CaTiO_3$–$CaFeO_{2.5}$. *Phase Transitions, 69,* 133–146.

Belonoshko, A. B., Dubrovinsky, L. S., & Dubrovinsky, N. A. (1996). A new high-pressure silica phase obtained by molecular dynamics. *American Mineralogist, 81,* 785–788.

Bina, C. R. (2010). Scale limits of free-silica seismic scatterers in the lower mantle. *Physics of the Earth and Planetary Interiors, 183,* 110–114.

Bindi, L., Dymshits, A. M., Bobrov, A. V., Litasov, K. D., Shatsky, A. F., Ohtani, E., et al. (2011). Crystal chemistry of sodium in the Earth's interior: The structure of $Na_2MgSi_5O_{12}$ synthesized at 17.5 GPa and 1700 °C. *American Mineralogist, 96*(2–3), 447–450.

Bindi, L., Tamarova, A., Bobrov, A. V., Sirotkina, E. A., Tschauner, O., Walter, M. J., et al. (2016). Incorporation of high amounts of Na in ringwoodite: Possible implications for transport of alkali into lower mantle. *American Mineralogist, 101,* 483–486.

Bläß, U. W., Langenhorst, F., Frost, D. J., & Seifert, F. (2007). Oxygen deficient perovskites in the system $CaSiO_3$–$CaAlO_{2.5}$ and implications for the Earth's interior. *Physics and Chemistry of Minerals, 34,* 363–376.

Bobrov, A. V., Kojitani, H., Akaogi, M., & Litvin, Yu A. (2008a). Phase relations on the diopside-hedenbergite-jadeite join up to 24 GPa and stability of Na-bearing majorite garnet. *Geochimica et Cosmochimica Acta, 72*(9), 2392–2408.

Bobrov, A. V., Litvin, Yu A, Bindi, L., & Dymshits, A. M. (2008b). Phase relations and formation of sodium-rich majorite garnet in the system $Mg_3Al_2Si_3O_{12}$–$Na_2MgSi_5O_{12}$ at 7.0 and 8.5 GPa. *Contributions to Mineralogy and Petrology, 156*(2), 243–257.

Bolfan-Casanova, N., Andrault, D., Amiguet, E., & Guignot, N. (2009). Equation of state and post-stishovite transformation of Al-bearing silica up to 100 GPa and 3000 K. *Physics of the Earth and Planetary Interiors, 174,* 70–77. doi:10.1016/j.pepi.2008.06.024

Boujibar, A., Bolfan-Casanova, N., Andrault, D., Bouhifd, M. A., & Trcera, N. (2016). Incorporation of Fe^{2+} and Fe^{3+} in bridgmanite during magma ocean crystallization. *American Mineralogist, 101,* 1560–1570. doi:10.2138/am-2016-5561

Brenker, F. E., Stachel, T., & Harris, J. W. (2002). Exhumation of lower mantle inclusions in diamond: ATEM investigation of retrograde phase transitions, reactions and exsolution. *Earth and Planetary Science Letters, 198,* 1–9.

Brodholt, J. P. (2000). Pressure-induced changes in the compression mechanism of aluminous perovskite in the Earth's mantle. *Nature, 407,* 620–622.

Bulanova, G. P., Smith, C. B., Kohn, S. C., Walter, M. J., Gobbo, L., & Kearns, S. (2008). Machado River, Brazil—A newly recognised ultradeep diamond occurrence. In *9th International Kimberlite Conference*. Extended Abstract No. 9IKC-A-00233.

Bulanova, G. P., Walter, M. J., Smith, C. B., Kohn, S. C., Armstrong, L. S., Blundy, J., et al. (2010). Mineral inclusions in sublithospheric diamonds from Collier 4 kimberlite pipe, Juina, Brazil: Subducted protoliths, carbonated melts and primary kimberlite magmatism. *Contributions to Mineralogy and Petrology, 159*(4), 489–510. doi:10.1007/s00410-010-0490-6

Burnham, A. D., Bulanova, G. P., Smith, C. B., Whitehead, S. C., Kohn, S. C., Gobbo, L., et al. (2016). Diamonds from the Machado River alluvial deposit, Rondônia, Brazil, derived from both lithospheric and sublithospheric mantle. *Lithos, 265,* 199–213. doi:10.1016/j.lithos.2016.05.022

Cammarano, F., & Romanowicz, B. (2007). Insights into the nature of the transition zone from physically constrained inversion of long-period seismic data. *Proceedings of the National Academy of the U.S.A., 104*(22), 9139–9144.

Caracas, R., Wentzcovitch, R., Price, G. D., & Brodholt, J. (2005). $CaSiO_3$ perovskite at lower mantle pressures. *Geophysical Research Letters, 32,* L06306. doi:10.1029/2004GL022144

Carpenter, M. A. (2006). Elastic properties of minerals and the influence of phase transitions. *American Mineralogist, 91,* 229–246.

Carpenter, M. A., Hemley, R. J., & Mao, H.-K. (2000). High-pressure elasticity of stishovite and the $P4_2/mnm \leftrightarrow Pnnm$ phase transition. *Journal of Geophysical Research, 105,* 10807–10816.

Carpenter, M. A., & Salje, E. K. H. (1998). Elastic anomalies in minerals due to structural phase transitions. *European Journal of Mineralogy, 10,* 693–812.

Cassidy, K. F., Groves, D. I., & Binns, R. A. (1988). Manganoan ilmenite formed during regional metamorphism of Archean mafic and ultramafic rocks from Western Australia. *Canadian Mineralogist, 26*(4), 999–1012.

Chao, E. C. T., Fahey, J. J., Littler, J., & Milton, D. J. (1962). Stishovite, SiO_2, a very high pressure new mineral from Meteor Crater, Arizona. *Geophysical Research, 67,* 419–421.

Chen, M., Shu, J., & Mao, H.-K. (2008). Xieite, a new mineral of high-pressure $FeCr_2O_4$ polymorph. *Chinese Science Bulletin, 53,* 3341–3345.

Chen, M., Shu, J., Mao, H.-K., Xie, X., & Hemley, R. J. (2003a). Natural occurrence and synthesis of two new postspinel polymorphs of chromite. *Proceedings of the National Academy of the U.S.A., 100,* 14651–14654.

Chen, M., Shu, J., Xie, X., & Mao, H.-K. (2003b). Natural $CaTi_2O_4$-structured $FeCr_2O_4$ polymorph in the Suizhou meteorite and its significance in mantle mineralogy. *Geochimica et Cosmochimica Acta, 67,* 3937–3942.

Chinn, J. L., Milledge, H. J., & Gurney, J. J. (1998). Diamonds and inclusions from the Jagersfontein kimberlite. In *Seventh International Kimberlite Conference Extended Abstracts* (pp. 156–157), Cape Town.

Chizmeshya, A. V. G., Wolf, G. H., & McMillan, P. F. (1996). First-principles calculation of the equation-of-state, stability, and polar optic modes of $CaSiO_3$ perovskite. *Geophysical Research Letters, 23*(20), 2725–2728.

Chizmeshya, A. V. G., Wolf, G. H., & McMillan, P. F. (1998). Correction to "First-principles calculation of the equation-of-state, stability, and polar optic modes of $CaSiO_3$ perovskite". *Geophysical Research Letters, 25*(5), 711.

Chung, J. I., & Kagi, H. (2002). High concentration of water in stishovite in the MORB system. *Geophysical Research Letters, 29*(21), 2020. doi:10.1029/2002GL015579

Civet, F., Thëbault, E., Verhoeven, O., Langlais, B., & Saturnino, D. (2015). Electrical conductivity of the Earth's mantle from the first Swarm magnetic field measurements. *Geophysical Research Letters, 42,* 3338–3346.

Cohen, R. E. (1987). Calculation of elasticity and high pressure instabilities in corundum and stishovite with the potential induced breathing model. *Geophysical Research Letters, 14*(1), 37–40.

Cohen, R. E. (1991). Bonding and elasticity of stishovite SiO_2 at high pressure: Linearized augmented plane wave calculations. *American Mineralogist, 76,* 733–742.

Coppari, F., Smith, R. F., Eggert, J. H., Wang, J., Rygg, J. R., Lazicki, A., et al. (2013). Experimental evidence for a phase transition in magnesium oxide at exoplanet pressures. *Nature Geoscience, 6,* 926–929.

Corgne, A., Liebske, C., Wood, B. J., Rubie, D. C., & Frost, D. J. (2005). Silicate perovskite-melt partitioning of trace elements and geochemical signature of a deep perovskitic reservoir. *Geochimica et Cosmochimica Acta, 146,* 249–260.

Corgne, A., & Wood, B. J. (2002). $CaSiO_3$ and $CaTiO_3$ perovskite-melt partitioning of trace elements: Implications for gross mantle differentiation. *Geophysical Research Letters, 29.* doi:10.1029/2001GL014398

Crowhurst, J., Brown, J., Goncharov, A., & Jacobsen, S. (2008). Elasticity of (Mg, Fe)O through the spin transition of iron in the lower mantle. *Science, 319,* 451–453. doi:10.1126/science.1149606

Daniel, I., Bass, J. D., Fiquet, G., Cardon, H., Zhang, J. Z., & Hanfland, M. (2004). Effect of aluminium on the compressibility of silicate perovskite. *Geophysical Research Letters, 31,* L15608.

Daniel, I., Cardon, H., Fiquet, G., Guyot, F., & Mezouar, M. (2001). Equation of state of Al-bearing perovskite to lower mantle pressure conditions. *Geophysical Research Letters, 28* (19), 3789–3792.

Davies, R. M., Griffin, W. L., O'Reilly, S. Y., & Doyle, B. J. (2004). Mineral inclusions and geochemical characteristics of microdiamonds from the DO27, A154, A21, A418, DO18, DD17 and Ranch Lake kimberlites at Lac de Gras, Slave Craton, Canada. *Lithos, 77*(1–4), 39–55. doi:10.1016/j.lithos.2004.04.016

Dera, P., Prewitt, C. T., Boctor, N. Z., & Hemley, R. J. (2002). Characterization of a high-pressure phase of silica from the Martian meteorite Shergotty. *American Mineralogist, 87,* 1018–1023.

Deschamps, F., & Khan, A. (2016). Electrical conductivity as a constraint on lower mantle thermo-chemical structure. *Earth and Planetary Science Letters, 450*(2016), 108–119. doi:10.1016/j.epsl.2016.06.027

Dorfman, S. M., Meng, Y., Prakapenka, V. B., & Duffy, T. S. (2013). Effects of Fe-enrichment on the equation of state and stability of (Mg, Fe)SiO_3 perovskite. *Earth and Planetary Science Letters, 361,* 249–257. doi:10.2138/am-2015-5190

Dorfman, S. M., Prakapenka, V. B., Meng, Y., & Duffy, T. S. (2012). Intercomparison of pressure standards (Au, Pt, Mo, MgO, NaCl and Ne) to 2.5 Mbar. *Journal of Geophysical Research, 117,* B08210.

Driver, K. P., Cohen, R. E., Wu, Z., Militzer, B., Ríos, P. L., Towler, M. D., et al. (2010). Quantum Monte Carlo for minerals at high pressures: Phase stability, equations of state, and elasticity of silica. *Proceedings of the National Academy of the U.S.A., 107,* 9519–9524. doi:10.1073/pnas.0912130107

Dubrovinsky, L. S., Dubrovinskaia, N. A., Annersten, H., Halenius, E., & Harryson, H. (2001a). Stability of $(Mg_{0.5}Fe_{0.5})O$ and $(Mg_{0.8}\,Fe_{0.2})O$ magnesiowüstites in the lower mantle. *European Journal of Mineralogy, 13*(5), 857–861.

Dubrovinsky, L. S., Dubrovinskaia, N. A., Prakapenka, V., Seifert, F., Langenhorst, F., Dmitriev, V., et al. (2003). High-pressure and high-temperature polymorphism in silica. *High Pressure Research, 23*(1–2), 35–39.

Dubrovinsky, L. S., Dubrovinskaia, N. A., Saxena, S. K., Annersten, H., Halenius, E., Harryson, H., et al. (2000). Stability of ferropericlase in the lower mantle. *Science, 289*(5478), 430–432.

Dubrovinsky, L. S., Dubrovinskaia, N. A., Saxena, S. K., Tutti, F., Rekhi, S., Le Bihan, T., et al. (2001b). Pressure-induced transformation of cristobalite. *Chemical Physics Letters, 333*(3–4), 264–270.

Dubrovinsky, L. S., Saxena, S. K., Lazor, P., Ahuja, R., Eriksson, O., Wills, J. M., et al. (1997). Experimental and theoretical identification of a new high-pressure phase of silica. *Nature, 388,* 362–365.

Duffy, T. S., Hemley, R. J., & Mao, H.-K. (1995). Equation of state and shear strength at multimegabar pressures: Magnesium oxide to 227 GPa. *Physical Review Letters, 74,* 1371–1375.

Dymshits, A. M., Bobrov, A. V., Litasov, K. D., Shatsky, A. F., Ohtani, E., & Litvin, Yu A. (2010). Experimental study of the pyroxene-garnet phase transition in the $Na_2MgSi_5O_{12}$ system at pressures of 13–20 GPa: First syntheses of sodium majorite. *Doklady Earth Sciences, 434*(1), 1263–1266.

Dziewonski, A. M., & Anderson, D. L. (1981). Preliminary reference Earthmodel. *Physics of the Earth and Planetary Interiors, 25,* 297–356.

El Goresy, A., Dera, P., Sharp, T. G., Prewitt, C. T., Chen, M., Dubrovinsky, L., et al. (2008). Seifertite, a dense or-thorhombic polymorph of silica from the Martian meteorites Shergotty and Zagami. *European Journal of Mineralogy, 20,* 523–528.

El Goresy, A., Dubrovinsky, L., Sharp, T. G., Saxena, S. K., & Chen, M. (2000). A monoclinic post-stishovite polymorph of silica in the Shergotty meteorite. *Science, 288*(5471), 1632–1634.

Elsdon, R. (1975). Manganoan ilmenite from the Leinster granite, Ireland. *Mineralogical Magazine, 40*(312), 419–421.

Fei, Y. (1996). *Crystal chemistry of FeO at high pressure and temperature* (pp. 243–254). Geochemical Society Special Publication 5.

Fei, Y., & Bertka, C. M. (1999). Phase transitions in the Earth's mantle and mantle mineralogy. In Y. Fei, C. M. Bertka, & B. O. Mysen (Eds.), *Mantle petrology: Field observations and high pressure experimentation: A tribute to Francis R. (Joe) Boyd* (pp. 189–207). Geochemical Society Special Publication No. 6.

Fei, Y., & Mao, H. K. (1994). In situ determination of the NiAs phase of FeO at high pressure and temperature. *Science, 266*(5191), 1678–1680.

Fei, Y. W., Wang, Y. B., & Finger, L. W. (1996). Maximum solubility of FeO in (Mg, Fe)SiO_3 perovskite as a function of temperature at 26 GPa: Implication for FeO content in the lower mantle. *Journal of Geophysical Research, 101,* 11525–11530.

Finger, L. W., & Conrad, P. G. (2000). The crystal structure of "Tetragonal Almandine-Pyrope Phase" (TAPP): A reexamination. *American Mineralogist, 85,* 1804–1807.

Fiquet, G., Dewaele, A., Andrault, D., Kunz, M., & Le Bihan, T. (2000). Thermoelastic properties and crystal structure of $MgSiO_3$ perovskite at lower mantle pressure and temperature conditions. *Geophysical Research Letters, 27,* 21–24.

Fischer, R. A., & Campbell, A. J. (2010). High pressure melting of wüstite. *American Mineralogist, 95,* 1473–1477.

Fischer, R. A., Campbell, A. J., Lord, O. T., Shofner, G. A., Dera, P., & Prakapenka, V. B. (2011a). Phase transition and metallization of FeO at high pressures and temperatures. *Geophysical Research Letters, 38,* L24301. doi:10.1029/2011GL049800

Fischer, R. A., Campbell, A. J., Shofner, G. A., Lord, O. T., Dera, P., & Prakapenka, V. P. (2011b). Equation of state and phase diagram of FeO. *Earth and Planetary Science Letters, 304,* 496–502.

Flynn, G. J., Nittler, L. R., & Engrand, C. (2016). Composition of cosmic dust: Sources and implications for the early Solar system. *Elements, 12*(6), 177–183.

Frost, D. J., Liebske, C., Langenhorst, F., McCammon, C. A., Trønnes, R. G., & Rubie, D. C. (2004). Experimental evidence for the existence of iron-rich metal in the Earth's lower mantle. *Nature, 428,* 409–412.

Frost, D. J., & McCammon, C. A. (2008). The redox state of Earth's mantle. *Annual Review of Earth and Planetary Sciences, 36,* 389–420. doi:10.1146/annurev.earth.36.031207.124322

Frost, D. J., & Myhill, R. (2016). Chemistry of the lower mantle. In H. Terasaki & R. A. Fischer (Eds.), *Deep Earth; Physics and Chemistry of the Lower Mantle and Core: Vol. 217. Geophysical monograph* (pp. 225–240).

Fukui, H., Yoneda, A., Nakatsuka, A., Tsujino, N., Kamada, S., Ohtani, E., et al. (2016). Effect of cation substitution on bridgmanite elasticity: A key to interpret seismic anomalies in the lower mantle. *Scientific Reports, 6,* 33337. doi:10.1038/srep33337

Funamori, N., Jeanloz, R., Miyajima, N., & Fujino, K. (2000). Mineral assemblages of basalt in the lower mantle. *Journal of Geophysical Research, 105*(B11), 26037–26043.

Funamori, N., Jeanloz, R., Nguyen, J. H., Kavner, A., Caldwell, W. A., Fiujino, K., et al. (1998). High-pressure transformations in $MgAl_2O_4$. *Journal of Geophysical Research, 103*(B9), 20813–20818.

Gaspar, J. C., & Wyllie, P. J. (1984). The alleged kimberlite-carbonatite relationship: Evidence from ilmenite and spinel from Premier and Wesselton mines and the Bentfontein sill, South Africa. *Contributions Mineralogy and Petrology, 85*(2), 133–140.

Gasparik, T., Wolf, K., & Smith, C. M. (1994). Experimental determination of phase relations in the $CaSiO_3$ system from 8 to 15 GPa. *American Mineralogist, 79*(11–12), 1219–1222.

Grocholski, B., Shim, S. H., & Prakapenka, V. B. (2013). Stability, metastability, and elastic properties of a dense silica polymorph, seifertite. *Journal of Geophysical Research: Solid Earth, 118*(9), 4745–4757. doi:10.1002/jgrb.50360

Gu, T., Li, M., McCammon, C., & Lee, K. K. M. (2016). Redox-induced lower mantle density contrast and effect on mantle structure and primitive oxygen. *Nature Geoscience, 9,* 723–729. doi:10.1038/NGEO2772

Guyot, F., Madon, M., & Poirier, J.-P. (1988). X-ray microanalysis of high pressure/high temperature phases synthesized from natural olivine in the diamond–anvil cell. *Earth and Planetary Science Letters, 90,* 52–64.

Harris, J. W., Hutchison, M. T., Hursthouse, M., Light, M., & Harte, B. (1997). A new tetragonal silicate mineral occurring as inclusions in lower mantle diamonds. *Nature, 387*(6632), 486–488.

Harte, B. (2010). Diamond formation in the deep mantle: The record of mineral inclusions and their distribution in relation to mantle dehydration zones. *Mineralogical Magazine, 74*(2), 189–215.

Harte, B., & Cayzer, N. (2007). Decompression and unmixing of crystals included in diamonds from the mantle transition zone. *Physics and Chemistry of Minerals, 34,* 647–656.

Harte, B., & Harris, J. W. (1994). Lower mantle mineral association preserved in diamonds. *Mineralogical Magazine, 58A,* 384–385.

Harte, B., Harris, J. W., Hutchison, M. T., Watt, G. R., & Wilding, M. C. (1999). Lower mantle mineral associations in diamonds from Sao Luiz, Brazil. In Y. Fei, C. M. Bertka, & B. O. Mysen (Eds.), *Mantle petrology: Field observations and high pressure experimentation: A tribute to Francis R. (Joe) Boyd* (pp. 125–153). Geochemical Society Special Publication No. 6.

Harte, B., & Hudson, N. F. C. (2013). Mineral associations in diamonds from the lowermost upper mantle and uppermost lower mantle. In *Proceedings of the 10th International Kimberlite Conference* (Vol. 1, pp. 235–254). Special Issue of the Journal of the Geological Society of India.

Hayman, P. C., Kopylova, M. G., & Kaminsky, F. V. (2005). Lower mantle diamonds from Rio Soriso (Juina, Brazil). *Contributions to Mineralogy and Petrology, 149*(4), 430–445. doi:10.1007/s00410-005-0657-8

Hemley, R. J., Prewitt, C. T., & Kingma, K. J. (1994) High-pressure behavior of silica. In P. J. Heaney & C. T. Prewitt (Eds.), *Silica: Physical behavior, geochemistry and materials applications. Reviews in mineralogy* (Vol. 29, pp. 41–82).

Hemley, R. J., Shu, J., Carpenter, M. A., Hu, J., Mao, H. K., & Kingma, K. J. (2000). Strain/order parameter coupling in the ferroelastic transition in dense SiO_2. *Solid State Communications, 114*(10), 527–532.

Hernández, E. R., Alfe, D., & Brodholt, J. (2013). The incorporation of water into lowermantle perovskites: A first-principles study. *Earth and Planetary Science Letters, 364,* 37–43. doi:10.1016/j.epsl.2013.01.005

Hirose, K., Fei, Y. W., Ma, Y. Z., & Mao, H. K. (1999). The fate of subducted basaltic crust in the Earth's lower mantle. *Nature, 397*(6714), 53–56.

Hirose, K., Takafuji, N., Sata, N., & Ohishi, Y. (2005). Phase transition and density of subducted MORB crust in the lower mantle. *Earth and Planetary Science Letters, 237,* 239–251.

Horiuchi, H., Ito, E., & Weidner, D. J. (1987). Perovskite-type $MgSiO_3$: Single-crystal X-ray diffraction. *American Mineralogist, 72,* 357–360.

Hsu, H., Yu, Y. G., & Wentzcovitch, R. M. (2012). Spin crossover of iron in aluminous $MgSiO_3$ perovskite and post-perovskite. *Earth and Planetary Science Letters, 359–360,* 34–39.

Hummer, D. R., & Fei, Y. (2012). Synthesis and crystal chemistry of Fe^{3+}-bearing (Mg, Fe^{3+})(Si, Fe^{3+})O_3 perovskite. *American Mineralogist, 97,* 1915–2012.

Hutchison, M. T. (1997). *Constitution of the deep transition zone and lower mantle shown by diamonds and their inclusions* (Unpublished Ph.D. thesis). University of Edinburgh, UK. Vol. 1, 340 pp., Vol. 2 (Tables and Appendices), 306 pp.

Hutchison, M. T., Hurtshouse, M. B., & Light, M. E. (2001). Mineral inclusions in diamonds: Associations and chemical distinctions around the 670-km discontinuity. *Contributions to Mineralogy and Petrology, 142*(2), 119–126.

Irifune, T. (1994). Absence of an aluminous phase in the upper part of the Earth's lower mantle. *Nature, 370,* 131–133.

Irifune, T., Koizumi, T., & Ando, J. I. (1996). An experimental study of the garnet-perovskite transformation in the system $MgSiO_3$-$Mg_3Al_2Si_3O_{12}$. *Physics of the Earth and Planet Interior, 96*(3–4), 147–157.

Irifune, T., & Ringwood, A. E. (1993). Phase transformations in subducted oceanic crust and buoyancy relationships at depths of 600–800 km in the mantle. *Earth and Planetary Science Letters, 117*(1–2), 101–110.

Irifune, T., Shinmei, T., McCammon, C. A., Miyajima, N., Rubie, D. C., & Frost, D. J. (2010). Iron partitioning and density changes of pyrolite in Earth's lower mantle. *Science, 327*(5962), 193–195. doi:10.1126/science.1181443

Irifune, T., & Tsuchiya, T. (2007). Mineralogy of the Earth—Phase transitions and mineralogy of the lower mantle. In G. D. Price & G. Schubert (Eds.), *Treatise on geophysics. Mineral physicas* (pp. 33–62). Elsevier.

Ishii, T., Hiroshi Kojitani, H., Tsukamoto, S., Fujino, K., Mori, D., Inaguma, Y., et al. (2014). High-pressure phase transitions in $FeCr_2O_4$ and structure analysis of new postspinel $FeCr_2O_4$ and $Fe_2Cr_2O_5$ phases with meteoritical and petrological implications. *American Mineralogist, 99,* 1788–1797.

Ito, E., Akaogi, M., Topor, L., & Navrotsky, A. (1990). Negative pressure-temperature slopes for reactions forming $MgSiO_3$ from calorimetry. *Science, 249,* 1275–1278.

Ito, E., & Matsui, Y. (1978). Synthesis and crystal-chemical characterization of $MgSiO_3$ perovskite. *Earth and Planetary Science Letters, 38,* 443–450.

Ito, E., & Takahashi, E. (1989). Postspinel transformations in the system Mg_2SiO_4-Fe_2SiO_4 and some geophysical implications. *Journal of Geophysical Research, 94*(B8), 10637–10646.

Jackson, I., Khanna, S. K., Revcolevschi, A., & Berthon, J. (1990). Elasticity, shear-mode softening and high-pressure polymorphism of wüstite ($Fe_{1-x}O$). *Journal of Geophysical Research, 95,* 21671–21685.

Jackson, J. M., Sinogeikin, S. V., Jacobsen, S. D., Reichmann, H. J., Mackwell, S. J., & Bass, J. D. (2006). Single-crystal elasticity and sound velocities of ($Mg_{0.94}Fe_{0.06}$)O ferropericlase to 20 GPa. *Journal of Geophysical Research, 111,* B09203.

Jackson, J. M., Zhang, J., & Bass, J. D. (2004). Sound velocities and elasticity of aluminous $MgSiO_3$ perovskite: Implications for aluminum heterogeneity in Earth's lower mantle. *Geophysical Research Letters, 31*(10), L10614.

Jacobsen, S. D., Reichmann, H. J., Spetzler, H., Mackwell, S. J., Smyth, J. R., Angel, R. J., et al. (2002). Structure and elasticity of single-crystal (Mg, Fe)O and a new method of generating shear waves for gigahertz ultrasonic interferometry. *Journal of Geophysical Research, 107,* 5867–5871.

Javoy, M. (1995). The integral enstatite chondrite model of the Earth. *Geophysical Research Letters, 22,* 2219–2222.

Javoy, M., Kaminski, E., Guyot, F., Andrault, D., Sanloup, C., Moreira, M., et al. (2010). The chemical composition of the Earth: Enstatite chondrite models. *Earth and Planetary Science Letters, 293*(3–4), 259–268.

Jeanloz, R., & Ahrens, T. J. (1980). Equations of state of FeO and CaO. *Geophysical Journal of the Royal Astronomical Society, 62,* 505–528.

Jiang, F., Gwanmesia, G. D., Dyuzheva, T. I., & Duffy, T. S. (2009). Elasticity of stishovite and acoustic mode softening under high pressure by Brillouin scattering. *Physics of the Earth and Planetary Interiors, 172*(3–4), 235–240. doi:10.1016/j.pepi.2008.09.017

Joswig, W., Stachel, T., Harris, J. W., Baur, W. H., & Brey, G. P. (1999). New Ca-silicate inclusions in diamonds—Tracers from the lower mantle. *Earth and Planetary Science Letters, 173*(1–2), 1–6.

Jung, D. Y., & Oganov, A. R. (2005). Ab initio study of the high-pressure behavior of $CaSiO_3$ perovskite. *Physics and Chemistry of Minerals, 32,* 146–153. doi:10.1007/s00269-005-0453-z

Kaercher, P., Speziale, S., Miyagi, L., Kanitpanyacharoen, W., & Wenk, H.-R. (2012). Crystallographic preferred orientation in wüstite (FeO) through the cubic-to-rhombohedral phase transition. *Physical Chemistry Minerals, 39,* 613–626. doi:10.1007/s00269-012-0516-x

Kaminski, E., & Javoy, M. (2013). A two-stage scenario for the formation of the Earth's mantle and core. *Earth and Planetary Science Letters, 365,* 97–107.

Kaminski, E., & Javoy, M. (2015). The composition of the Deep Earth. In A. Khan & F. Deschamps (Eds.), *The Earth's heterogeneous mantle* (pp. 303–328). Berlin: Springer.

Kaminsky, F. V. (2012). Mineralogy of the lower mantle: A review of 'super-deep' mineral inclusions in diamond. *Earth-Science Reviews, 110*(1–4), 127–147.

Kaminsky, F. V., & Belousova, E. A. (2009). Manganoan ilmenite as kimberlite/diamond indicator mineral. *Russian Geology and Geophysics, 50*(10), 1212–1220.

Kaminsky, F. V., Khachatryan, G. K., Andreazza, P., Araujo, D., & Griffin, W. L. (2009a). Super-deep diamonds from kimberlites in the Juina area, Mato Grosso State, Brazil. *Lithos, 112S*(2), 833–842.

Kaminsky, F. V., & Lin, J.-F. (2017). Iron partitioning in natural lower-mantle minerals: Toward a chemically heterogeneous lower mantle. *American Mineralogist, 102*(4), 824–832. doi:10.2138/am-2017-5949.

Kaminsky, F. V., Ryabchikov, I. D., McCammon, C., Longo, M., Abakumov, A. M., Turner, S., et al. (2015a). Oxidation potential in the Earth's lower mantle as recorded from ferropericlase inclusions in diamond. *Earth and Planetary Science Letters, 417,* 49–56.

Kaminsky, F. V., Ryabchikov, I. D., & Wirth, R. (2016). A primary natrocarbonatitic association in the Deep Earth. *Mineralogy and Petrology, 110*(2–3), 387–398. doi:10.1007/s00710-015-0368-4

Kaminsky, F., & Wirth, R. (2017). Nitride, carbonitride and nitrocarbide inclusions in lower-mantle diamonds: A key to the balance of nitrogen in the Earth. *European Geosciences Union General Assembly Abstract No. EGU2017-1751*, Vienna, Austria.

Kaminsky, F., Wirth, R., Matsyuk, S., Schreiber, A., & Thomas, R. (2009b). Nyerereite and nahcolite inclusions in diamond: Evidence for lower-mantle carbonatitic magmas. *Mineralogical Magazine, 73*(5), 797–816.

Kaminsky, F. V., Wirth, R., & Schreiber, A. (2015b). A microinclusion of lower-mantle rock and some other lower-mantle inclusions in diamond. *Canadian Mineralogist, 53*(1), 83–104. doi:10.3749/canmin.1400070

Kaminsky, F. V., Zakharchenko, O. D., Channer, D. M. D., Blinova, G. K., & Bulanova, G. P. (1997). Diamonds from the Guaniamo area, Bolivar state, Venezuela. In *Memorias del VIII Congreso Geologico Venezolana, Soc Venezolana de Geologia* (pp. 427–430).

Kaminsky, F. V., Zakharchenko, O. D., Davies, R., Griffin, W. L., Khachatryan-Blinova, G. K., & Shiryaev, A. A. (2001). Superdeep diamonds from the Juina area, Mato Grosso State, Brazil. *Contributions to Mineralogy and Petrology, 140*(6), 734–753.

Kaminsky, F. V., Zakharchenko, O. D., Griffin, W. L., Channer, D. M. D., & Khachatryan-Blinova, G. K. (2000). Diamond from the Guaniamo area, Venezuela. *Canadian Mineralogist, 38*(6), 1347–1370.

Kantor, A. P., Jacobsen, S. D., Kantor, I Yu., Dubrovinsky, L. S., McCammon, C. A., Reichmann, H. J., et al. (2004). Pressure-induced magnetization in FeO: Evidence from elasticity and Mössbauer spectroscopy. *Physical Review Letters, 93,* 215502.

Karki, B., Stixrude, L., & Crain, J. (1997a). Ab initio elasticity of three high-pressure polymorphs of silica. *Geophysical Research Letters, 24*(24), 3269–3272.

Karki, B. B., Warren, M. C., Stixrude, L., Ackland, C. J., & Crain, J. (1997b). *Ab initio* studies of high-pressure structural transformations in silica. *Physical Review B, 55,* 3465–3471.

Karki, B. B., Wentzcovitch, R. M., de Gironcoli, S., & Baroni, S. (2001). First principles thermoelasticity of $MgSiO_3$-perovskite; consequences for the inferred properties of the lower mantle. *Geophysical Research Letters, 28*(14), 2699–2702.

Katsura, T., & Ito, E. (1996). Determination of Fe–Mg partitioning between perovskite and magnesiowüstite. *Geophysical Research Letters, 23*(16), 2005–2008.

Katsura, T., Yoneda, A., Yamazaki, D., Yoshino, T., & Ito, E. (2010). Adiabatic temperature profile in the mantle. *Physics of the Earth and Planetary Interiors, 183,* 212–218. doi:10.1016/j.pepi.2010.07.001

Kesson, S. E., & Fitz Gerald, J. D. (1991). Partitioning of MgO, FeO, NiO, MnO and Cr_2O_3 between magnesian silicate perovskite and magnesiowustite: Implications for the origin of inclusions in diamond and the composition of the lower mantle. *Earth and Planetary Science Letters, 111,* 229–240.

Kesson, S. E., Fitz Gerald, J. D., O'Neill, H., St, C., & Shelley, J. M. G. (2002). Partitioning of iron between magnesian silicate perovskite and magnesiowüstite at about 1 Mbar. *Physics of the Earth and Planetary Interiors, 131,* 295–310.

Kesson, S. E., Fitz Gerald, J. D., & Shelley, J. M. (1998). Mineralogy and dynamics of a pyrolite lower mantle. *Nature, 393*(6682), 252–255.

Kesson, S. E., Fitz Gerald, J. D., Shelley, J. M. G., & Withers, R. L. (1995). Phase relations, structure and crystal chemistry of some aluminous silicate perovskite. *Earth and Planetary Science Letters, 134,* 187–201.

Kind, R., & Li, X. (2007). Deep Earth structure—Transition zone and mantle discontinuities. In G. Schubert, B. Romanowicz, & A. Dziewonski (Eds.), *Treatise on Geophysics: Vol. 1. Seismology and the structure of the Earth* (pp. 591–612). Amsterdam: Elsevier.

Kingma, K. J., Cohen, R. E., Hemley, R. J., & Mao, H.-K. (1995). Transformation of stishovite to a denser phase at lower mantle pressures. *Nature, 374,* 243–245.

Kingma, J. K., Mao, H. K., & Hemley, R. J. (1996). Synchrotron XRD of SiO_2 to multimegabar pressures. *High Pressure Research, 14,* 363.

Klug, D. D., Rousseau, R., Uehara, K., Bernasconi, M., Le Page, Y., & Tse, J. S. (2001). Ab initio molecular dynamics study of the pressure-induced phase transformations in cristobalite. *Physical Review B, 63,* 104106.

Knittle, E., & Jeanloz, R. (1987). Synthesis and equation of state of (Mg, Fe)SiO_3 perovskite to over 100 gigapascals. *Science, 235,* 668–670.

Knittle, E., & Jeanloz, R. (1991). Earth's core-mantle boundary: Results of experiments at high pressures and temperatures. *Science, 251,* 1438–1443.

Kobayashi, Y., Kondo, T., Ohtani, E., Hirao, N., Miyajima, N., Yagi, T., et al. (2005). Fe–Mg partitioning between (Mg, Fe)SiO_3 post-perovskite, perovskite, and magnesiowüstite in the Earth's lower mantle. *Geophysical Research Letters, 32,* L19301. doi:10.1029/2005GL023257

Kojitani, H., Katsura, T., & Akaogi, M. (2007). Aluminum substitution mechanisms in perovskite-type $MgSiO_3$: An investigation by Rietveld analysis. *Physics and Chemistry of Minerals, 34*(4), 257–267. doi:10.1007/s00269-007-0144-z

Komabayashi, T., Hirose, K., Sata, N., Ohishi, Y., & Dubrovinsky, L. S. (2007). Phase transition in $CaSiO_3$ perovskite. *Earth and Planetary Science Letters, 260,* 564–569.

Kondo, T., Ohtani, E., Hirao, N., Yagi, T., & Kikegawa, T. (2004). Phase transitions of (Mg, Fe)O at megabar pressures. *Physics of the Earth and Planetary Interiors, 143–144,* 201–213.

Kubo, T., Suzuki, T., & Akaogi, M. (1997). High pressure phase equilibria in the system $CaTiO_3$–$CaSiO_3$: Stability of perovskite solid solutions. *Physics and Chemistry of Minerals, 24,* 488–494.

Kubo, A., Yagi, T., Ono, S., & Akaogi, M. (2000). Compressibility of $Mg_{0.9}Al_{0.2}Si_{0.9}O_3$ perovskite. *Proceedings of the Japan Academy Series B, 76*(8), 103–107.

Kudoh, Y., Prewitt, C. T., Finger, L. W., Darovskikh, A., & Ito, E. (1990). Effect of iron on the crystal structure of (Mg, Fe)SiO_3 perovskite. *Geophysical Research Letters, 17*(10), 1481–1484. doi:10.1029/GL017i010p01481

Kurashina, T., Hirose, K., Ono, S., Sata, N., & Ohishi, Y. (2004). Phase transition in Albearing $CaSiO_3$ perovskite: Implications for seismic discontinuities in the lower mantle. *Physics of the Earth and Planetary Interiors, 145,* 67–74.

Kuwayama, Y., Hirose, K., Satam, N., & Ohishi, Y. (2005). The pyrite-type high-pressure form of silica. *Science, 309,* 923–925.

Lakshtanov, D. L., Sinogeikin, S. V., Konstantin, D., Litasov, K. D., Vitali, B., Prakapenka, V. B., et al. (2007). The post-stishovite phase transition in hydrous alumina-bearing SiO_2 in the lower mantle of the Earth. *Proceedings of the National Academy of Sciences of the U.S.A., 104,* 13588–13590.

Lauterbach, S., McCammon, C. A., Aken, P. V., Langenhorst, F., & Seifert, F. (2000). Mossbauer and ELNES spectroscopy of (Mg, Fe)(Si, Al)O_3 perovskite: A highly oxidized component of the lower mantle. *Contributions to Mineralogy and Petrology, 138,* 17–26.

Lee, Y. M., & Nassaralla, C. L. (1997). Minimization of hexavalent chromium in magnesite-chrome refractory. *Metallurgical and Materials Transactions B, 28,* 855–859.

Lee, K. K. M., O'Neill, B., Panero, W. R., Shim, S. H., Benedetti, L. R., & Jeanloz, R. (2004). Equations of state of the high-pressure phases of a natural peridotite and implications for the Earth's lower mantle. *Earth and Planetary Science Letters, 223*(3–4), 381–393. doi:10.1016/j.epsl.2004.04.033

Li, L., Nagai, T., Seto, Y., Fujino, K., Kawano, J., & Itoh, S. (2015). Superior solid solubility of $MnSiO_3$ in $CaSiO_3$ perovskite. *Physics and Chemistry of Minerals 42*(2), 123–129.

Li, L., Weidner, D. J., Brodholt, J., Alfe, D., Price, G. D., Caracas, R., et al. (2006). Phase stability of $CaSiO_3$ perovskite at high pressure and temperature: Insights from ab initio molecular dynamics. *Physics of the Earth and Planetary Interiors, 155,* 260–268.

Lin, J.-F., Alp, E. E., Mao, Z., Inoue, T., McCammon, C., Xiao, Y., et al. (2012). Electronic spin states of ferric and ferrous iron in the lower-mantle silicate perovskite. *American Mineralogist, 97,* 592–597.

Lin, J.-F., Heinz, D. L., Mao, H.-K., Hemley, R. J., Devine, J. M., Li, J., et al. (2003). Stability of magnesiowüstite in Earth's lower mantle. *Proceedings of the National Academy of Sciences of the U.S.A., 100*(8), 4405–4408.

Lin, J.-F., Speziale, S., Mao, Z., & Marquardt, H. (2013). Effects of the electronic spin transitions of iron in lower-mantle minerals: Implications to deep-mantle geophysics and geochemistry. *Reviews of Geophysics, 51,* 244–275. doi:10.1002/rog.20010

Lin, J.-F., Struzhkin, V. V., Jacobsen, S. D., Hu, M. Y., Chow, P., King, J., et al. (2005). Spin transition of iron in magnesiowüstite in Earth's lower mantle. *Nature, 436,* 377–380.

Litasov, K. D., Kagi, H., Shatskiy, A., Ohtani, E., Lakshtanov, D. L., Bass, J. D., et al. (2007). High hydrogen solubility in Al-rich stishovite and water transport in the lower mantle. *Earth and Planetary Science Letters, 262,* 620–634. doi:10.1016/j.epsl.2007.08.015

Litasov, K., Ohtani, E., Langenhorst, F., Yurimoto, H., Kubo, H., & Kondo, T. (2003). Water solubility in Mg-perovskites and water storage capacity in the lower mantle. *Earth and Planetary Science Letters, 211,* 189–203.

Litvin, Y. A. (2014). The stishovite paradox in the genesis of superdeep diamonds. *Doklady Earth Sciences, 455*(1), 274–278.

Litvin, Y. A., Spivak, A. V., & Dubrovinsky, L. S. (2016). Magmatic evolution of the material of the Earth's lower mantle: Stishovite paradox and origin of superdeep diamonds (experiments at 24–26 GPa). *Geochemistry International, 54*(11), 936–947. doi:10.1134/S0016702916090032

Liu, L.-G. (1974). Silicate perovskite from phase transformations of pyrope-garnet at high pressure and temperature. *Geophysical Research Letters, 1,* 277–280.

Liu, L.-G. (1975). Post oxide phases of forsterite and enstatite. *Geophysical Research Letters, 2,* 417–419.

Liu, L. (1977). Ilmenite-type solid solutions between $MgSiO_3$ and Al_2O_3 and some structural systematics among ilmenite compounds. *Geochimica et Cosmochimica Acta, 41,* 1355–1361.

Liu, L.-G. (2002). An alternative interpretation of lower mantle mineral associations in diamonds. *Contributions to Mineralogy and Petrology, 144*(1), 16–21.

Liu, Z., Irifune, T., Nishi, M., Tange, Y., Arimoto, T., & Shinmei, T. (2016). Phase relations in the system $MgSiO_3$–Al_2O_3 up to 52 GPa and 2000 K. *Physics of the Earth and Planetary Interiors, 257,* 18–27. doi:10.1016/j.pepi.2016.05.006

Liu, L.-G., & Ringwood, A. E. (1975). Synthesis of a perovskite-type polymorph of $CaSiO_3$. *Earth and Planetary Science Letters, 28,* 209–211.

Lobanov, S. S., Zhu, Q., Holtgrewe, N., Prescher, C., Prakapenka, V. B., Oganov, A. R., et al. (2015). Stable magnesium peroxide at high pressure. *Scientific Review, 5,* 13582. doi:10.1038/srep13582

Longo, M., McCammon, C. A., & Jacobsen, S. D. (2011). Microanalysis of the iron oxidation state in (Mg, Fe)O and application to the study of microscale processes. *Contributions to Mineralogy and Petrology, 162,* 1249–1257.

Magyari-Köpe, B., Vitos, L., Grimvall, G., Johansson, B., & Kollar, J. (2002a). Low-temperature crystal structure of $CaSiO_3$ perovskite: An *ab initio* total energy study. *Physical Review B, 65,* 193107. doi:10.1103/PhysRevB.65.193107

Magyari-Köpe, B., Vitos, L., Johansson, B., & Kollar, J. (2002b). Model structure of perovskites: Cubic-orthorhombic phase transition. *Computational Material Science, 25*(4), 615–621.

Mao, H. K., Chen, L. C., Hemley, R. J., Jephcoat, A. P., Wu, Y., & Bassett, W. A. (1989). Stability and equation of state of $CaSiO_3$-perovskite to 134 GPa. *Journal of Geophysical Research, 94,* 17889–17894.

Mao, Z., Lin, J. F., Scott, H. P., Watson, H. C., Prakapenka, V. B., Xiao, Y., et al. (2011). Iron-rich perovskite in the Earth's lower mantle. *Earth and Planetary Science Letters, 309,* 179–184. doi:10.1016/j.epsl.2011.06.030

Mao, Z., Lin, J.-F., Yang, J., Inoue, T., & Prakapenka, V. B. (2015). Effects of the Fe^{3+} spin transition on the equation of state of bridgmanite. *Geophysical Research Letters, 42,* 4335–4342. doi:10.1002/2015GL064400

Mao, H.-K., Shu, J., Fei, Y., Hu, J., & Hemley, R. J. (1996). The wüstite enigma. *Physics of the Earth and Planetary Interiors, 96,* 135–145.

Mao, W., Shu, J., Hu, J., Hemley, R., & Mao, H.-K. (2002). Displacive transition in magnesiowüstite. *Journal of Physics: Condensed Matter, 14,* 11349–11354.

Mao, Z., Wang, F., Lin, J.-F., Fu, S., Yang, J., Wu, X., et al. (2017). Equation of state of high-spin bridgmanite in the Earth's lower mantle by synchrotron X-ray diffraction and Mössbauer spectroscopy. *American Mineralogist, 102*(2), 357–368. doi:10.2138/am-2017-5770

Marquardt, H., Speziale, S., Reichmann, H. J., Frost, D. J., & Schilling, F. R. (2009a). Single-crystal elasticity of $(Mg_{0.9}Fe_{0.1})O$ to 81 GPa. *Earth and Planetary Science Letters, 287,* 345–352.

Marquardt, H., Speziale, S., Reichmann, H. J., Frost, D. J., Schilling, F. R., & Garnero, E. J. (2009b). Elastic shear anisotropy of ferropericlase in Earth's lower mantle. *Science, 324* (6289), 224–226.

Martirosyan, N.S., Yoshino, T., Shatskiy, A., Chanyshev, A.D., & Litasov, K.D. (2016). The $CaCO_3$–Fe interaction: Kinetic approach for carbonate subduction to the deep Earth's mantle. *Physics of the Earth and Planetary Interiors, 259*, 1–9.

Matas, J., Bass, J. D., Ricard, Y., Mattern, E., & Bukowinsky, M. S. (2007). On the bulk composition of the lower mantle: predictions and limitations from generalized inversion of radial seismic profiles. *Geophysical Journal International, 170,* 764–780.

McCammon, C. A. (1993). Effect of pressure on the composition of the Lower Mantle End Member Fe_xO. *Science, 259,* 66–68.

McCammon, C. A. (1997). Perovskite as a possible sink for ferric iron in the lower mantle. *Nature, 387,* 694–696.

McCammon, C. A. (2005). Mantle oxidation state and oxygen fugacity: Constraints on mantle chemistry, structure, and dynamics. In *Earth's Deep Mantle: Structure, Composition, and Evolution: Vol. 160. Geophysical monograph* (pp. 219–240).

McCammon, C., Hutchison, M. T., & Harris, J. W. (1997). Ferric iron content of mineral inclusions in diamonds from São Luiz: A view into the lower mantle. *Science, 278*(5337), 434–436.

McCammon, C. A., Ringwood, A. E., & Jackson, I. (1983). Thermodynamics of the system Fe-FeO-MgO at high pressure and temperature and a model for formation of the Earth's core. *Geophysical Journal of the Royal Astronomical Society, 72,* 577–595.

McDonough, W. F., & Sun, S.-S. (1995). The composition of the Earth. *Chemical Geology, 120* (3–4), 223–253.

McWilliams, R. S., Spaulding, D. K., Eggert, J. H., Celliers, P. M., Hicks, D. G., Smith, R. F., et al. (2012). Phase transformations and metallization of magnesium oxide at high pressure and temperature. *Science, 338,* 1330–1333.

Meade, C., Mao, H. K., & Hu, J. (1995). High-temperature phase transition and dissociation of (Mg, Fe)SiO_3 perovskite at lower mantle pressures. *Science, 268*(5218), 1743–1745.

Merli, M., Bonadiman, C., Diella, V., & Pavese, A. (2016). Lower mantle hydrogen partitioning between periclase and perovskite: A quantum chemical modelling. *Geochimica et Cosmochimica Acta, 173,* 304–318.

Metsue, A., & Tsuchiya, T. (2011). Lattice dynamics and thermodynamic properties of (Mg, Fe^{2+}) SiO_3 postperovskite. *Journal of Geophysical Research, 116,* JB008018. doi:10.1029/2010JB008018

Meyer, H. O. A., & Svisero, D. P. (1975). Mineral inclusions in Brazilian diamonds. *Physics and Chemistry of the Earth, 9,* 785–795.

Mitchell, R. H. (1978). Manganoan magnesian ilmenite and titanian clinohumite from the Jacupiranga carbonatite, Sao Paulo, Brazil. *American Mineralogist, 63*(5–6), 544–547.

Miyahara, M., Kaneko, S., Ohtani, E., Sakai, T., Nagase, T., Kayama, M., et al. (2013). Discovery of seifertite in a shocked lunar meteorite. *Nature Communications, 4,* 1737. doi:10.1038/ncomms2733

Moore, R. O., & Gurney, J. J. (1985). Pyroxene solid solution in garnets included in diamonds. *Nature, 318*(6046), 553–555.

Moore, R. O., Otter, M. L., Rickard, R. S., Harris, J. W., & Gurney, J. J. (1986). The occurrence of moissanite and ferro-periclase as inclusions in diamond. In *4th International Kimberlite Conference Extended Abstracts, Perth* (Vol. 16, pp. 409–411). Geological Society of Australia Abstracts.

Moran, E., Blesa, M. C., Medina, M.-E., Tornero, J.-D., Menendez, N., & Amador, U. (2002). Nonstoichiometric spinel ferrites obtained from α-$NaFeO_2$ via molten media reactions. *Inorganic Chemistry, 41*(23), 5961–5967.

Muir, J. M. R., & Brodholt, J. P. (2015). Elastic properties of ferrous bearing $MgSiO_3$ and their relevance to ULVZs. *Geophysical Journal International, 201*(1), 496–504. doi:10.1093/gji/ggv045

Muir, J. M. R., & Brodholt, J. P. (2016). Ferrous iron partitioning in the lower mantle. *Physics of the Earth and Planetary Interiors, 257,* 12–17. doi:10.1016/j.pepi.2016.05.008

Murakami, M., Hirose, K., Kawamura, K., Sata, N., & Ohishi, Y. (2004a). Post-perovskite phase transition in $MgSiO_3$. *Science, 304,* 855–858.

Murakami, M., Hirose, K., Ono, S., & Ohishi, Y. (2003). Stability of $CaCl_2$-type and α-PbO_2-type SiO_2 at high pressure and temperature determined by in situ X-ray measurements. *Geophysical Research Letters, 30,* 1207. doi:10.1029/2002GL016722

Murakami, M., Hirose, K., Ono, S., Tsuchiya, T., Isshiki, M., & Watanuki, T. (2004b). High pressure and high temperature phase transitions of FeO. *Physics of the Earth and Planetary Interiors, 146,* 273–282.

Murakami, M., Hirose, K., Sata, N., & Ohishi, Y. (2005). Post-perovskite phase transition and mineral chemistry in the pyrolitic lowermost mantle. *Geophysical Research Letters, 32,* L03304. doi:10.1029/2004GL021956

Murakami, M., Hirose, K., Yurimoto, H., Nakashima, S., & Takafuji, N. (2002). Water in Earth's Lower Mantle. *Science, 295,* 1885–1887.

Murakami, M., Ohishi, Y., Hirao, N., & Hirose, K. (2012). A perovskitic lower mantle inferred from high-pressure, high-temperature sound velocity data. *Nature, 485*(7396), 90–94. doi:10.1038/nature11004

Nakajima, Y., Frost, D. J., & Rubie, D. C. (2012). Ferrous iron partitioning between magnesium silicate perovskite and ferropericlase and the composition of perovskite in the Earth's lower mantle. *Journal of Geophysical Research, 117,* B08201. doi:10.1029/2012JB009151

Navrotsky, A., Schoenitz, M., Kojitani, H., Xu, H., Zhang, J., Weidner, D. J., et al. (2003). Aluminum in magnesium silicate perovskite: Formation, structure, and energetics of magnesium-rich defect solid solutions. *Journal of Geophysical Research, 108*(B7). doi:10.1029/2002JB002055

Nestola, F., Burnham, A. D., Peruzza, L., Tauro, L., Alvaro, M., Walter, M. J., et al. (2016a). Tetragonal almandine-pyrope phase, TAPP: Finally a name for it, the new mineral jeffbenite. *Mineralogical Magazine, 80*(7), 1219–1232. doi:10.1180/minmag.2016.080.059

Nestola, F., Burnham, A., Peruzzo, L., Tauro, L., Alvaro, M., Walter, M. J., et al. (2015) Jeffbenite, IMA 2014-097. CNMNC Newsletter No. 24, April 2015, page 250. *Mineralogical Magazine, 79,* 247–251.

Nestola, F., Cerantola, V., Milani, S., Anzolini, C., McCammon, C., Novella, D., et al. (2016b). Synchrotron Mössbauer Source technique for *in situ* measurement of iron-bearing inclusions in natural diamonds. *Lithos, 265,* 328–333. doi:10.1016/j.lithos.2016.06.016

Nomura, R., Hirose, K., Sata, N., & Ohishi, Y. (2010). Precise determination of post-stishovite phase transition boundary and implications for seismic heterogeneities in the mid-lower mantle. *Physics of the Earth and Planetary Interiors, 183,* 104–109. doi:10.1016/j.pepi.2010.08.004

Nomura, R., Ozawa, H., Tateno, S., Hirose, K., Hernlund, H., Muto, S., et al. (2011). Spin crossover and iron-rich silicate melt in the Earth's deep mantle. *Nature, 473,* 199–203.

Oganov, A. R., Brodholt, J. P., & Price, G. D. (2001). Ab initio elasticity and thermal equation of state of $MgSiO_3$ perovskite. *Earth and Planetary Science Letters, 184*(3–4), 555–560.

Oganov, A. R., Gillan, M. J., & Price, G. D. (2005). Structural stability of silica at high pressures and temperatures. *Physical Review B, 71,* 64104.

Oganov, A. R., & Ono, S. (2004). Theoretical and experimental evidence for a post-perovskite phase of $MgSiO_3$ in Earth's D″ layer. *Nature, 430,* 445–448.

Ohta, K., Cohen, R. E., Hirose, K., Haule, K., Shimizu, K., & Ohishi, Y. (2012). Experimental and theoretical evidence for pressure-induced metallization in FeO with rocksalt-type structure. *Physical Review Letters, 108,* 026403. doi:10.1103/PhysRevLett.108.026403

Ohta, K., Fujino, K., Kuwayama, Y., Kondo, T., Shimizu, K., & Ohishi, Y. (2014). Highly conductive iron rich (Mg, Fe)O magnesiowustite and its stability in the Earth's lower mantle. *Journal of Geophysical Research: Solid Earth, 119*(6), 4656–4665.

Ohta, K., Hirose, K., Shimizu, K., & Ohishi, Y. (2010). High-pressure experimental evidence for metal FeO with normal NiAs-type structure. *Physical Review B, 82*(174), 120. doi:10.1103/PhysRevB.82.174120

Ohta. K., Yagi, T., Hirose, K., & OhishI, Y. (2017). Thermal conductivity of ferropericlase in the Earth's lower mantle. *Earth and Planetary Science Letters, 465*, 29–37. doi: 10.1016/j.epsi.2017.02.030

Omori, K., & Hasegawa, S. (1955). Chemical composition of perthite, ilmenite, allanite and pyroxmangite occurred in pegmatites of vicinity of Iwaizami Town, Iwate Prefecture. *Journal of Japanese Association of Mineralogy, Petrography, and Economic Geology, 39,* 89–98.

Ono, S., Hirose, K., Murakami, M., & Isshiki, M. (2002). Post-stishovite phase boundary in SiO_2 determined by in situ X-ray observations. *Earth and Planetary Science Letters, 197,* 187–192.

Ono, S., Ohishi, Y., Isshiki, M., & Watanuki, T. (2005). In situ X-ray observations of phase assemblages in peridotite and basalt compositions at lower mantle conditions: Implications for density of subducted oceanic plate. *Journal of Geophysical Research, 110,* B02208.

Ono, S., Ohishi, Y., & Mibe, K. (2004). Phase transition in CaSi-perovskite and stability of Al-bearing Mg-perovskite in the lower mantle. *American Mineralogist, 89,* 1480–1485.

Ozawa, H., Hirose, K., Ohta, K., Ishii, H., Hiraoka, N., Ohishi, Y., et al. (2011). Spin crossover, structural change, and metallization in NiAs-type FeO at high pressure. *Physical Review B, 84,* 134417.

Ozawa, H., Hirose, K., Tateno, S., Sata, N., & Ohishi, Y. (2010). Phase transition boundary between B1 and B8 structures of FeO up to 210 GPa. *Physics of the Earth and Planetary Interiors, 179,* 157–163.

Ozawa, H., Hirose, K., Yonemitsu, K., & Ohishi, Y. (2016). High-pressure melting experiments on Fe–Si alloys and implications for silicon as a light element in the core. *Earth and Planetary Science Letters, 456,* 47–54. doi:10.1016/j.epsl.2016.08.042

Palot, M., Jacobsen, S. D., Townsend, J. P., Nestola, F., Marquardt, K., Harris, J. W., et al. (2016). Evidence for H_2O-bearing fluids in the lower mantle from diamond inclusion. *Lithos, 265,* 237–243. doi:10.1016/j.lithos.2016.06.023

Panero, W. R., & Stixrude, L. P. (2004). Hydrogen incorporation in stishovite at high pressure and symmetric bonding in δ-AlOOH. *Earth and Planetary Science Letters, 221,* 421–431.

Park, K. T., Terakura, K., & Matsui, Y. (1988). Theoretical evidence for a new ultra-high-pressure phase of SiO_2. *Nature, 336,* 670–672. doi:10.1038/336670a0

Perrillat, J. P., Ricolleau, A., Daniel, I., Fiquet, G., Mezouar, M., Guignot, N., et al. (2006). Phase transformations of subducted basaltic crust in the upmost lower mantle. *Physics of the Earth and Planetary Interiors, 157*(1–2), 139–149.

Perry, S.N., Pigott, J.S., & Panero, W.R. (2017). Ab initio calculations of uranium and thorium storage in $CaSiO_3$-perovskite in the Earth's lower mantle. *American Mineralogist, 102,* 321–326. doi: 10.2138/am-2017-5816

Reid, A. F., & Ringwood, A. E. (1969). Newly observed high pressure transformations in Mn_3O_4, $CaAl_2O_4$, and $ZrSiO_4$. *Earth and Planetary Science Letters, 6,* 205–208.

Reid, A.F., & Ringwood, A.E. (1970). The crystal chemistry of dense M_3O_4 polymorphs: High pressure Ca_2GeO_4 of K_2NiF_4 structure type. *Journal of Solid State Chemistry, 1,* 557–565.

Reid, A. F., & Ringwood, A. E. (1975). High pressure modification of $ScAlO_3$ and some geophysical implications. *Journal of Geophysical Research, 80,* 3363–3369.

Richet, P., Mao, H.-K., & Bell, P. M. (1989). Bulk moduli of magnesiowiistites from static compression measurements. *Journal of Geophysical Research, 94*(B3), 3037–3045.

Richmond, N. C., & Brodholt, J. P. (1998). Calculated role of aluminum in the incorporation of ferric iron into magnesium silicate perovskite. *American Mineralogist, 83,* 947–951.

Ricolleau, A., Fiquet, G., Addad, A., Menguy, N., Vanni, C., Perrillat, J.-P., et al. (2008). Analytical transmission electron microscopy study of a natural MORB sample assemblage transformed at high pressure and high temperature. *American Mineralogist, 93,* 144–153. doi:10.2138/am.2008.2532

Righter, K., Danielson, L., Drake, M. J., & Domanik, K. (2014). Partition coefficients at high pressure and temperature. In R. W. Carlson (Ed.), *Treatise on geochemistry* (2nd ed., Vol. 3, pp. 449–477). Elsevier.

Ringwood, A. E. (1962). Mineralogical constitution of the deep mantle. *Journal of Geophysical Research, 67*(10), 4005–4010.

Ringwood, A. E. (1975). *Composition and petrology of the Earth's mantle* (p. 618). New York: McGraw-Hill.

Ringwood, A. E. (1977). Composition of core and implications for origin of Earth. *Geochemical Journal, 11*(3), 111–135.

Róg, G., Kozlowska-Róg, A., & Dudek, M. (2007). The standard Gibbs free energy of formation of calcium chromium (III) oxide in the temperature range (1073 to 1273 K). *The Journal of Chemical Thermodynamics, 39,* 275–278.

Ross, N. L., Angel, R. J., & Seifert, F. (2002). Compressibility of brownmillerite ($Ca_2Fe_2O_5$): Effect of vacancies on the elastic properties of perovskite. *Physics of the Earth and Planetary Interiors, 129,* 145–151.

Ryabchikov, I. D. (2011). Conditions of diamond formation in the Earth's lower mantle. *Doklady Earth Sciences, 438*(2), 788–791.

Ryabchikov, I. D., & Kaminsky, F. V. (2013a). The composition of the lower mantle: Evidence from mineral inclusions in diamonds. *Doklady Earth Sciences, 453*(2), 1246–1249.

Ryabchikov, I. D., & Kaminsky, F. V. (2013b). Oxygen potential of diamond formation in the lower mantle. *Geology of Ore Deposits, 55*(1), 1–12. doi:10.1134/S1075701513010066

Ryabchikov, I. D., & Kaminsky, F. V. (2014). Physico-chemical parameters of material in mantle plumes: Evidence from the thermodynamic analysis of mineral inclusions in sublithospheric diamonds. *Geochemistry International, 52*(11), 903–911. doi:10.1134/S001670291411007X

Sakai, T., Ohtani, E., Terasaki, H., Sawada, N., Kobayashi, Y., Miyahara, M., et al. (2009). Fe–Mg partitioning between perovskite and ferropericlase in the lower mantle. *American Mineralogist, 94,* 921–925. doi:10.2138/am.2009.3123

Saxena, S. K., Dubrovinsky, L. S., Lazor, P., Cerenius, Y., Haggkvist, P., Hanfland, M., et al. (1996). Stability of perovskite ($MgSiO_3$) in the Earth's mantle. *Science, 274*(5291), 1357–1359.

Saxena, S. K., Dubrovinsky, L. S., Lazor, P., & Hu, J. Z. (1998). In situ X-ray study of perovskite ($MgSiO_3$): Phase transition and dissociation at mantle conditions. *European Journal of Mineralogy, 10,* 1275–1281.

Scott Smith, B. H., Danchin, R. V., Harris, J. W., & Stracke, K. J. (1984). Kimberlites near Orroroo, South Australia. In J. Kornprobst (Ed.), *Kimberlites I: Kimberlites and related rocks* (pp. 121–142). Amsterdam: Elsevier.

Seagle, C. T., Heinz, D. L., Campbell, A. J., Prakapenka, V. B., & Wanless, S. T. (2008). Melting and thermal expansion in the Fe–FeO system at high pressure. *Earth and Planetary Science Letters, 265,* 655–665.

Serghiou, G., Zerr, A., & Boehler, R. (1998). (Mg, Fe)SiO_3-perovskite stability under lower mantle conditions. *Science, 280*(5372), 2093–2095.

Shannon, R. D., & Prewitt, C. T. (1969). Effective ionic radii in oxides and fluorides. *Acta Crystallographica, B25,* 925–946.

Sharp, T. G., El Goresy, A., Wopenka, B., & Chen, M. (1999). A post-stishovite SiO_2 polymorph in the meteorite Shergotty: Implications for impact events. *Science, 284,* 1511–1513.

Sharygin, I. S., Litasov, K. D., Shatskiy, A., Safonov, O. G., Ohtani, E., & Pokhilenko, N. P. (2012) Interaction of orthopyroxene with carbonatite melts at 3 and 6.5 GPa: Implication for evolution of kimberlite magma. In *G–COE Symposium 2012. Achievement of G–COE Program for Earth and Planetary Science, Sendai, Japan, 2012* (pp. 146–149).

Sherman, D. M. (1989). The nature of pressure-induced metallization of FeO and its implications to the core-mantle boundary. *Geophysical Research Letters, 16,* 515–518.

Sherman, D. M., & Jansen, H. J. F. (1995). First-principles predictions of the high pressure phase transition and electronic structure of FeO: Implications for the chemistry of the lower mantle and core. *Geophysical Research Letters, 22,* 1001–1004.

Shieh, S. R., Duffy, T. S., & Shen, G. (2005). XRD study of phase stability in SiO_2 at deep mantle conditions. *Earth and Planetary Science Letters, 235,* 273–282.

Shim, S.-H., Duffy, T. S., & Shen, G. (2000a). The stability and P–V–T equation of state of $CaSiO_3$ perovskite in the Earth's lower mantle. *Journal of Geophysical Research, 106*(B11), 25955–25968.

Shim, S. H., Duffy, T., & Shen, G. (2000b). The equation of state of $CaSiO_3$ perovskite to 108 GPa at 300 K. *Physics of the Earth and Planetary Interiors, 120,* 327–338.

Shim, S. H., Duffy, T. S., & Shen, G. (2001). Stability and structure of $MgSiO_3$ perovskite to 2300-kilometer depth in Earth's mantle. *Science, 293*(5539), 2437–2440.

Shim, S.-H., Jeanloz, R., & Duffy, T. S. (2002). Tetragonal structure of $CaSiO_3$ perovskite above 20 GPa. *Geophysical Research Letters, 29*(24), 2466. doi:10.1029/2002GL016148

Shu, J., Mao, H. K., Hu, J., Fei, Y., & Hemley, R. J. (1998). Single-crystal XRD of wüstite to 30 GPa hydrostatic pressure. *Neues Jahrbuch für Mineralogie Abhandlungen, 172,* 309–323.

Shukla, G., Wu, Z., Hsu, H., Floris, A., Cococcioni, M., & Wentzcovitch, R. M. (2015). Thermoelasticity of Fe^{2+}-bearing bridgmanite. *Geophysical Research Letters, 42,* 1741–1749.

Sidorin, I., Michael, G., & Helmberger, D. V. (1999). Evidence for a ubiquitous seismic discontinuity at the base of the mantle. *Science, 286,* 1326–1331.

Simpson, E. S. (1929). Contributions to the mineralogy of Western Australia. *Journal of the Royal Society of Western Australia, 15,* 99–113.

Sinmyo, R., Bykova, E., McCammon, C., Kupenko, I., Potapkin, V., & Dubrovinsky, L. (2014). Crystal chemistry of Fe^{3+}-bearing (Mg, Fe)SiO_3 perovskite: a single-crystal X-ray diffraction study. *Physics and Chemistry of Minerals, 41,* 409–417.

Sinmyo, R., & Hirose, K. (2013). Iron partitioning in pyrolitic lower mantle. *Physics and Chemistry of Minerals, 40,* 107–113. doi:10.1007/s00269-012-0551-7

Sinogeikin, S. V., Zhang, J., & Bass, J. D. (2004). Elasticity of single crystal and polycrystalline $MgSiO_3$ perovskite by Brillouin spectroscopy. *Geophysical Research Letters, 31,* L06620. doi:10.1029/2004GL019559

Snetsinger, K. G. (1969). Manganoan ilmenite from a Sierran adamellite. *American Mineralogist, 54*(4), 431–435.

Sobolev, N. V., Yefimova, E. S., Channer, D. M. D., Anderson, P. F. N., & Barron, K. M. (1998). Unusual upper mantle beneath Guaniamo, Guyana shield, Venezuela: Evidence from diamond inclusions. *Geology, 26*(11), 971–974.

Speziale, S., Milner, A., Lee, V. E., Clark, S. M., Pasternak, M. P., & Jeanloz, R. (2005). Iron spin transition in Earth's mantle. *Proceedings of the National Academy of the U.S.A., 102,* 17918–17922. doi:10.1073/pnas.0508919102

Stachel, T., Harris, J. W., Brey, G. P., & Joswig, W. (2000). Kankan diamonds (Guinea) II: Lower mantle inclusion parageneses. *Contributions to Mineralogy and Petrology, 140*(1), 16–27.

Stebbins, J. F., Kroeker, S., & Andrault, D. (2001). The mechanism of solution of aluminum oxide in $MgSiO_3$ perovskite. *Geophysical Research Letters, 28,* 615–618.

Stishov, S. M., & Belov, N. V. (1962). Crystal structure of a new dense modification of silica SiO_2. *Doklady Akademii Nauk SSSR, 143*(4), 951.

Stishov, S. M., & Popova, S. V. (1961). A new dense modification of silica. *Geochemistry (USSR), 10,* 923–926.

Stixrude, L., & Cohen, R. E. (1993). Stability of orthorhombic $MgSiO_3$ perovskite in the Earth's lower mantle. *Nature, 364*(6438), 613–616.

Stixrude, L., Cohen, R. E., Yu, R., & Krakauer, H. (1996). Prediction of phase transition in $CaSiO_3$ perovskite and implications for lower mantle structure. *American Mineralogist, 81,* 1293–1296.

Stixrude, L., Lithgow-Bertelloni, C., Kiefer, B., & Fumagalli, P. (2007). Phase stability and shear softening in $CaSiO_3$ perovskite at high pressure. *Physical Review B, 75,* 024108.

Sugahara, M., Yoshiasa, A., Komatsu, Y., Yamanaka, T., Bolfan-Kasanova, N., Nakatsuka, A., et al. (2006). Reinvestigation of the $MgSiO_3$ perovskite structure at high pressure. *American Mineralogist, 91,* 533–536.

Sun, N., Mao, Z., Yan, S., Lin, J. F., Wu, X., & Prakapenka, V. B. (2016). Confirming a pyrolitic lower mantle using self-consistent pressure scales and new constraints on $CaSiO_3$-perovskite. *Journal of Geophysical Research, 121*(7), 4876–4892. doi:10.1002/2016JB013062

Takafuji, N., Yagi, T., Miyajima, N., & Sumita, T. (2002). Study on Al_2O_3 content and phase stability of aluminous-$CaSiO_3$ perovskite at high pressure and temperature. *Physics and Chemistry of Minerals, 29,* 532–537. doi:10.1007/s00269-002-0271-5

Tange, Y., Takahashi, E., Nishihara, Y., Funakoshi, K., & Sata, N. (2009). Phase relations in the system MgO-FeO-SiO_2 to 50 GPa and 2000°C: An application of experimental techniques using multianvil apparatus with sintered diamond anvils. *Journal of Geophysical Research, 114*(B02), 214.

Tappert, R., Foden, J., Stachel, T., Muehlenbachs, K., Tappert, M., & Wills, K. (2009a). The diamonds of South Australia. *Lithos, 112S,* 806–821.

Tappert, R., Foden, J., Stachel, T., Muehlenbachs, K., Tappert, M., & Wills, K. (2009b). Deep mantle diamonds from South Australia: A record of Pacific subduction at the Gondwanan margin. *Geology, 37*(1), 43–46. doi:10.1130/G25055A.1

Tappert, R., Stachel, T., Harris, J. W., Muehlenbachs, K., Ludwig, T., & Brey, G. (2005a). Diamonds from Jagersfontein (South Africa): Messengers from the sublithospheric mantle. *Contributions to Mineralogy and Petrology, 150*(5), 505–522.

Tappert, R., Stachel, T., Harris, J. W., Shimizu, N., & Brey, G. P. (2005b). Mineral inclusions in diamonds from the Slave Province, Canada. *European Journal of Mineralogy, 17*(3), 423–440.

Tarrida, M., & Richet, P. (1989). Equation of state of $CaSiO_3$ perovskite to 96 GPa. *Geophysical Research Letters, 16,* 1351–1354.

Tateno, S., Hirose, K., & Ohishi, Y. (2014). Melting experiments on peridotite to lowermost mantle conditions. *Journal of Geophysical Research: Solid Earth, 119,* 4684–4694. doi:10.1002/2013JB010616

Teter, D. M., Hemley, R. J., Kresse, G., & Hafner, J. (1998). High-pressure polymorphism in silica. *Physical Review Letters, 80,* 2145–2148.

Thomson, A. R., Kohn, S. C., Bulanova, G. P., Smith, C. B., Araujo, D., & Walter, M. J. (2014). Origin of sub-lithospheric diamonds from the Juina-5 kimberlite (Brazil): Constraints from carbon isotopes and inclusion compositions. *Contributions to Mineralogy and Petrology, 168,* 1081.

Thomson, A. R., Kohn, S. C., Bulanova, G. P., Smith, C. B., Araujo, D., & Walter, M. J. (2016). Trace element composition of silicate inclusions in sub-lithospheric diamonds from the Juina-5 kimberlite: Evidence for diamond growth from slab melts. *Lithos, 265,* 108–124. doi:10.1016/j.lithos.2016.08.035

Tomioka, N., & Fujino, K. (1997). Natural (Mg, Fe)SiO_3-ilmenite and perovskite in the Tenham meteorite. *Science, 277,* 1084–1086.

Tomioka, N., & Fujino, K. (1999). Akimotoite, (Mg, Fe)SiO_3, a new silicate mineral of the ilmenite group in the Tenham chondrite. *American Mineralogist, 84,* 267–271.

Townsend, J. P., Tsuchiya, J., Bina, C. R., & Jacobsen, S. D. (2016). Water partitioning between bridgmanite and postperovskite in the lowermost mantle. *Earth and Planetary Science Letters, 454,* 20–27. doi:10.1016/j.epsl.2016.08.009

Tschauner, O., Ma, Ch., Beckett, J. R., Prescher, C., Prakapenka, V. B., & Rossman, G. R. (2014). Discovery of bridgmanite, the most abundant mineral in Earth, in a shocked meteorite. *Science, 346*(6213), 1100–1102.

Tsuchida, Y., & Yagi, T. (1989). A new, post-stishovite high-pressure polymorph of silica. *Nature, 340,* 217–220.

Tsuchiya, T., & Kawai, K. (2013). Ab initio mineralogical model of the Earth's lower mantle. In: S.-i. Karato (Ed.), *Physics and Chemistry of the Deep Earth*. John Wiley, Sons, pp. 213–243.

Tsuchiya, T., Caracas, R., & Tsuchiya, J. (2004a). First principles determination of the phase boundaries of high-pressure polymorphs of silica. *Geophysical Research Letters, 31,* L11610. doi:10.1029/2004GL019649

Tsuchiya, T., Tsuchiya, J., Umemoto, K., & Wentzcovitch, R. M. (2004b). Phase transition in $MgSiO_3$ perovskite in the earth's lower mantle. *Earth and Planetary Science Letters, 224,* 241–248.

Tsuchiya, T., & Wang, X. (2013). Ab initio investigation on the high-temperature thermodynamic properties of Fe^{3+}-bearing $MgSiO_3$ perovskite. *Journal of Geophysical Research, 118,* 83–91. doi:10.1029/2012JB009696

Umemoto, K., Kawamura, K., Hirose, K., & Wentzcovitch, R. M. (2016). Post-stishovite transition in hydrous aluminous SiO_2. *Physics of the Earth and Planetary Interiors, 255,* 18–26. doi:10.1016/j.pepi.2016.03.008

Van Aken, P. A., & Liebscher, B. (2002). Quantification of ferrous/ferric ratios in minerals: New evaluation schemes of Fe L_{23} electron energy-loss near-edge spectra. *Physics and Chemistry of Minerals, 29*(3), 188–200.

Van Rythoven, A. D., & Schulze, D. J. (2009). In-situ analysis of diamonds and their inclusions from the Diavik Mine, Northwest Territories, Canada: Mapping diamond growth. *Lithos, 112S,* 870–879.

Vanpeteghem, C., Angel, R., Ross, N., Jacobsen, S., Dobson, D., Litasov, K., et al. (2006). Al, Fe substitution in the $MgSiO_3$ perovskite structure: A single-crystal X-ray diffraction study. *Physics of the Earth and Planetary Interiors, 155*(1–2), 96–103.

Vilella, K., Shim, S.-H., Farnetani, C. G., & Badro, J. (2015). Spin state transition and partitioning of iron: Effects on mantle dynamics. *Earth and Planetary Science Letters, 417,* 57–66. doi:10.1016/j.epsl.2015.02.009

Vincent, E. A., & Phillips, R. (1954). Iron-titanium oxide minerals in layered gabbros of the Skaergaard intrusion, East Greenland. *Geochimica et Cosmochimica Acta, 6*(1), 1–34.

Wadhawan, V. K. (1982). Ferroelasticitay and related properties of crystals. *Phase Transitions, 3,* 3–103.

Walter, M. J., Kubo, A., Yoshino, T., Brodholt, J., Koga, K. T., & Ohishi, Y. (2004a). Phase relations and equation-of-state of aluminous Mg-silicate perovskite and implications for Earth's lower mantle. *Earth and Planetary Science Letters, 222*(2), 501–516.

Walter, M. J., Nakamura, E., Tronnes, R. G., & Frost, D. J. (2004b). Experimental constraints on crystallization differentiation in a deep magma ocean. *Geochimica et Cosmochimica Acta, 68,* 4267–4284.

Wang, W., Gasparik, T., & Rapp, R. (2000). Partitioning of rare earth elements between $CaSiO_3$ perovskite and coexisting phases: Constraints on the formation of $CaSiO_3$ inclusions in diamond. *Earth and Planetary Science Letters, 181*(3), 291–300.

Wang, Y. B., & Weidner, D. J. (1994). Thermoelasticity of $CaSiO_3$ perovskite and implications for the lower mantle. *Geophysical Research Letters, 21,* 895–898.

Warren, M. C., Ackland, G. J., Karki, B. B., & Clark, S. J. (1998). Phase transitions in silicate perovskites from first principles. *Mineralogical Magazine, 62*(5), 585–598.

Wentzcovitch, R. M., Ross, N. L., & Price, G. D. (1995). Ab initio study of $MgSiO_3$ and $CaSiO_3$ perovskites at lower-mantle pressures. *Physics of the Earth and Planetary Interiors, 90,* 101–112.

Wicks, J. K., & Duffy, T. S. (2016). Crystal structures of minerals in the lower mantle. In H. Terasaki & R. A. Fischer (Eds.), *Deep Earth: Physics and Chemistry of the Lower Mantle and Core: Vol. 217. Geophysical monograph* (pp. 69–87).

Wicks, J. K., Jackson, J. M., Sturhahn, W., Zhuravlev, K. K., Tkachev, S. N., & Prakapenka, V. B. (2015). Thermal equation of state and stability of $(Mg_{0.06}Fe_{0.94})O$. *Physics of the Earth and Planetary Interiors, 249,* 28–42.

Wilding, M. C. (1990). *A study of diamonds with syngenetic inclusions* (Unpublished Ph.D. thesis). University of Edinburgh, UK, 281 pp.

Wilding, M. C., Harte, B., & Harris, J. W. (1991). Evidence for a deep origin for the Sao Luiz diamonds. In *Fifth International Kimberlite Conference Extended Abstracts, Araxa, June 1991*, pp. 456–458.

Williams, Q., & Knittle, E. (2005). The highly uncertain bulk composition of Earth's mantle. In R. van der Hilst, J. Trampert, J. Bass, & J. Matas (Eds.), *Structure, dynamics and properties of Earth's Mantle* (pp. 187–200). Washington, D.C.: AGU Press.

Wirth, R., Dobrzhinetskaya, L., Harte, B., Schreiber, A., & Green, H. W. (2014). High-Fe (Mg, Fe)O inclusion in diamond apparently from the lowermost mantle. *Earth and Planetary Science Letters, 404,* 365–376.

Wirth, R., Vollmer, C., Brenker, F., Matsyuk, S., & Kaminsky, F. (2007). Nanocrystalline hydrous aluminium silicate in superdeep diamonds from Juina (Mato Grosso State, Brazil). *Earth and Planetary Science Letters, 259*(3–4), 384–399.

Wolf, G. H., & Jeanloz, R. (1985). Lattice dynamics and structural distortions of $CaSiO_3$ and $MgSiO_3$ perovskites. *Geophysical Research Letters, 12*(7), 413–416.

Wood, B. J. (2000). Phase transformations and partitioning relations in peridotite under lower mantle conditions. *Earth and Planetary Science Letters, 174*(3–4), 341–354.

Wood, B. J., & Corgne, A. (2009). Mineralogy of the Earth—Trace elements and hydrogen in the Earth's transition zone and lower mantle. In G. D. Price (Ed.), *Treatise on geophysics. Mineral physics* (pp. 63–89.). Elesevier.

Wu, Z., João, F., Justo, F., & Wentzcovitch, R. M. (2013). Elastic anomalies in a spin-crossover system: Ferropericlase at lower mantle conditions. *Physical Review Letters, 110,* 228501.

Wu, Z., & Wentzcovitch, R. M. (2014). Spin crossover in ferropericlase and velocity heterogeneities in the lower mantle. *Proceedings of the National Academy of the U.S.A., 111,* 10468–10472.

Xu, S., Lin, J.-F., Morgan, D. (2017). Iron speciation induced chemical and seismic heterogeneities in the lower mantle. *Journal of Geophysical Research. Solid Earth, 122.* doi: 10.1002/2016JB013543.

Xu, Y., McCammon, C., & Poe, B. T. (1998). The effect of alumina on the electrical conductivity of silicate perovskite. *Science, 282,* 922–924.

Xu, S., Shim, S.-H., & Morgan, D. (2015). Origin of Fe^{3+} in Fe-containing, Al-free mantle silicate perovskite. *Earth and Planetary Science Letters, 409,* 319–328.

Yagi, T., Okabe, K., Nishiyama, N., Kubo, A., & Kikegawa, T. (2004). Complicated effects of aluminum on the compressibility of silicate perovskite. *Physics of the Earth and Planetary Interiors, 143–144,* 81–91.

Yagi, T., Suzuki, T., & Akimoto, S. (1985). Static compression of wüstite $(Fe_{0.98})O$ to 120 GPa. *Journal of Geophysical Research, 90,* 8784–8788.

Yamamoto, T., Yuen, D. A., & Ebisuzaki, T. (2003). Substitution mechanism of Al ions in $MgSiO_3$ perovskite under high pressure conditions from first-principles calculations. *Earth and Planetary Science Letters, 206,* 617–625.

Yamanaka, T., Fukuda, T., & Tsuchiya, J. (2002). Bonding character of SiO_2 stishovite under high pressures up to 30 GPa. *Physics and Chemistry of Minerals, 29,* 633–641. doi:10.1007/s00269-002-0257-3

Yamanaka, T., Uchida, A., & Nakamoto, Y. (2008). Structural transition of post-spinel phases $CaMn_2O_4$, $CaFe_2O_4$, and $CaTi_2O_4$ under high pressures up to 80 GPa. *American Mineralogist, 93,* 1874–1881. doi:10.2138/am.2008.2934

Yamazaki, D., Ito, E., Yoshino, T., Tsujino, N., Yoneda, A., Guo, X., et al. (2014). Over 1 Mbar generation in the Kawai-type multianvil apparatus and its application to compression of $(Mg_{0.92}Fe_{0.08})SiO_3$ perovskite and stishovite. *Physics of the Earth and Planetary Interiors, 228,* 262–267. doi:10.1016/j.pepi.2014.01.013

Yang, J., Tong, X., Lin, J.-F., Okuchi, T., & Tomioka, N. (2015). Elasticity of ferropericlase across the spin crossover in the Earth's lower mantle. *Scientific Reports, 5,* 17188. doi:10.1038/srep17188

Yeganeh-Haeri, A., Weidner, D. J., & Ito, E. (1989). Elasticity of $MgSiO_3$ in the perovskite structure. *Science, 243,* 787–789.

Yoshino, T., Kamada, S., Zhao, C., Ohtani, E., & Naohisa, H. (2016). Electrical conductivity model of Al-bearing bridgmanite with implications for the electrical structure of the Earth's lower mantle. *Earth and Planetary Science Letters, 434,* 208–219.

Yu, Y. G., Wentzcovitch, R. M., Vinograd, V. L., & Angel, R. J. (2011). Thermodynamic properties of $MgSiO_3$ majorite and phase transitions near 660 km depth in $MgSiO_3$ and Mg_2SiO_4: A first principles study. *Journal of Geophysical Research, 116,* B02208. doi:10.1029/2010JB007912

Zedgenizov, D. A., Kagi, H., Shatsky, V. S., & Ragozin, A. L. (2014a). Local variations of carbon isotope composition in diamonds from Sao-Luis (Brazil): Evidence for heterogenous carbon reservoir in sublithospheric mantle. *Chemical Geology, 240*(1–2), 114–124. doi:10.1016/j.chemgeo.2013.10.033

Zedgenizov, D. A., Ragozin, A. L., Kalininaa, V. V., & Kagi, H. (2016). The mineralogy of Ca-rich inclusions in sublithospheric diamonds. *Geochemistry International, 54*(10), 890–900. doi:10.1134/S0016702916100116

Zedgenizov, D. A., Shatskiy, A., Ragozin, A. L., Kagi, H., & Shatsky, V. S. (2014b). Merwinite in diamond from São Luiz, Brazil: A new mineral of the Ca-rich mantle environment. *American Mineralogist, 99,* 547–550.

Zedgenizov, D. A., Shatsky, V. S., Panin, A. V., Evtushenko, O. V., Ragozin, A. L., & Kagi, H. (2015). Evidence for phase transitions in mineral inclusions in superdeep diamonds of the Sao Luiz deposit, Brazil. *Russian Geology and Geophysics, 56*(1), 296–305.

Zerr, A., Serghiou, G., & Boehler, R. (1997). Melting of $CaSiO_3$ perovskite to 430 kbar and first in situ measurements of lower mantle eutectic temperatures. *Geophysical Research Letters, 12* (24), 909–912.

Zhai, S., Yin, Y., Shieh, S. R., Shan, S., Xue, W., Wang, C.-P., et al. (2016). High-pressure X-ray diffraction and Raman spectroscopy of $CaFe_2O_4$-type β-$CaCr_2O_4$. *Physics and Chemistry of Minerals, 43,* 307–314. doi:10.1007/s00269-015-0795-0

Zhang, S., Cottaar, S., Liu, T., Stackhouse, S., & Militzera, B. (2016a). High-pressure, temperature elasticity of Fe-and Al-bearing $MgSiO_3$: Implications for the Earth's lower mantle. *Earth and Planetary Science Letters, 434,* 264–273.

Zhang, J., Li, B., Utsumi, W., & Liebermann, R. C. (1996). In situ X-ray observations of the coesite-stishovite transition: Reversed phase boundary and kinetics. *Physics and Chemistry of Minerals, 23,* 1–10.

Zhang, L., Meng, Y., Yang, W., Wang, L., Mao, W. L., Zeng, Q. S., et al. (2014). Disproportionation of (Mg, Fe)SiO_3 perovskite in Earth's deep lower mantle. *Science, 344* (6186), 877–882. doi:10.1126/science.1250274

Zhang, L., Popov, D., Meng, Y., Wang, J., Ji, C., Li, B., et al. (2016b). In situ crystal structure determination of seifertite SiO_2 at 129 GPa: Studying a minor phase near Earth's core–mantle boundary. *American Mineralogist, 101,* 231–234.

Zhang, J., & Weidner, D. J. (1999). Thermal equation of state of aluminum-enriched silicate perovskite. *Science, 284,* 782–784.

Zhang, J., & Zhao, Y. (2005). Effects of defect and pressure on the thermal expansivity of Fe_XO. *Physics and Chemistry of Minerals, 32,* 241–247.

Zhu, Q., Oganov, A. R., & Lyakhov, A. O. (2013). Novel stable compounds in the Mg–O system under high pressure. *Physical Chemistry Chemical Physics, 15,* 7696–7700.

Zou, G. T., Mao, H. K., Bell, P. M., & Virgo, D. (1980). High-pressure experiments on the iron oxide wüstite ($Fe_{1-x}O$). *Yearbook Carnegie Institution of Washington, 79,* 374–376.

Chapter 5
Mafic Lower-Mantle Mineral Association

Abstract Mafic mineral association in the lower mantle is subordinate to the ultramafic one. It includes bridgmanite, CaSi-perovskite, SiO_2 and anhydrous aluminous phases. The former three are the same as that observed in the ultramafic association; but their chemical compositions differ from those in the ultramafic association, mainly in the significant enrichment of Al. Among aluminous phases the NAL phase occurs at low-pressure conditions and is replaced by a CF phase at a depth of 800–1200 km depth. NAL phase is also concentrated in Na and K, while CF phase does not contain K. The partition coefficient of aluminium between bridgmanite and the NAL phase vary from 0.10 to 0.26, demonstrating that the Al enrichment in bridgmanite occurs at the expense of the Al decrease in the NAL phases. The Al concentration in the CF phase remains constant and the Al concentration in bridgmanite, after reaching maximal concentrations (24–25 wt%) with disintegration of the NAL phase, remains constant as well. In addition to the major minerals, phase Egg, δ-AlOOH, and a series of dense hydrous magnesium silicates (DHMS) are expected to be present in the mafic association. Among these DHMS, Phase D and Phase H are most likely to occur in the subducting slabs within the lower mantle. Some of these minerals (phase Egg and δ-AlOOH) are observed in natural geological materials; the others have only been synthesized in laboratory experiments.

5.1 General

Mafic association in the lower mantle appears as a consequence of recrystallization from subducting slab materials, predominantly of the middle-ocean ridge basalt (MORB) composition during the course of the subduction of lithospheric plates through the 660 km discontinuity. MORB contains more silica (48–51 wt% SiO_2) and aluminium (15–16 wt% Al_2O_3) than the bulk mantle, and therefore is composed of a different mineral association than the juvenile ultramafic composition. Four major mineral phases complete the mafic mineral association in the lower mantle: bridgmanite, CaSi-perovskite, SiO_2, and anhydrous aluminous phase (NAL, CF and/or CAS) (Funamori et al. 2000; Hirose et al. 2005; Akaogi 2007; Ricolleau et al.

F.V. Kaminsky, *The Earth's Lower Mantle*, Springer Geology,
DOI 10.1007/978-3-319-55684-0_5

2010). The first three minerals above are the same as those observed in the ultramafic association; they experienced similar phase transformations with pressure. However, the mafic association is not as commonly observed as the ultramafic counterparts; only a few samples have been identified as inclusions in lower-mantle diamond to date (Wirth et al. 2007; Walter et al. 2011; Thomson et al. 2014).

The chemical compositions of the mafic association, established from high pressure–temperature (*P*–*T*) experiments on MORB materials, differ from those in the ultramafic association, mainly in the significant enrichment in Al (Kesson et al. 1994; Hirose et al. 1999, 2005; Ono et al. 2001; Hirose and Fei 2002; Litasov et al. 2004; Litasov and Ohtani 2005; Ricolleau et al. 2010). In addition to the major mineral phases mentioned above, high *P*–*T* experimental results have showed the occurrence of various minor hydrous minerals at relevant *P*–*T* conditions of the subducting slabs, such as hydrous aluminium phases (Phase Egg and δ-AlOOH) and a series of DHMS. Phase Egg and δ-AlOOH have been observed in geological environments, but other hydrous minerals are known only from experiments and theoretical calculations.

5.2 Major Minerals

5.2.1 *Bridgmanite*

According to high-pressure experiments with dry MORB compositions, the occurrence of bridgmanite is associated with CaSi-perovskite, stishovite and Al-rich phase(s) at 27–80 GPa. It has a significantly lower magnesium index mg = 0.567–0.771 (average 0.658 $\pm$ 0.124), higher concentration of Al (Al_2O_3 = 11.1–24.0 wt %; average 16.23 $\pm$ 4.15 wt%), as well as higher concentrations of Ti and Ca than low-Al bridgmanite from the ultramafic association with an average mg = 0.916 and Al_2O_3 = 1.55 wt% (Kesson et al. 1994; Hirose et al. 1999, 2005; Ono et al. 2001; Hirose and Fei 2002; Litasov et al. 2004; Litasov and Ohtani 2005; Ricolleau et al. 2010) (Table 5.1; Fig. 5.1). Experimental synthesis of bridgmanite using a hydrated mafic composition at 25–28 GPa and ~1500 °C shows that bridgmanite has a similar composition with mg = 0.630 and may be associated with Al-rich Phase D (Litasov and Ohtani 2005; Pamato et al. 2015).

Bridgmanite from the natural mafic association identified in Juina-5 kimberlite pipe, Brazil, has similar compositions with fe = 0.346–0.401 and Al_2O_3 = 16.26–18.37 wt% (Table 5.1) (Thomson et al. 2014).

The concentration of Al in bridgmanite of the mafic association depends on the pressure at which it forms. The Al concentration is doubled from 11 to 13 wt% Al_2O_3 at 26 GPa to 24 wt% Al_2O_3 at 100 GPa, while the mg value remains almost constant (Fig. 5.2). Considering this correlation as a possible geobarometer, natural 'mafic' bridgmanite from Juina-5 pipe with 16.26–18.37 wt% Al_2O_3 have been formed at ~65–75 GPa. Such a drastic increase of wt% Al_2O_3 with pressure is most likely correlated to the redistribution of Al in bridgmanite from an Al-rich phase with increasing pressure. On the other hand, while Ca and Na concentrations in bridgmanite show a positive correlation with Al concentration (Fig. 5.3a,b) that is similar to that in

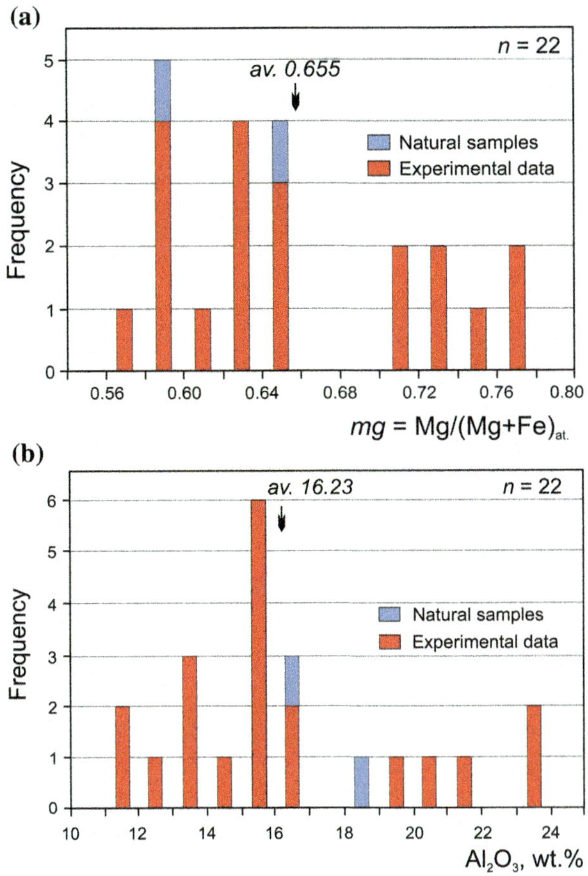

Fig. 5.1 Distribution of *mg* index (**a**) and Al_2O_3 (**b**) in bridgmanite from the mafic association. Data from Kesson et al. (1994), Hirose et al. (1999, 2005), Ono et al. (2001), Hirose and Fei (2002), Litasov et al. (2004), Litasov and Ohtani (2005), Ricolleau et al. (2010), Thomson et al. (2014)

bridgmanite of the ultramafic association, the correlation of Al with iron index *fe* is negative (Fig. 5.3c). That is, the most Al-rich bridgmanite are more Mg-rich.

Among bridgmanite varieties of the ultramafic association, there is also a group with a high Al concentration (see Sect. 4.2.2). However, those high-Al 'ultramafic' bridgmanite samples, in comparison with bridgmanite from the mafic association, are significantly richer in Mg, Ca, Na and Cr (0.64–5.32 wt% against 0.02–1.19 wt % CaO; 0.82–6.21 wt% against 0–0.11 wt% Na_2O; 1.19–3.14 wt% against 0.04–0.45 wt% Cr_2O_3, respectively) (Fig. 5.3).

Table 5.1 Representative compositions of bridgmanite synthesised from MORB materials at high pressures and temperatures and from natural samples (wt%)

Starting material	Synthetic MORB glass			Fresh natural MORB		Synthetic MORB glass		Juina-5 pipe, Brazil	
P (Gpa)	29.2	28	28	44	55	80	100	N/A	N/A
T (K)	1943	1873	1873	2391	2550	N/A	N/A	N/A	N/A
Sample No.	mbk-2	K-244dry	K-244wet	MORB#4	MORB#5			Ju5-31a	Ju5-117b
Mineral association	Brd + CaSiPrv + Sti + CF	Brd + CaSiPrv + Sti + NAL		Brd + CaSiPrv + Sti + NAL + CF	Brd + CaSiPrv + Sti + CF			Brd	Brd
SiO_2	41.5	38.6	35.2	42.3	40.9	43.0	44.7	35.64	36.14
TiO_2	3.27	4.38	5.12	2.5	2.1	n.a.	n.a.	5.48	3.77
Al_2O_3	11.4	13.4	15.3	21.2	23.6	20.5	24.0	16.26	18.37
Cr_2O_3	n.a.	n.a.	n.a.	n.a.	n.a.	n.a.	n.a.	0.05	0.04
FeO	21.0	21.2	23.2	11.3	10.1	13.9	11.7	21.79	18.39
MnO	n.a.	n.a.	n.a.	n.a.	n.a.	n.a.	n.a.	0.68	0.56
MgO	20.3	20.2	19.3	18.7	19.1	20.2	15.9	18.24	19.45
CaO	1.13	0.57	0.51	3.5	3.3	0.6	1.4	0.12	0.11
Na_2O	0.74	0.35	0.39	0.5	0.8	1.9	2.2	0.11	0.16
K_2O	0.02	0.01	0.01	n.a.	n.a.	n.a.	n.a.	0.01	0.00
Total	99.4	98.7	99.0	100.0	99.9	100.1	99.9	98.38	97.00
fe	0.367	0.370	0.403	0.253	0.229	0.278	0.292	0.401	0.346
mg	0.633	0.630	0.597	0.747	0.771	0.722	0.708	0.599	0.654
Reference	Litasov et al. (2004)	Litasov and Ohtani (2005)		Ricolleau et al. (2010)		Kesson et al. (1994)		Thomson et al. (2014)	

Note n.a. not analysed

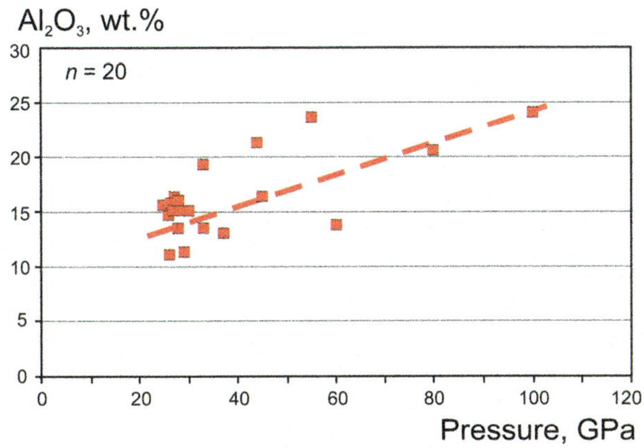

Fig. 5.2 Al_2O_3 concentration as a function of pressure in bridgmanite synthesised using MORB compositions at high pressures and temperatures. Data from Kesson et al. (1994), Hirose et al. (1999, 2005), Ono et al. (2001), Hirose and Fei (2002), Litasov et al. (2004), Litasov and Ohtani (2005), Ricolleau et al. (2010)

5.2.2 *$CaSiO_3$*

Representative chemical compositions of CaSi-perovskite from experimental studies on mafic associations are shown in Table 5.2. Similar to the compositions of bridgmanite discussed above, it is enriched in Al: 0.89–4.49 wt% Al_2O_3 (average: 2.49 wt%) as compared with the Al concentration of 0.1–1.52 wt% (average: 0.29 wt%) in CaSi-perovskite from the ultramafic association (Fig. 5.4a). The Al concentration in CaSi-perovskite synthesized from MORB materials remains almost constant within the entire pressure range of 25–113 GPa where the mineral was synthesized. CaSi-perovskite from natural association identified in Juina-5 pipe (Brazil), by contrast, has very low concentrations of Al_2O_3 (0.14–0.20 wt%; Thomson et al. 2014), similar to concentrations of Al_2O_3 in CaSi-perovskite from the ultramafic association (see Figs. 4.21 and 5.4b).

Although Fe and Mg concentrations in CaSi-perovskite from the mafic association are significantly higher than in the ultramafic association, the magnesium indices in CaSi-perovskites from the both associations are similar (averages: 0.519 and 0.521, respectively) (Fig. 5.4b).

There are, among natural CaSi-perovskite from the mafic association, high-Ti varieties with TiO_2 = 6.64–23.09 wt% TiO_2 (Thomson et al. 2014). Similar to the ultramafic association, Ca(Si,Ti)O_3 grains have dissolution structures (Fig. 4e in Thomson et al. 2014).

Concentrations of Fe, Mg and minor elements, such as Na, are significantly greater in CaSi-perovskite from the mafic association in comparison to this mineral from the ultramafic association (Fig. 5.5). Fe and Mg show positive correlation with Al.

The Al-concentration in $CaSiO_3$ is found to influence the transition boundary of the tetragonal to cubic structure transition, increasing it from 550 to 1900 K at

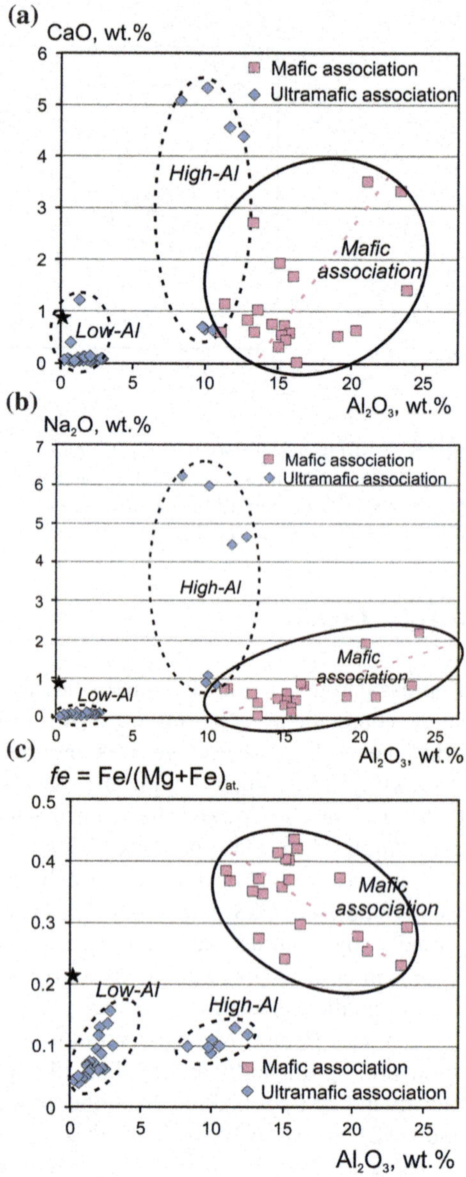

Fig. 5.3 Chemical characteristics of bridgmanite from the mafic association ($n = 20$) in comparison to bridgmanite from the ultramafic association ($n = 34$). **a** CaO versus Al_2O_3. **b** Na_2O versus Al_2O_3. **c** Iron index *fe* versus Al_2O_3. Data from Kesson et al. (1994), Hutchison (1997), Hirose et al. (1999, 2005), Stachel et al. (2000), Kaminsky et al. (2001), Ono et al. (2001), Hirose and Fei (2002), Davies et al. (2004), Litasov et al. (2004), Hayman et al. (2005), Litasov and Ohtani (2005), Tappert et al. (2009), Ricolleau et al. (2010), Zedgenizov et al. (2014, 2015). Star shows the composition of bridgmanite from the Tenham meteorite after Tschauner et al. (2014)

60 GPa with increasing Al_2O_3 concentration from 0 to 5.9 wt% (Kurashina et al. 2004). The tetragonal → cubic transition may thus occur in relatively cold subducted lithospheric slabs where $CaSiO_3$ perovskite can contain a higher proportion of Al_2O_3 (Hirose et al. 2005).

Table 5.2 Representative compositions of CaSi-perovskite synthesized from MORB materials and from natural samples (wt%)

Starting material	Synthetic MORB glass			Fresh natural MORB		Synthetic MORB glass		Juina-5 pipe, Brazil	
P (Gpa)	29.2	28	28	44	55	80	100	N/A	N/A
T (K)	1943	1873	1873	2391	2550	N/A	N/A	N/A	N/A
Sample No.	mbk-2	K-244wet	K-244dry	MORB#4	MORB#5			Ju5-52	Ju5-82
Mineral association	Brd + CaSiPrv + Sti + CF			Brd + CaSiPrv + Sti + NAL + CF	Brd + CaSiPrv + Sti + CF			CaSiPrv + ZrO_2	Ca(Si,Ti) Prv
SiO_2	47.9	48.9	49.3	51.9	52.1	56.7	55.0	51.27	37.48
TiO_2	2.22	2.71	2.58	1.3	1.2	n.a.	n.a.	0.02	23.09
Al_2O_3	4.06	1.64	2.08	3.5	2.5	2.1	2.2	0.14	2.93
FeO	5.13	0.97	1.46	4.4	3.5	2.6	1.6	0.24	0.14
MnO	n.a.	n.a.	n.a.	n.a.	n.a.	n.a.	n.a.	0.06	0.00
MgO	3.10	0.61	0.71	3.2	3.9	1.4	0.9	0.05	0.00
CaO	36.4	44.4	43.3	35.7	36.8	36.7	39.2	47.38	28.74
Na_2O	0.75	0.31	0.32	0.1	0.1	<0.5	1.2	0.16	0.06
K_2O	0.17	0.04	0.06	n.a.	n.a.	n.a.	n.a.	0.11	0.02
Total	99.7	99.6	99.8	100.1	100.1	99.5	100.1	99.43	93.80
fe	0.481	0.471	0.535	0.435	0.335	0.510	0.499	0.729	1.000
mg	0.519	0.529	0.465	0.565	0.665	0.490	0.501	0.271	0.000
Reference	Litasov et al. (2004)	Litasov and Ohtani (2005)		Ricolleau et al. (2010)		Kesson et al. (1994)		Thomson et al. (2014)	

Note n.a. not analysed

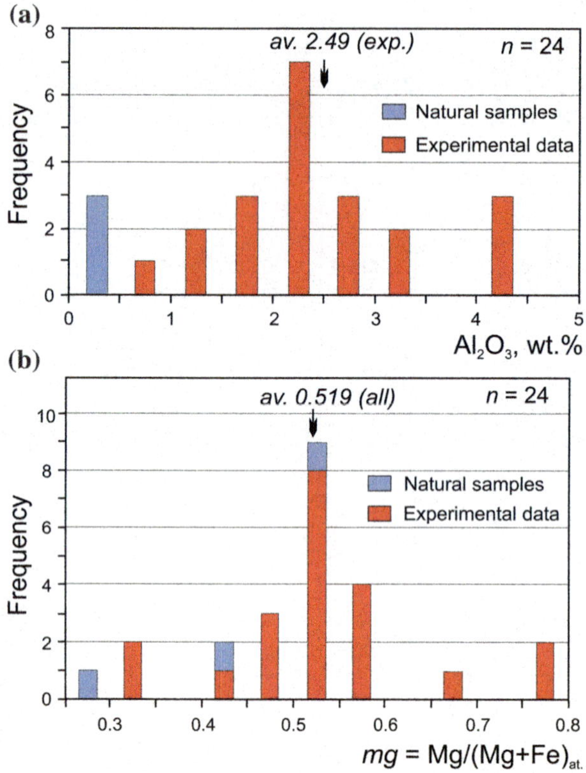

Fig. 5.4 Distribution of Al_2O_3 (**a**) and *mg* index (**b**) in CaSi-perovskite from the mafic association. Data from Kesson et al. (1994), Hirose et al. (1999, 2005), Ono et al. (2001), Hirose and Fei (2002), Litasov et al. (2004), Litasov and Ohtani (2005), Ricolleau et al. (2010)

5.2.3 SiO_2

SiO_2 from the mafic association is synthesized within a range of 25–100 GPa and, like perovskitic minerals, also has an elevated Al concentration: 0.35–4.2 wt% Al_2O_3 with an average of 2.59 wt% Al_2O_3; 10.3 wt% Al_2O_3 as was reported for one case (Table 4.3). It increases up to 12.6 wt% at 113 GPa in seifertite (Hirose et al. 2005). The only known natural sample of SiO_2 from the mafic association in Juina-5 pipe, Brazil, does not have high Al_2O_3 (0.09 wt%; Thomson et al. 2014). This may be caused by its association with kyanite, which is probably formed as a result of dissociation and collecting the entire amount of Al from the initial high-pressure composition.

Stishovite reported in experimental works under pressure values of 80–100 GPa (Kesson et al. 1994) is doubtful because at such *P–T* conditions $CaCl_2$-structured SiO_2 is stable. Moreover, experiments by Hirose et al. (2005) with MORB compositions demonstrated that the phase transition boundary between Al-bearing stishovite and $CaCl_2$-structured SiO_2 is lower than that of pure SiO_2 by ~25 GPa: it is located at ~62 GPa under 2000 K, which corresponds to a depth of ~1500 km (Table 5.3).

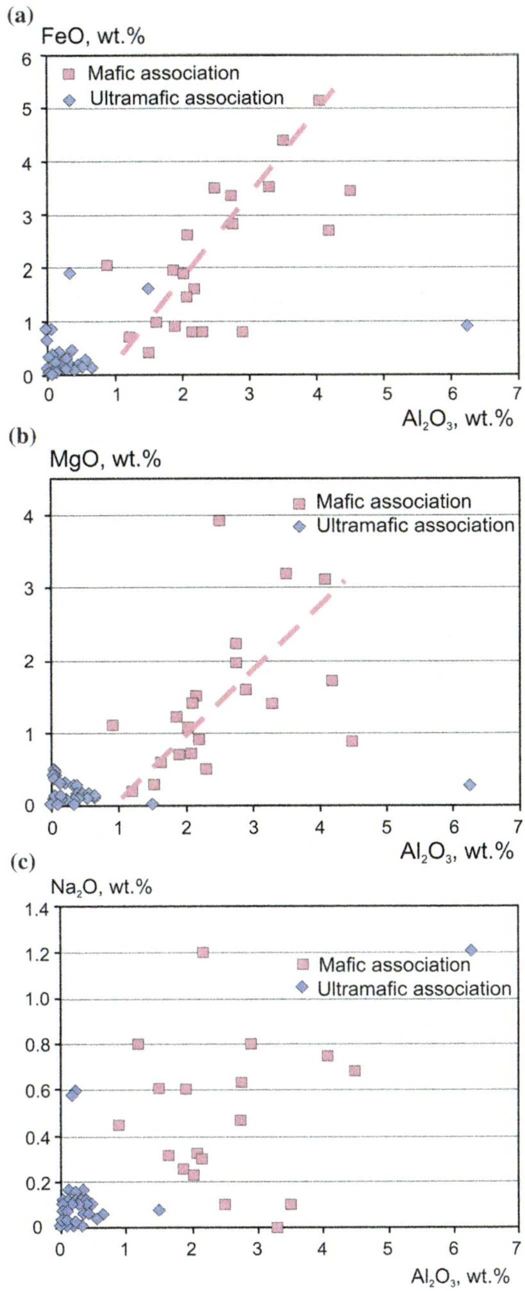

Fig. 5.5 Chemical characteristics of CaSi-perovskite from the mafic association. **a** *fe* versus Al_2O_3; **b** *mg* versus Al_2O_3; **c** Na versus Al_2O_3

Table 5.3 Representative compositions of SiO_2 from experimental and natural mafic associations (wt%)

Starting material	Synthetic MORB glass			Fresh natural MORB		Synthetic MORB glass		Juina-5 pipe, Brazil
P (Gpa)	29.2	28	28	44	55	80	100	N/A
T (K)	1943	1873	1873	2391	2550	N/A	N/A	N/A
Sample No.	mbk-2	K-244wet	K-244dry	MORB#4	MORB#5			Ju5-90
Mineral association	Brd + CaSiPrv + Sti + CF			Brd + CaSiPrv + Sti + NAL + CF	Brd + CaSiPrv + Sti + CF			SiO_2 + Ky
SiO_2	98.5	96.5	96.4	98.7	98.5	88.5	96.1	100.07
TiO_2	n.a.	0.01	0.04	0.1	0.1	n.a.	n.a.	0.03
Al_2O_3	1.15	3.04	2.41	0.8	0.9	10.3	2.7	0.09
FeO	0.14	0.13	0.23	0.1	0.1	<0.5	<0.5	0.01
NiO	n.a.	n.a.	n.a.	n.a.	n.a.	n.a.	n.a.	0.02
MgO	n.a.	0.08	0.09	0	0.1	<0.5	<0.5	0.01
CaO	n.a.	0.06	0.05	0.1	0.4	<0.5	<0.5	0.02
Na_2O	0.34	0.04	0.03	0.0	0.0	<0.5	<0.5	0.03
Total	100.1	99.9	99.3	99.8	100.1	98.8	98.8	100.29
Reference	Litasov et al. (2004)	Litasov and Ohtani (2005)		Ricolleau et al. (2010)		Kesson et al. (1994)		Thomson et al. (2014)

Note n.a. not analysed

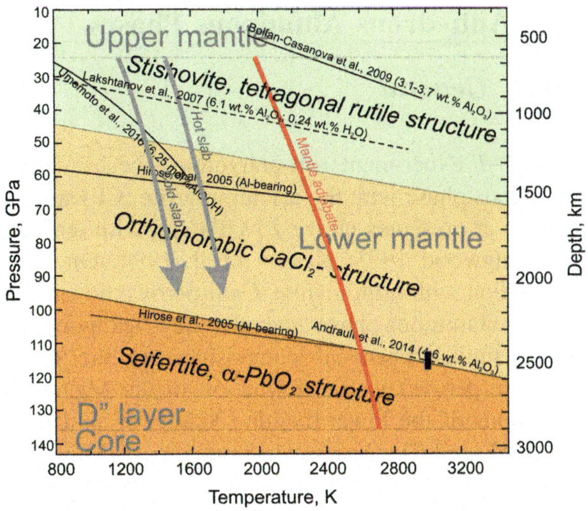

Fig. 5.6 Phase diagram for SiO_2 in subducted slabs. Temperature profiles in slabs after Kirby et al. (1996). Data from Hirose et al. (2005), Lakshtanov et al. (2007), Bolfan-Casanova et al. (2009), Andrault et al. (2014), and Umemoto et al. (2016). Mantle adiabate after Katsura et al. (2010)

Concentration of H_2O in stishovite has been shown to be related to the Al admixture in silica. While Al-free stishovite contains only 16–30 wt. ppm H_2O, synthetic stishovite with 4.4 wt% Al_2O_3 contains ~3010 wt. ppm H_2O (Litasov et al. 2007a). Most hydrogen in stishovite is associated with Al^{3+} substitutional defects on the octahedral (Si^{4+}) site. The hydrogen can occupy 40% of vacancies created by the incorporation of Al^{3+} at 20 GPa (Litasov et al. 2007a). It is possible that Al actually forms an AlOOH phase with H_2O, which is highly soluble in stishovite (e.g., Chung and Kagi 2002; Panero and Stixrude 2004). The incorporation of Al and H_2O into SiO_2 significantly reduces the transition pressure for the stishovite to post-stishovite transition boundary (Fig. 5.6). Specifically, hydrous Al-bearing stishovite transforms to post-stishovite at 24 GPa and 298 K, which is far lower than the transition pressure of 50–60 GPa for pure silica (Lakshtanov et al. 2007). First-principles static calculations and molecular dynamics demonstrated that hydrogen bonds and hydrogen mobility play a crucial role in lowering the post-stishovite transition pressure (Umemoto et al. 2016). A co-operative redistribution of hydrogen atoms is the main mechanism responsible for the reduction of the transition pressure in hydrous Al-bearing stishovite. Stishovite, containing 6.25 mol% AlOOH, transforms into $CaCl_2$-structured phase at pressures from 30 to 32 GPa (1000 K) to 50 GPa (1300 K), i.e., within the range of 800–1200 km depth. Since SiO_2 is a candidate composition in the subducted slabs, it can undergo a series of the aforementioned phase transitions that are influenced by their Al concentrations, causing stratification in the subducting slabs in the lower mantle: at 1200–1300 km depth with stishovite → $CaCl_2$-structure transition and at 2200–2300 km with $CaCl_2$-structure → seifertite transition (Fig. 5.6). This is similar to the stratification in juvenile ultramafic mantle, except for the presence of the aluminous mineral phase. In addition, phase boundaries in subducted slabs are located higher compared with the juvenile ultramafic host.

5.3 Anhydrous Aluminous Phases

5.3.1 General

High *P–T* experiments on anhydrous oceanic basalts have demonstrated that an aluminous phase with the calcium–ferrite (CF) structure (space group *Pbnm*, $Z = 4$) appears at pressures above 25 GPa and with temperatures above 1200 °C (Irifune and Ringwood 1993; Kesson et al. 1994; Ono 2001). The CF phase occurs in association with bridgmanite, CaSi-perovskite and SiO_2 at high *P–T*. Subsequently, another aluminous phase, named a 'new hexagonal aluminous phase' (NAL), which has a similar chemical composition to the CF phase, was observed in the garnet-to-perovskite transformation in the $MgAl_2O_4$–$CaAl_2O_4$ system at relevant conditions of the lower mantle (Akaogi et al. 1999; Miyajima et al. 1999, 2001). The NAL and CF phases can comprise 10–30% of the subducted MORB materials in the lower mantle (Ricolleau et al. 2010; Mookherjee et al. 2012).

Studies of diamonds from the Juina-5 kimberlite in Brazil indicated that their mineral inclusions have compositions consistent with the phase assemblage expected to crystallize from basaltic materials at lower-mantle conditions: bridgmanite, CaSi-perovskite, SiO_2, and NAL and CF phases (Walter et al. 2011). These results provide geological evidence for the possible existence of the NAL and CF phases in the lower mantle.

5.3.2 New Hexagonal Aluminous Phase (NAL)

The NAL phase has a general formula of ${}^{IX}A{}^{VIII}B_2^{VI}C_6O_{12}$, where ${}^{IX}A$ sites are comprised by large monovalent or divalent cations (Na^+, K^+ and/or Ca^{2+}), ${}^{VIII}B$ sites are represented by smaller cations (Mg^{2+} and Fe^{2+}), and ${}^{VI}C$ octahedral sites are occupied by Al^{3+} (mostly), Si^{4+} and Fe^3. The full chemical formula of NAL is $(Na, K, Ca)(Mg, Fe^{2+})_2(Al, Fe^{3+}, Si)_6O_{12}$. Both Na- and K-rich NAL phases were synthesized at pressures of 22–25 GPa and a temperature of 1500 °C (Kojitani et al. 2011). Hence, the NAL phase concentrates not only Al but also alkalis, while the CF phase does not accommodate potassium (Miyajima et al. 2001; Guignot and Andrault 2004).

Representative compositions of NAL are shown in Table 5.4. Concentrations of both Na (4.93–5.9 wt% Na_2O; average 5.46 wt%) and K (0.54–1.94 wt% K_2O; average 1.00 wt%) are quite constant, and are independent of other components and pressure conditions. Al_2O_3 has a strong linear negative correlation with SiO_2, demonstrating their substitution in the C site (Fig. 5.7a). The Al/Si_{at} strongly decreases from 4.7 at 25 GPa to 1.73 at 33 GPa (Fig. 5.7b), which coincides with the increase of the Al concentration with increasing pressure in the associated bridgmanite.

Table 5.4 Representative compositions of the NAL phase from experimental mafic associations (wt%)

Starting material	Synthetic MORB glass					Fresh natural MORB	
P (Gpa)	25	26	28	28	28	33	44
T (K)	1273	1773	1873	1873	2573		
Sample No.	K-156	K-206wet	K-244wet	K-244dry	K-254	MORB#1	MORB#2
Mineral association	Brd + CaSiPrv + Sti + NAL						
Brd + CaSiPrv + Sti + NAL + CF							
SiO_2	13.5	15.0	15.9	20.2	22.4	26.5	23.3
TiO_2	0.40	1.21	1.11	1.06	1.14	0.3	0.3
Al_2O_3	53.8	52.1	51.0	47.3	44.2	38.9	45.0
FeO	11.9	11.7	11.4	12.4	11.4	8.7	4.7
MgO	12.6	11.9	12.2	10.7	12.6	17.7	18.9
CaO	1.16	1.39	1.80	1.60	2.01	2.2	1.9
Na_2O	5.79	5.67	5.40	5.68	5.12	5.9	5.9
K_2O	0.72	0.79	0.89	0.54	0.96	n.a.	n.a.
Total	99.9	99.8	99.7	99.5	99.8	100.2	100.0
fe	0.346	0.355	0.344	0.394	0.336	0.216	0.122
mg	0.654	0.645	0.656	0.606	0.664	0.784	0.878
Al/Si (at.)	4.698	4.094	3.781	2.760	2.326	1.730	2.277
Reference	Litasov and Ohtani (2005)					Ricolleau et al. (2010)	

Note n.a. not analysed

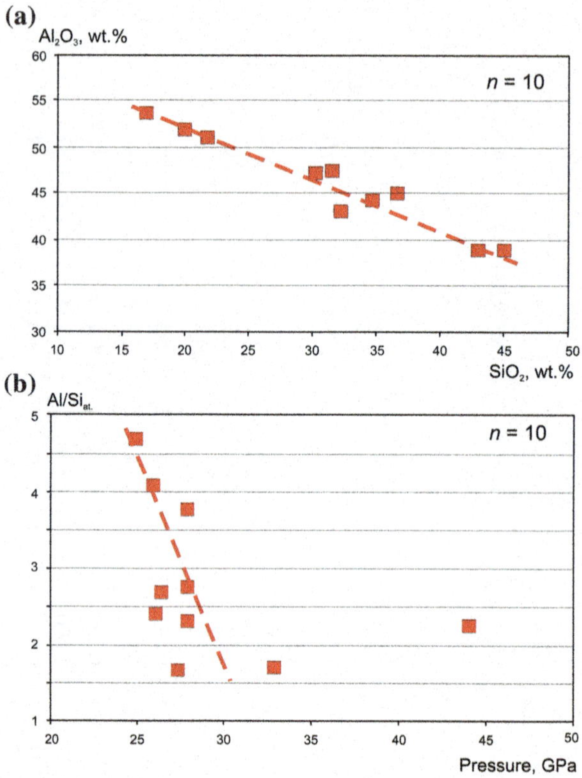

Fig. 5.7 Chemical characteristics of the NAL phase. **a** Al_2O_3 versus SiO_2. **b** Al/Si versus pressure. Data from Hirose and Fei (2002), Litasov et al. (2004), Litasov and Ohtani (2005), Ricolleau et al. (2010)

The NAL phase has a hexagonal crystal structure with the space group of *P*6$_3$/*m* (Gasparik et al. 2000; Miura et al. 2000; Miyajima et al. 2001; Kojitani et al. 2011; Mookherjee et al. 2012; Wicks and Duffy 2016; Wu et al. 2016a) (Fig. 5.8). The ^{IX}A site is a nine-fold coordinated tunnel with a hexagonal cross-section and is occupied by a large monovalent (Na^+, K^+) or divalent cation (Ca^{2+}). Cations in the ^{IX}A site are likely to be disordered owing to its Wyckoff symmetry being 2*a* and the fact that only one atom occupies either of the two equivalent sites, i.e., there is half occupancy (Mookherjee et al. 2012). The B site is six-fold co-ordinated but trigonal prismatic rather than octahedral and is typically occupied by Mg^{2+} (or replacing with Fe^{2+} in a Fe-rich variety). Each Mg^{2+} ion is surrounded by three double chains of AlO_6 (SiO_6) octahedra connected by corner-sharing oxygen ions. They form the C site. Large channels are formed by these octahedra in three double chains and are occupied typically by Na^+ or K^+ ions at half occupancy. The distorted C octahedra form edge-sharing double chains extending along the *c* axis that are corner linked to form tunnels. Within the tunnels lie the larger A sites while the smaller B sites are

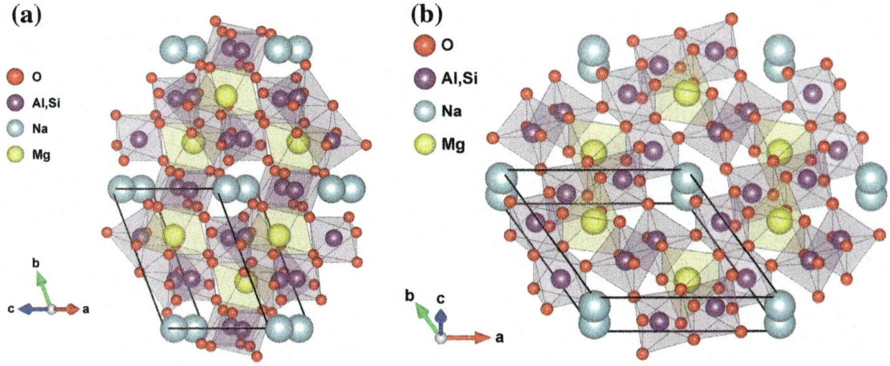

Fig. 5.8 Hexagonal crystal structure of Fe-free NAL (space group $P6_3/m$) in different axis orientations. Lattice parameters from Pamato et al. (2014)

surrounded by three double chains. With multiple sites that can each accept more than one cation, the NAL structure can accommodate a wide range of compositional variation. The large cation sites in the NAL structure allow them to be potential hosts for the alkali elements (Wicks and Duffy 2016).

The hexagonal NAL structure is stable under pressure conditions of up to 40–50 GPa, after which it transforms to the orthorhombic CF phase (Ono et al. 2009; Ricolleau et al. 2010; Imada et al. 2011) in accordance with the later theoretical results obtained using the first-principles computational method (Kawai and Tsuchiya 2012). The enrichment of the NAL phase in potassium significantly increases its stability up to 141 GPa (Guignot and Andrault 2004; Kato et al. 2013).

Density functional theory calculations showed that the NAL phase has lower seismic velocities compared with the CF phase, and NAL could contribute to low-velocity provinces in the deep mantle (Mookherjee et al. 2012). However, calculations performed by Kawai and Tsuchiya (2012) concluded that NAL and CF phases would have comparable shear wave velocities, producing less than 1% velocity difference across the phase transition; this conclusion is in agreement with experiments carried out by Pamato et al. (2014).

Since the NAL phase probably contains up to 12 mol% of iron (Litasov and Ohtani 2005; Ricolleau et al. 2010; see Table 5.4), the valence and spin states of iron in the NAL phase can potentially affect its elasticity, which in turn can affect our understanding of the mineral physics of the MORB assemblage in the lower mantle (Lin et al. 2013; Wu et al. 2016b). Recently, two single-crystal samples (iron-free $Na_{1.14}Mg_{1.83}Al_{4.74}Si_{1.23}O_{12}$ and iron-bearing $Na_{0.71}Mg_{2.05}Al_{4.62}Si_{1.16}Fe^{2+}_{0.09}Fe^{3+}_{0.17}O_{12}$) were studied by Wu et al. (2016a) using synchrotron nuclear forward scattering (NFS) and X-ray diffraction (XRD) combined with diamond-anvil cells (DACs) up to 86 GPa at room temperature to investigate the potential effects of the spin transition on its equation of state at high pressures. Based on the evaluation of the hyperfine quadrupole splitting from NFS results, a pressure-induced high-spin (HS) to low-spin (LS) transition of the octahedral Fe^{3+} in the iron-bearing NAL was observed at approximately 30 GPa. Compared with the equation of state for the iron-free NAL phase using XRD data, the iron-bearing NAL undergoes a volume reduction of 1.0% (~ 1.2 Å^3) at 33–47 GPa, which is associated with the spin transition of the octahedral Fe^{3+}. The bulk sound velocities (V_{Φ}) of the iron-free and iron-bearing NAL phase are approximately 6% larger than those of Al, Fe-bearing bridgmanite and CaSi-perovskite in the lower mantle, except for the spin transition region where a notable softening of V_{Φ} with a maximum reduction of 9.4% occurs in the iron-bearing NAL phase at 41 GPa. Considering the high volume proportion of the NAL phase in subducted MORB, the distinct elastic properties of the iron-bearing NAL phase across the spin transition may provide an explanation for the observed seismic heterogeneities of subducted slabs in the lower mantle at depths below 1200 km.

5.3.3 Aluminous Phase with the Calcium–Ferrite Structure (CF Phase)

The CF phase, which is considered as a high-pressure modification of NAL, has a general formula on a 12 oxygen basis to make it better comparable with the NAL phase, of $^{VIII}B_3^{VI}C_6O_{12}$, where ^{VIII}B sites are occupied by Na^+, Ca^{2+} and Mg^{2+} cations and the ^{VI}C sites are framework-forming cations Al^{3+} (mostly) and Si^{4+}, similar to that in the NAL phase. However, unlike the NAL phase, it is significantly enriched (almost doubled) in sodium (average 10.55 wt% Na_2O) and does not contain potassium. Its full chemical formula is $(Na, Ca, Mg)_3(Al, Si)_6O_{12}$ (Irifune and Ringwood 1993; Kesson et al. 1994; Ono et al. 2001; Hirose and Fei 2002). Al concentration in the CF phase varies from 32.8 to 45.1 wt% Al_2O_3 (av.: 39.4 ± 3.6 wt% Al_2O_3), independently of Si and pressure values. Representative compositions of the CF phase from experimental data are presented in Table 5.5.

The CF phase has an orthorhombic space group *Pbnm* and four formula units ($Z = 4$ on a four oxygen primitive basis) in the unit cell (Decker and Kasper 1957). Such CF structure is a common high-pressure structure type (Wicks and Duffy 2016) experimentally adopted by several compositions including $MgAl_2O_4$ (Irifune et al. 1991), $CaAl_2O_4$ (Reid and Ringwood 1969), and $NaAlSiO_4$ (Liu 1977). It is one of a series of similar structures ($CaFe_2O_4$, $CaTi_2O_4$, $CaMn_2O_4$) that are common high-pressure polymorphs of spinel-structured phases (Yamanaka et al. 2008). Single-crystal diffraction experiments have shown that forsterite, Mg_2SiO_4, metastably adopts a related structure upon room temperature compression above 58 GPa (Finkelstein et al. 2014). CF structure was established for $FeCr_2O_4$ composition in meteorites (Chen et al. 2003) and suggested a new $CaCr_2O_4$ oxide found in ultramafic lower-mantle microxenolite (Kaminsky et al. 2015; see Sect. 4.7.3).

The CF phase has a chain framework structure formed by edge-sharing AlO_6 octahedra along the *c*-axis direction similar to the NAL phase, but it only forms one kind of 'tunnel' formed by the linked double chains being in eight-fold coordination, where sodium atoms are located (Yamada et al. 1983) (Fig. 5.9). This model is

Table 5.5 Representative compositions of the CF phase from experimental mafic associations (wt%)

Starting material	Powdered MORB composition			Fresh natural MORB		Synthetic MORB glass	
P (Gpa)	30	33	37	44	55	80	100
T (K)				2391	2550		
Sample No.				MORB#4	MORB#5		
Mineral association	Brd + CaSiPrv + Sti + CF			Brd + CaSiPrv + Sti + NAL + CF	Brd + CaSiPrv + Sti + CF		
SiO_2	27.5	27.9	28.9	28.5	26.7	26.0	32.4
TiO_2	0.7	0.7	0.7	0.5	0.2		
Al_2O_3	40.1	37.6	36.6	39.8	43.2	45.1	44.1
FeO	7.9	7.1	6.7	9.7	6.2	9.0	5.9
MgO	10.0	10.0	7.1	11.0	12.9	8.9	7.7
CaO	0.8	1.3	0.5	1.3	1.7	0.8	0.7
Na_2O	12.1	13.2	15.2	9.2	9.1	10.3	9.3
K_2O	0.0	0.0	0.0	n.a.	n.a.	n.a.	n.a.
Total	99.1	97.8	95.7	100.0	100.0	100.1	100.1
fe	0.307	0.285	0.346	0.331	0.212	0.362	0.300
mg	0.693	0.715	0.654	0.669	0.788	0.638	0.700
Al/Si (at.)	1.719	1.589	1.493	1.646	1.907	2.045	1.604
Reference	Ono et al. (2001)			Ricolleau et al. (2010)		Kesson et al. (1994)	

Note n.a. not analysed

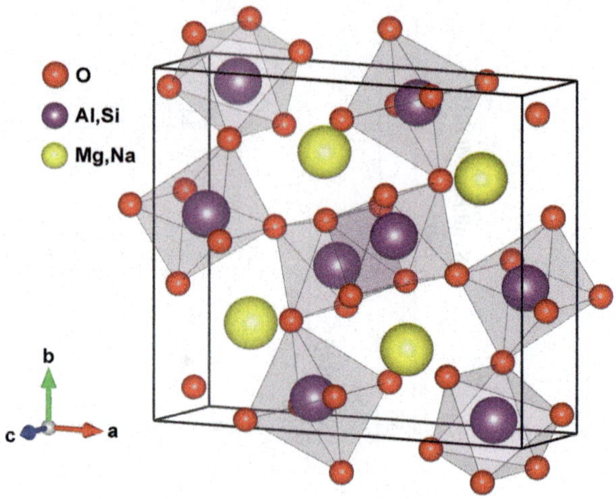

Fig. 5.9 Orthorhombic structure of the CF phase (space group *Pbnm*). Lattice parameters from Kojitani et al. (2007)

based on structural parameters for the $MgAl_2O_4$ end-member (Kojitani et al. 2007), whereas the CF-type phase in basaltic compositions of the lower mantle can exist within the $NaAlSiO_4$–$MgAl_2O_4$ system.

Experimental studies on the $NaAlSiO_4$–$MgAl_2O_4$ system in a laser-heated DAC coupled with in situ synchrotron XRD measurements demonstrated that the CF phase is the high pressure form of the NAL phase, with a transition pressure between 34 and 43 GPa (Ono et al. 2009; Imada et al. 2011). On the other hand, the CF phase exists in association with the NAL phase at pressures of up to 50 GPa (Ono et al. 2009; Ricolleau et al. 2010; Imada et al. 2011), and at much higher pressures of up to 120 GPa and 2300 K, in the potassium-rich system. The calcium-bearing CF phase is also found to coexist with the NAL phase (Kato et al. 2013). Thus far, these studies have shown that the relative stability of NAL and CF phases is a complex, sensitive function of *P*–*T* and composition. While pressure tends to favour stability of the CF phase, enrichment in Mg or alkalis tends to favour the NAL phase (Mookherjee et al. 2012).

The NAL to CF phase transition causes an increase in density and bulk modulus (Guignot and Andrault 2004) and a change in compressional wave velocity by −0.2% and shear wave velocity by +0.9% (Kawai and Tsuchiya 2012); the shear wave velocity increases by 2.5% across phase transition at ~40 GPa (Dai et al. 2013). The change in the elastic properties and shear wave anisotropy across the NAL to CF phase transition is proposed to be significant enough to be seismically detectable (Kawai and Tsuchiya 2012; Dai et al. 2013). A study on the physical properties of the aluminous phases performed by Mookherjee et al. (2012), demonstrated that the sound wave velocities of the NAL phase is significantly lower than the CF phase and other major lower mantle phases. The existence of such a lower-velocity NAL phase in the deeply subducted MORB materials can compensate for the faster sound velocity of silica (SiO_2) and its high-pressure polymorphic phase(s) in the lower mantle.

5.3.4 *Calcium Aluminium Silicate with Ba-Ferrite Structure (CAS Phase)*

A Ca–Al silicate (CAS) with an end-member chemical formula of $CaAl_4Si_2O_{11}$ containing some amount of water was first reported by Irifune et al. (1994) in experiments performed at 17.5–24 GPa and 1500–1900 °C, and later by Gautron et al. (1996) at 14 GPa and 1400–1500 °C. The CAS phase has a hexagonal unit cell with lattice parameters a = 5.4 Å and c = 12.7 Å and a space group, most likely, $P6_3/mmc$ with Z = 2, analogous to hexagonal Ba-ferrite $BaTi_2Fe_4O_{11}$. It has a six-layer, close-packed structure, reasonable for a phase stable under the conditions of the transition zone (Gautron et al. 1997).

Subsequently, the CAS phase was synthesized as an Al-rich phase using anhydrous MORB materials in multi-anvil apparatus at 25.5 GPa (Hirose and Fei 2002; Litasov and Ohtani 2005). In these experiments, the NAL and CF phases appeared at higher pressures, above 27–27.5 GPa, coexisting with bridgmanite, stishovite and CaSi-perovskite. The NAL phase synthesized at 26 GPa was stable below 2030 °C but transformed to the CAS phase at higher temperatures. The CAS phase does not appear below the solidus temperature in the presence of bridgmanite. It remains unclear whether the CAS phase can present in the lower mantle.

In the natural environment, a single sample of CAS in association with kyanite was reported from an inclusion in Juina-5 kimberlite pipe, Brazil (Thomson et al. 2014). The natural CAS is poorer in Mg and Al (with Al/Si = 1.027) than when obtained from experiments (Table 5.6).

5.4 Partitioning of Aluminium Between Bridgmanite and Aluminous Phases

Aluminium in the lower-mantle mafic association has been observed to be concentrated in three major phases: bridgmanite, NAL phas, and CF phase. The concentration of Al in bridgmanite of the mafic association varies from 11 to 13 wt% Al_2O_3 at 26 GPa to 24 wt% Al_2O_3 at 100 GPa (see Fig. 5.1), independently of the magnesium index value (mg), which is rather constant. Al-rich NAL phase with 38.9–53.8 wt% Al_2O_3 (average: 46.2 ± 4.9 wt% Al_2O_3) probably incorporated its aluminium from majorite in the lower mantle, and shows a strong correlation of its Al_2O_3 concentration with SiO_2 (Fig. 5.7a). The ratio Al/Si_{at} in the NAL phase strongly decreases from 4.7 at 25 GPa to 1.73 at 33 GPa (Fig. 5.7b). Al concentration in the CF phase varies from 32.8 to 45.1 wt% Al_2O_3 (average 39.4 ± 3.6 wt % Al_2O_3), independently of Si and pressure values. This implies that with redistribution of Al between bridgmanite and NAL occurs with increasing pressure, as subducting slabs penetrate deeper into the lower mantle.

Table 5.7 shows Si/Al ratio values in coexisting bridgmanite in the NAL phase from the experimental solidus associations, as well as Al/Si ratios and aluminium partitioning coefficient values. The partition coefficient of aluminium between bridgmanite and the

Table 5.6 Compositions of synthetic and natural CAS phase (wt%)

Starting material	Glass with a natural MORB composition	Synthetic MORB glass	Juina-5 pipe, Brazil
P (Gpa)	27.5	26.5	n/a
T (K)	2703	2523	n/a
Sample No.	LO#117	K-255	Ju5-85
Mineral association	Brd + CaSiPrv + Sti + NAL + CAS + Liq	Brd + CaSiPrv + Sti + NAL + CAS + Liq	CAS + Ky
SiO_2	41.99	37.50	40.19
TiO_2	0.26	0.46	0.79
Al_2O_3	39.65	44.40	35.03
Cr_2O_3	0.07	n.a.	0.07
FeO	1.80	1.43	1.50
NiO	n.a.	n.a.	0.05
MnO	0.03	n.a.	0.02
MgO	2.92	1.76	0.31
CaO	11.08	12.70	16.88
Na_2O	1.28	1.17	3.59
K_2O	1.07	0.32	0.80
Total	100.15	99.74	99.23
fe	0.257	0.313	0.731
mg	0.743	0.687	0.269
Al/Si (at.)	1.113	1.396	1.027
Reference	Hirose and Fei (2002)	Litasov and Ohtani (2005)	Thomson et al. (2014)

Note n.a. not analysed

Table 5.7 Si/Al ratio values (atomic) and partition coefficient of Al in coexisting bridgmanite and NAL

Sample No.	Mineral association*	Pressure, GPa	Al/Si in bridgmanite	Al/Si in NAL	$K_D^{Brd\text{–}NAL}$	Reference
K-156	Brd + CaSiPrv + Sti + NAL	25	0.463	4.698	0.099	Litasov and Ohtani (2005)
K-206wet	Brd + CaSiPrv + Sti + NAL	26	0.472	4.094	0.115	Litasov and Ohtani (2005)
mbk-1	Brd + CaSiPrv + Sti + NAL + CF + Maj	26.1	0.321	2.408	0.134	Litasov et al. (2004)
K-244wet	Brd + CaSiPrv + Sti + NAL	28	0.512	3.781	0.136	Litasov and Ohtani (2005)
K-244dry	Brd + CaSiPrv + Sti + NAL	28	0.409	2.760	0.148	Litasov and Ohtani (2005)
K-254	Brd + CaSiPrv + Sti + NAL	28	0.555	2.326	0.238	Litasov and Ohtani (2005)
–	Brd + CaSiPrv + Sti + NAL + CF	33	0.348	1.730	0.201	Ricolleau et al. (2010)
–	Brd + CaSiPrv + Sti + NAL + CF	44	0.591	2.277	0.260	Ricolleau et al. (2010)

Note *The associations with liquid are not included

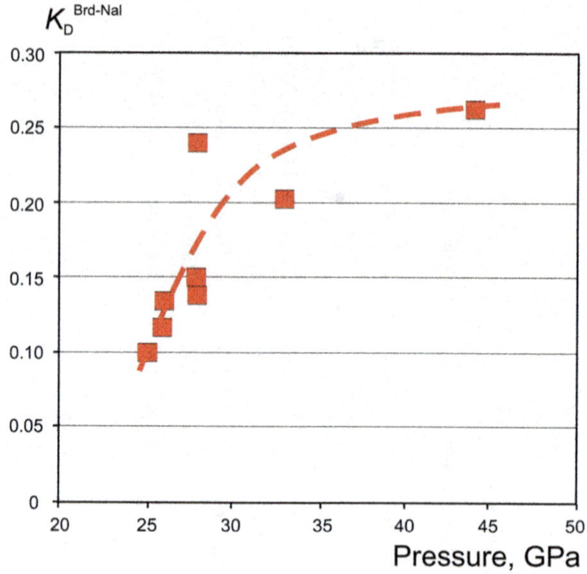

Fig. 5.10 Dependence of aluminium partition coefficient $K_D^{Brd-NAL}$ of pressure

NAL phase defined as $K_D^{Brd-NAL} = (Al/Si)^{Brd}/(Al/Si)_{at}^{NAL}$ varies from 0.10 to 0.26, asymptotically increasing within a pressure range of 25–44 GPa (Fig. 5.10). This confirms that the Al enrichment in bridgmanite with increasing pressure is a result of the dissolution of the NAL phase, while the Al concentration in the CF phase, which replaces the NAL phases at pressures of 40–50 GPa, remains constant. The Al concentration in bridgmanite, after reaching a maximal concentration of ~24–25 wt%, also remains constant.

5.5 Hydrous Aluminous Phases

5.5.1 General

Subducting slabs can contain significant amounts of water in hydrous minerals in hydrothermally altered basalts (e.g., serpentine) and sediments (e.g., clay minerals). High-pressure investigations of the Al_2O_3–SiO_2–H_2O system relevant to subducting slab materials have found that three hydrous mineral phases, phase Egg, δ-AlOOH and topaz-OH can possibly become stable at *P–T* conditions corresponding to the lower mantle. Initially, however, these three phases were reported to be stable at relatively lower *P–T* conditions at 5.5–21 GPa and temperatures up to 1500 °C (Wunder et al. 1993; Schmidt 1995; Ono 1999; Suzuki et al. 2000). Subsequently, they were experimentally proven to be stable at greater depths relevant to the lower-mantle conditions. Phase Egg and δ-AlOOH are found as inclusions in natural diamonds.

5.5.2 *Phase Egg*

A hydrous aluminous silicate $AlSiO_3OH$ was first observed in high-pressure experiments on hydrous basaltic systems at pressures exceeding 10 GPa. The first water-containing high-pressure mineral, initially suggested as $Al_5Si_5O_{17}OH$ (Egglton et al. 1978):

$$\begin{gathered} AlOOH + SiO_2 \rightarrow AlSiO_3OH. \\ \text{diaspore} + \text{stishovite} \rightarrow \text{phase Egg} \end{gathered}$$

This phase was subsequently synthesized from topaz-OH + stishovite mixture at 12 GPa, and has been later termed as phase Egg to give credit to Professor R.A. Egglton who first discovered the phase (Egglton et al. 1978; Schmidt 1995). High *P–T* stability of the phase was experimentally extended to 20 GPa and 1700 °C (Ono 1999), then to 23 GPa and 1200 °C (Sano et al. 2004), then to 25 GPa and ~1500 °C (Boffa Ballaran et al. 2010), and later to 26 GPa and 1460 °C (Pamato et al. 2015). Most recently Fukuyama et al. (2017) experimentally established the stability limit of phase Egg at 23.5 GPa, suggesting that differences in previous results were caused by the use of different pressure scales. They concluded that phase Egg is stable only within the transition zone (Fukuyama et al. 2017). However, other experiments and equation of state study of this phase demonstrate that, despite the presence of hydrogen in its composition, phase Egg is stable up to 40 GPa (Vanpeteghem et al. 2003). For this reason, the presence of phase Egg and stishovite in natural samples may be suggestive not only of the origin in the transition zone but in the lower mantle as well, and phase Egg may be considered as a major carrier of water in the deep Earth.

A natural association of stishovite and phase Egg was observed in a diamond from the Rio Soriso, Juina area, Brazil (Wirth et al. 2007) (Fig. 5.11). Both minerals were identified with the observation of in situ Raman spectroscopy and electron diffractometry, such that original mineral structures were not altered during the examination.

Phase Egg forms an aggregate of euhedral, cubic and cubo-octahedral grains that are 20–30 to 200–300 nm in size (Fig. 5.11b). The bulk Si/Al ratios of the two analyzed natural grains are approximately 1:1 (46.74–53.84 at.% and 45.63–52.60 at.%). A minor admixture of Fe (0.29–1.52 at.%) is present in its composition. The presence of OH in phase Egg was identified using the pre-peak at approximately 528 eV in the electron energy-loss spectrum and the OH-stretching band between 1800 and 3000 cm^{-1} in the Raman spectra. Numerous irregularly shaped pores are present in the stishovite matrix (Fig. 5.11a). They contained a fluid, which was released to the vacuum during the focused ion beam milling of the specimen for the transmission electron microscopy analysis. Quench products of the fluid contain minor amounts of K, Ca, Ba, P, S, Cl and F that were also detected along the walls of the pores. This porosity possibly reflects the initial stage of degassing of phase Egg under conditions close to its pressure-dependent disintegration. In some of the Raman spectra taken from phase Egg, characteristic spectral features of δ-AlOOH

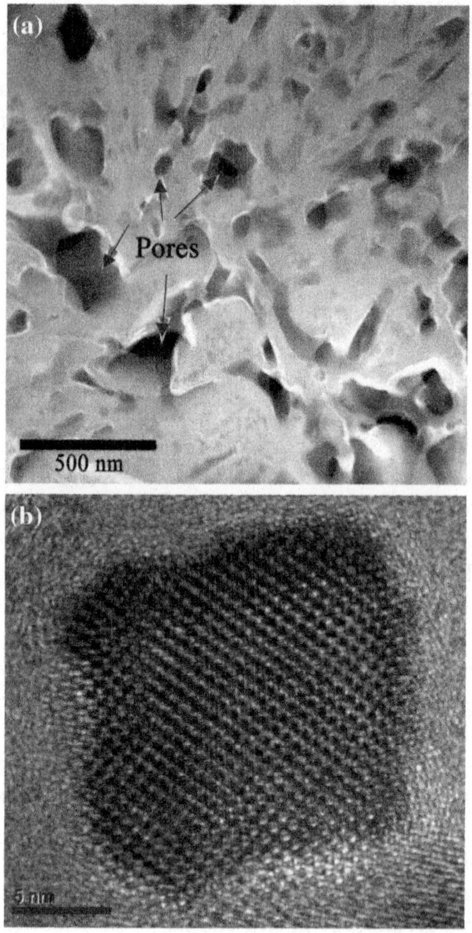

Fig. 5.11 Electron-microscope images of phase Egg. **a** Porous aggregate of stishovite and phase Egg. **b** Single cubic crystal of phase Egg. From Wirth et al. 2007. © Earth and Planetary Science Letters (Elsevier)

phase located near 2150 and 2500 cm^{-1} were detected, analogous to the spectral characteristics of experimentally produced δ-AlOOH phase (Ohtani et al. 2001b). The coexistence of δ-AlOOH phase with phase Egg indicates the most likely pressure interval for their joint formation to be 23–27 GPa, i.e., in the uppermost regions of the lower mantle (Sano et al. 2004).

The crystal structure of phase Egg, obtained in experimental studies, was determined, using powder XRD, to be monoclinic (Fig. 5.12) with a space group $P2_1/n$, a unit cell volume $V_0 = 212.99(1)$ Å^3, and zero-pressure density of 3.74 g/cm^3 (Schmidt et al. 1998). Most grains of natural phase Egg are monoclinic as well, but

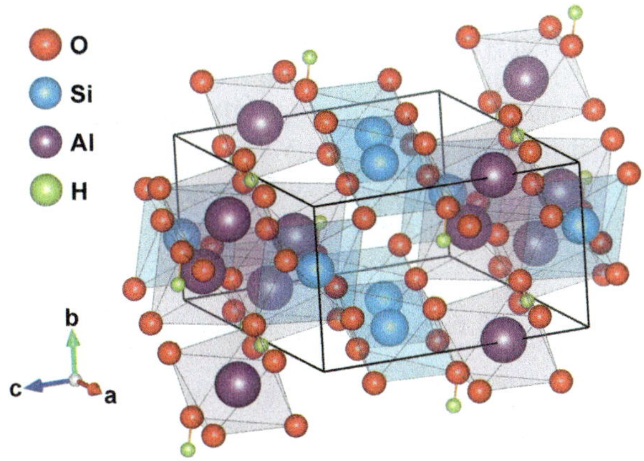

Fig. 5.12 Monoclinic crystal structure of phase Egg (space group $P2_1/n$) with lattice parameters from Schmidt et al. (1998)

some analyzed natural grains are tetragonal (a = 0.7435 nm and c = 0.706 nm), suggesting that the tetragonal polymorph may be a relic of the original phase Egg.

The monoclinic structure of phase Egg resembles that of stishovite and δ-AlOOH, containing an ordered arrangement of SiO_6 and AlO_6 octahedra linked by both edge- and corner-sharing (Fig. 5.12). Similar to δ-AlOOH (see Sect. 5.5.3), the crystal structure of phase Egg is characterized by strong hydrogen bonding, which can have great effects on many of the properties in the mineral including elasticity, rheology and electrical conductivity. Hydrogen bonding, which is the electrostatic attraction between two polar groups (in this case hydrogen and oxygen atoms), also plays an important role in stabilizing hydrous and nominally anhydrous phases by lowering the enthalpy, or increasing the entropy, or both (e.g., Prewitt and Parise 2000). The strong hydrogen bonding in phase Egg and δ-AlOOH (as well as in DHMSs) may be an important factor in stabilizing these structures at high pressures and temperatures.

5.5.3 Hydrous δ-AlOOH and ε-FeOOH Phases

Aluminium oxyhydroxide, the hydrous δ-AlOOH phase, which is a high-pressure polymorph of diaspore (α-AlOOH) and boehmite (γ-AlOOH), was initially synthesized in association with majorite and phase Egg at ~21 GPa and 1000 °C (Suzuki et al. 2000). Subsequently, it has been found to be stable at pressures above 30 GPa and temperatures of 1000–1200 °C, and thus it may serve (like phase Egg) as a water carrier in the cold subducted slabs, transporting water into the lower mantle (Ohtani et al. 2001a, b; Sano et al. 2004). Later, in situ XRD experiments in the diamond anvil cell were conducted to determine the stability field of δ-AlOOH over the entire P–T conditions of the lower mantle. δ-AlOOH was observed to exist at pressures from 33 to 134 GPa at 1350–2300 K, indicating that this hydrous mineral in the sediment layer of a subducting slab could transport water into the core-mantle boundary (Sano et al. 2008; Mashino et al. 2016). The experiments by Ohira et al. (2014) also confirmed that a hydrous δ–AlOOH phase could coexist with Al-depleted bridgmanite or post-perovskite under the *P–T* conditions relevant to the cold slab geotherm. δ-AlOOH was identified in geological samples to exist in association with phase Egg (Wirth et al. 2007).

Theoretical calculations further predicted the pressure-induced phase transition of δ-AlOOH into a new cubic phase at approximately 170 GPa (Tsuchiya and Tsuchiya 2011). Therefore, δ-AlOOH phase is likely to be the highest pressure phase among all hydrous mineral compounds, since these hydrous materials tend to dissociate and release H_2O at pressures of around a few tens of GPa (Ohtani et al. 2001b).

Pure δ-AlOOH has an orthorhombic symmetry with a non-centrosymmetric space group $P2_1nm$ that is very close to that of SiO_2 in the $CaCl_2$-type (Suzuki et al. 2000; Komatsu et al. 2006; Vanpeteghem et al. 2007) (Fig. 5.13a). The structure contains both AlO_6 and HO_6 octahedra, and the oxygen anions form a slightly distorted hexagonal close-packed arrangement perpendicular to the *b* direction

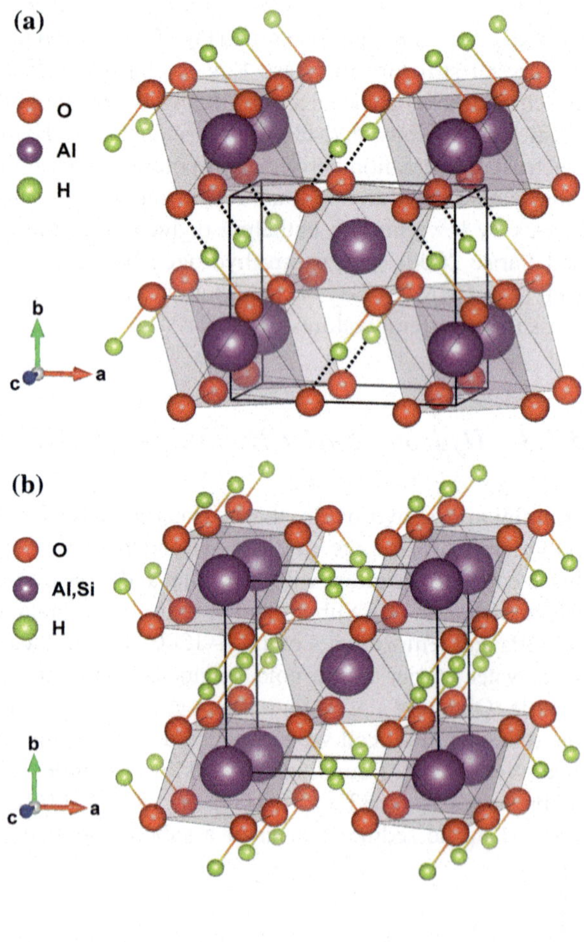

Fig. 5.13 Orthorhombic crystal structure models of δ-AlOOH based on $P2_1nm$ and *Pnnm* space groups. *Dotted lines* are partially ordered H bonds. Unit cell parameters from Komatsu et al. (2011)

(Wicks and Duffy 2016). A single-crystal XRD study indicates that δ-AlOOH transforms to the $CaCl_2$-type structure (space group *Pnnm*) at ~8 GPa (Kuribayashi et al. 2013) (Fig. 5.13). This structure consists of corner-sharing chains of edge-sharing octahedra (AlO_6, in this case). There is no discontinuity in unit cell volume across this displaceable transition. The main difference between the two structures lies in the disordering of hydrogen atoms in the high-pressure structure. The substitution of Mg^{2+} and Si^{4+} into δ-AlOOH also promotes the transformation of $P2_1nm$ structure (Suzuki et al. 2000; Kudoh et al. 2004; Komatsu et al. 2011). The disordered cations, Mg^{2+} and Si^{4+}, with different valences than Al^{3+} may fluctuate hydrogen positions, causing the disordered hydrogen to change its symmetry. As a result, Mg–Si-containing δ-AlOOH phase exhibits the *Pnnm* structure (Komatsu et al. 2011) or *Pnn*2 structure (Kudoh et al. 2004).

Pressure-induced hydrogen bond symmetrization starts to occur in δ-AlOOH at 28 GPa (Tsuchiya et al. 2002), similar to that reported for phase Egg. Since the crystal structure with the symmetric hydrogen bonding has a smaller volume than the asymmetric hydrogen bond counterpart, the occurrence of the hydrogen bond symmetrisation can further stabilize hydrous materials to higher-pressure conditions. It is conceivable that higher-pressure hydrous phases would also be expected to have symmetric hydrogen bonds. The details of the hydrogen bonding in δ-AlOOH phases has been extensively discussed, especially with regards to the formation of symmetric hydrogen bonds, but there is currently no consensus on this issue (Tsuchiya et al. 2002; Panero and Stixrude 2004; Li et al. 2006; Xue et al. 2006; Vanpeteghem et al. 2007; Sano-Furukawa et al. 2008).

The predicted ultrahigh-pressure cubic phase has a cubic pyrite-type AlO_2 framework with interstitial H atoms forming symmetric hydrogen bonds, whose symmetry is assigned to the space group $Pa\bar{3}$.

Other cation substitutions, such as Fe and Ti in the oxyhydroxide, have also been suggested based on high *P–T* experiments corresponding to the transition zone conditions: at pressures of 16 GPa and temperatures of 1000–1600 °C (Nishihara and Matsukage 2016). In the studied conditions, two stable iron-titanium oxyhydroxide phases were identified whose estimated composition is $(FeH)_{1-x}Ti_xO_2$. One is the Fe-rich solid solution ($x < 0.23$) with ε-FeOOH-type crystal structure (orthorhombic, *P*21 *nm*); and the other is the more Ti-rich solid solution ($x > 0.35$) with α-PbO_2-type structure (orthorhombic, *Pbcn*). The ε-FeOOH-type phase is stable up to ~1100 °C, whereas the Ti-rich α-PbO_2-type phase is stable up to 1500 °C for a composition of $x = 0.5$ and at least to 1600 °C for $x = 0.75$. These phases may be stable within subducting basaltic slabs into the transition zone, but their stability in the lower mantle remains uncertain and requires further clarifications in future studies.

In iron-rich compositions, a complete Fe–Al substitution may occur. ε-FeOOH, the high-pressure polymorph of goetite (α-FeOOH), is isostructural with δ-AlOOH, and has a slightly distorted rutile structure with corner-linked single bands of edge-shared octahedra parallel to the *c*-axis (Otte et al. 2009). ε-FeOOH undergoes a spin transition beginning at 49 GPa and completing at ~65 GPa, which is assisted by symmetrization of hydrogen bonds during the transition from $P2_1nm$ to *Pnnm*. (Gleason et al. 2008, 2013; Reagan et al. 2016).

5.5.4 Topaz-OH

Topaz-OH $Al_2SiO_4(OH)_2$, the OH end-member of the F-OH topaz, has been synthesized at a pressure of 6–10 GPa from diaspore (α-AlOOH) and phase Π $Al_3Si_2O_7(OH)_3$ starting materials as a result of their reaction with SiO_2. This phase is shown to be stable up to 13 GPa and 1500 °C (Wunder et al. 1993; Xue et al. 2006), after which it reacts with SiO_2 to form phase Egg $AlSiO_3OH$ (Egglton et al. 1978; Schmidt 1995, 1998; Ono 1999). At high temperatures, it undergoes a dehydration reaction to form kyanite (Wunder et al. 1993). At pressures above 14 GPa, a high-pressure topaz-OH II variety, with significant cation disorder occurs (Kanzaki 2010).

The structure of topaz-OH shows close agreement with that of F-rich topaz, containing tetrahedral SiO_4 and linked $AlO_4(OH)_2$ octahedra. It was first refined with the space group *Pbnm* from single-crystal XRD by Wunder et al. (1993), although the hydrogen positions could not be located. A subsequent single-crystal XRD study (Northrup et al. 1994) reported that the single H position found in F-rich topaz is split into two nonequivalent, half-occupied H positions in topaz-OH. They also suggested that the structure is non-centrosymmetric and might have a long-range ordered H distribution with a reduction of symmetry to $Pbn2_1$. Each H is associated with three O atoms in an irregular trifurcated H-bond arrangement. A lattice energy calculation and molecular dynamics simulation study (Churakov and Wunder 2004) predicted four nonequivalent crystallographic H positions in topaz-OH that are in dynamic exchange at ambient conditions and suggested instantaneous configuration of H violates in all symmetry elements, whereas the space- and time-averaged structure possesses *Pbnm* symmetry. Monoclinic structure $P2_1/c$ is also suggested for topaz-OH (Mookherjee et al. 2016).

The stability of topaz-OH under high *P*–*T* conditions has not been studied.

5.6 Dense Hydrous Magnesium Silicates (DHMSs)

5.6.1 General

Numerous experimental high-pressure works were carried out on DHMSs of relevant mantle compositions, showing that they can be stable with increasing pressure at characteristic slab temperatures. Ringwood and Major (1967) recognized the first of these DHMSs in experiments performed in the MgO–SiO–H_2O system under pressures of up to 18 GPa, and denoted these phases as A [$Mg_7Si_2O_8(OH)_6$], B [$Mg_{12}Si_4O_{19}(OH)_6$], and C, known also as superhydrous Phase B

[$Mg_{10}Si_3O_{14}(OH)_4$]. The naming system for the DHMS phases proposed by them was later adopted for naming a whole sequence of 'alphabetic' DHMS phases. Phases termed D (Yamamoto and Akimoto 1977; Liu 1986, 1987), E, F (Kanzaki 1989, 1991; Kudoh et al. 1993, 1995), and G (Ohtani et al. 1997) were subsequently reported; however the latter two, F and G, appear to be similar if not identical to Phase D (Yang et al. 1997; Frost and Fei 1998, 1999). These phases contain 2.14–18 wt% H_2O and may act as water carriers in subducting hydrated plates to >35 GPa (Yang et al. 1997; Ohtani et al. 2004), in addition to hydrous aluminium phases. Most of these phases appeared to be unstable under lower-mantle pressure conditions (e.g., Kanzaki 1991), with the exception of the following three: superhydrous Phase B, Phase D, and the recently discovered Phase H.

5.6.2 *Superhydrous Phase B*

'Superhydrous' Phase B [$Mg_{12}Si_4O_{19}(OH)_6$], containing as much as 5.8 wt% H_2O, is stable at *P–T* conditions relevant to the top of the lower mantle and decomposes into periclase, bridgmanite and Phase D at approximately 30 GPa and high temperatures (Shieh et al. 1998). On the other hand, the following reaction sequences were observed with increasing temperature from 1200 to 1600 °C in the pressure range from 18–20 GPa: superhydrous Phase B → Phase B + liquid → wadsleyite + periclase + liquid → anhydrous Phase B + periclase + liquid (Ohtani et al. 2001a).

5.6.3 *Phase D*

Initially, the term Phase D was used by Yamamoto and Akimoto (1977) to describe the hydroxyl chondrodite, which was synthesized at pressures between 3 and 12 GPa. However, Liu (1987) reapplied the name Phase D to describe a phase discovered as a dissociation product of $Mg_3Si_2O_5(OH)_4$ serpentine at pressures above ~20 GPa (an equivalent to phases F and G). It has the ideal formula $MgSi_2O_6H_2$ (Mg-Phase D), but is typically non-stoichiometric. Phase D contains nominally 10.1 wt% H_2O (varying from 10 to 18 wt%) and varies significantly in stoichiometry depending on synthesis conditions (Frost and Fei 1998) with Mg/Fe ratio from 0.56 to 0.71 within the range of 16–27 GPa and 1073–1573 K; the Mg/Si ratio and H_2O concentration decrease with increasing temperature and increase with increasing pressure (Shinmei et al. 2008). Along with most other DHMS phases, Phase D synthesized from the starting samples containing complex peridotitic bulk compositions have been found to contain up to 7 wt% Al_2O_3 and 5 wt% FeO (Frost 1999; Litasov et al. 2007b, 2008; Rosa et al. 2012; Ganskow and Langenhorst 2014), suggesting that natural Phase D samples contain not only water but also some Al and Fe. Al^{3+} substitutes for Si^{4+}, Fe^{2+} substitutes for Mg, and Fe^{3+} may occupy Si and Mg sites (Ganskow and Langenhorst

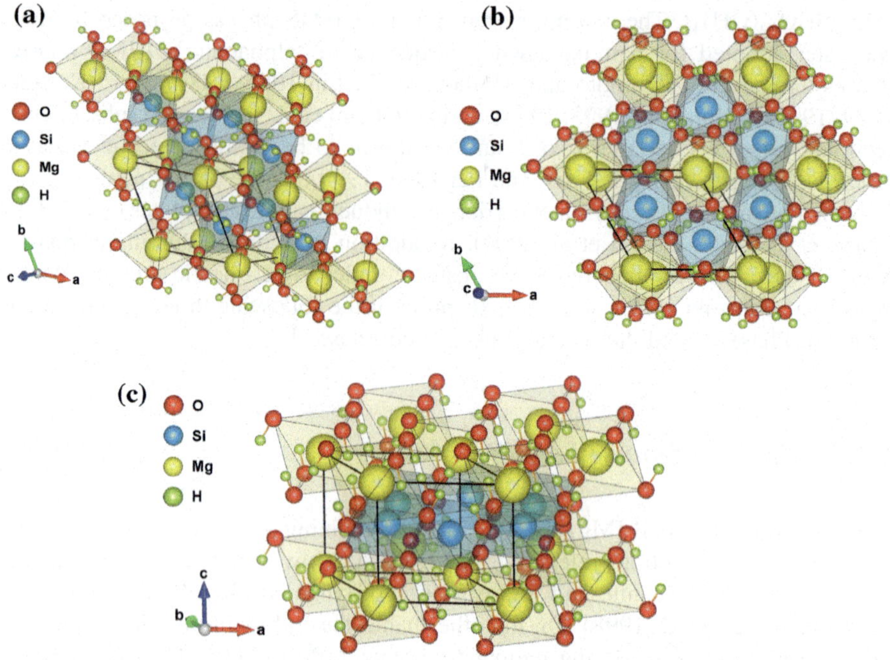

Fig. 5.14 Crystal structure of Al-free trigonal phase D (space group *P*-31*m*) in different axis orientations. Lattice parameters from Yang et al. (1997)

2014). Available data demonstrate that the $Fe^{3+}/\Sigma Fe$ ratio of Fe-bearing phase D determined by conventional Mössbauer spectroscopy and EEL spectroscopy is 0.60–0.94, and iron is not involved in water-substitution mechanisms (Chang et al. 2013; Ganskow and Langenhorst 2014).

Phase D has a trigonal structure with the space group *P*-31 *m*, based on a hexagonal close-packed array of oxygen atoms (Kudoh et al. 1997; Yang et al. 1997; Xue et al. 2008), similar to stishovite (Fig. 5.14). All cations except H are in the octahedrally coordinated sites. The MgO_6 and SiO_6 octahedra occur in separated, alternating layers stacked along the *c* direction (Fig. 5.14c), with Mg and Si at the M1 and M2 sites, respectively, and with the octahedral sites corresponding to the 2*c* (M3) and 1*b* Wyckoff positions being vacant (Yang et al. 1997). The MgO_6 and SiO_6 octahedra share edges with neighbours to form rings in layers perpendicular to the *c* axis (Fig. 5.14b). The SiO_6 octahedra form gibbsite-like layers, in which each octahedron shares three edges, leaving one third of the octahedral sites vacant. The MgO_6 octahedra lie above and below vacant sites in the SiO_6 layer and are corner-linked to the Si octahedra, and so two thirds of the Mg octahedral sites are empty (Wicks and Duffy 2016). Charge balance is achieved by hydrogen anions. The hydrogen positions are located in the MgO_6 layers, with O–H bonds facing away from SiO_6 octahedra.

Phase D is the only DHMS structure reported to date that contains Si exclusively in octahedral co-ordination. Owing to this, the calculated density of Phase D is 3.50 g/cm^3, which is substantially greater than any other DHMS found thus far, including Phase A (2.96 g/cm^3), Phase B (3.37 g/cm^3), superhydrous Phase B (3.21 g/cm^3), Phase E (2.82 g/cm^3), and Phase F (Yang et al. 1997 and references therein). As a result of the layered nature of the structure, with strong SiO_6 layers alternating with weaker MgO_6 octahedra resulting in enhanced compressibility along the *c* axis, Phase D exhibits anisotropic compression (Frost and Fei 1999). Theoretical calculations of the elastic constants indicate that the anisotropy decreases with depth but significant anisotropy in seismic wave velocities is retained to the highest pressures (Mainprice et al. 2007).

At pressures exceeding 35–40 GPa Phase D, like phase Egg and δ-AlOOH (see above), experiences hydrogen bond symmetrization (Tsuchiya et al. 2005; Shinmei et al. 2008). A symmetric hydrogen bond is one in which the hydrogen atom is located at the midpoint between the two neighbouring oxygen atoms, rather than the asymmetric O–H distances that characterize a conventional hydrogen bond. This serves stabilizing hydrous Phase D (as well as phase Egg and δ-AlOOH) under ultrahigh-pressure conditions. This pressure-induced hydrogen bond symmetrization has a significant effect on the compression behaviour of Phase D. The bulk modulus increases by about 20% with this structural change (Tsuchiya et al. 2005). A powder XRD study showed evidence of such an anomaly (Hushur et al. 2011), but a more recent high-resolution single-crystal XRD study to 65 GPa found no evidence for a bulk modulus anomaly, at least to this pressure (Rosa et al. 2013a, b). No evidence for hydrogen bond symmetrization was found in an infrared spectroscopic study (Shieh et al. 2009) or in other theoretical calculations (Mainprice et al. 2007).

Fe-bearing Phase D was reported to undergo the electron spin transition of Fe^{3+} at high pressure, which is shown to cause a softening in the bulk modulus and bulk sound velocity of the crystal (Chang et al. 2013). Recently, a two-stage HS-LS spin transition involving both Fe^{2+} and Fe^{3+} was reported to occur in Phase D at high pressures (Wu et al. 2016b). This transition of iron occurs for Fe^{2+} at 37–41 GPa and for Fe^{3+} at 64–68 GPa. These transitions are accompanied by an increase in density and a significant softening in the bulk modulus and bulk velocity at their respective pressure range. The occurrence of the dense low-spin FeAl-bearing phase D with relatively high velocity anisotropies in deep-subducted slabs can potentially contribute to small-scale seismic heterogeneities. However, hydrogen symmetrization and spin transitions in Phase D may overlap with each other, and their interplay at high pressures is not yet fully understood (Wu et al. 2016b).

The stability field of Phase D extends from 20 to 50 GPa, independent of temperature from ~1300–1700 K (Shieh et al. 1998, 2000; Frost and Fei 1998, 1999; Tsuchiya et al. 2005; Litasov et al. 2008; Shinmei et al. 2008; Hushur et al. 2011; Ghosh and Schmidt 2014; Nishi et al. 2014; Walter et al. 2015). Pure Phase D is stable only at relatively lower temperatures and can break down at temperatures lower than those of the typical mantle geotherm (e.g., Nishi et al. 2014). In experiments, Phase D decomposes at 44–50 GPa, corresponding to a depth of 1250–1400 km, into an assemblage of nominally anhydrous phases consisting of low-Al bridgmanite and H_2O (Shieh et al. 1998; Walter et al. 2015).

The more stable, Al-rich variety of Phase D with up to 50 wt% Al_2O_3 was synthesized in association with phase Egg in a simplified basaltic bulk composition at lower mantle conditions at 25 GPa and ~1500 °C, which is significantly higher than temperature stability for pure Phase D, allowing the Al-rich Phase D to remain stable and host H_2O in the subducting oceanic crust in the Earth's lower mantle (Boffa Ballaran et al. 2010). This idea was confirmed by experiments that demonstrated that Al-rich Phase D $Al_2SiO_6H_2$ in association with phase Egg and stishovite, has very high thermal stability extending to over 2000 °C at 26 GPa, i.e., higher than the expected geotherm of the normal lower mantle (Pamato et al. 2015). In the subsolidus, the Al-rich Phase D is stable to ~55 GPa (Walter et al. 2015). The Al-rich Phase D may have an admixture of Fe, which actually decreases its stability (Ghosh and Schmidt 2014). According to experimental data, Al/Fe ratios in Phase D and in coexisting bridgmanite are roughly equal (Pamato et al. 2015).

The Al-rich Phase D was suggested initially to have the same structure as Mg-Phase D (space group *P*-31*m*), but with a more disordered cation distribution (Boffa Ballaran et al. 2010). A more recent study by Pamato et al. (2015) attributes its structure to hexagonal with the space group $P6_3/mcm$, where Al replaces Mg and all six octahedral sites become partially occupied by a random and disordered distribution of Si and Al, such that M1 and M4 become equivalent (as do M2 and M3), resulting in an increase in symmetry. One of the main differences between the structures is that the Al–Si disorder in Al-rich Phase D results in essentially an undistorted octahedra of similar size, whereas in Mg-Phase D, large MgO distances cause the octahedra to be strongly distorted. The lesser extent of the octahedral distortion is likely to stabilize the Al-rich Phase D relative to its Mg-bearing counterpart. On the other hand, Bindi et al. (2015) reported no change in the unit-cell volume in Al-rich Phase D with respect to pure $MgSi_2O_6H_2$. Earlier, Boffa Ballaran et al. (2010) and Komatsu et al. (2011) observed that the Tscher-mak's $Si^{4+} + Mg^{2+} \rightarrow 2Al^{3+}$ substitution, in hydrous silicates, does not induce either an enlargement of the unit cell or a lengthening of the octahedral bond distances.

An important factor in the charge distribution in the Al-rich Phase D is the large degree of Al–Si disorder. Although the contribution to the configurational entropy due to Al–Si disorder has a stabilizing influence on low-pressure minerals, such as feldspar, the preference of Al for the octahedral site over the tetrahedral coordination at pressures of a few gigapascals means that this effect plays only a minor role in much of the upper mantle and transition zone, such disorder seems to become important again in the lower mantle, where Si and Al both exist exclusively in the octahedral coordination. The Al-rich Phase D may be the first member of a new class of completely disordered hydrous aluminosilicates that consequently have high thermal stabilities (Pamato et al. 2015) (Fig. 5.15).

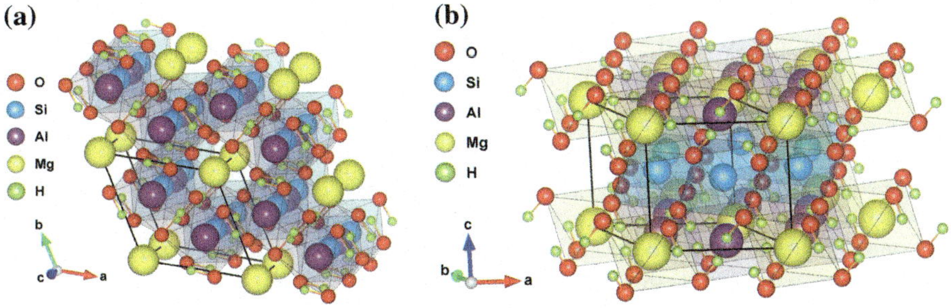

Fig. 5.15 Crystal structure of the Al-rich hexagonal phase D (space group *P*6_3*/mcm*) in different axis orientations. Lattice parameters from Pamato et al. (2015)

5.6.4 Phase H

Using theoretical calculations, Tsuchiya (2013) predicted that Phase D should undergo a phase transition to a new hydrous (with ~15% of water) Phase H with an ideal formula of $MgSiO_4H_2$, in association with SiO_2 ($MgSi_2O_6H_2 \rightarrow MgSiO_4H_2 + SiO_2$) at pressures exceeding ~40 GPa. Subsequently, multi-anvil apparatus experiments using sintered diamond anvils and in situ X-ray measurements showed that Phase D transforms to an assemblage containing Phase Hat pressures above 44–50 GPa (Nishi et al. 2014; Ohtani et al. 2014; Walter et al. 2015).

The crystal structure of Phase H was initially suggested as non-centrosymmetric orthorhombic with a space group *P*2_1*nm* similar to the one of the high-pressure hydrous phase δ-AlOOH (Suzuki et al. 2000). Later, Tsuchiya (2013) suggested that it should have a monoclinic symmetry with a space group *P2/m*. However, single-crystal XRD study on Phase D synthesized at 45 GPa unequivocally showed the orthorhombic symmetry with the $CaCl_2$-structure (space group *Pnnm*) (Bindi

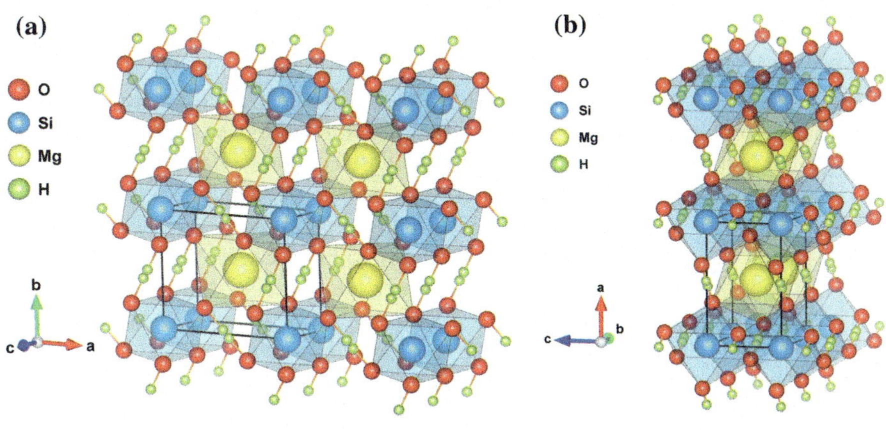

Fig. 5.16 Orthorhombic crystal structure model of phase H (space group *Pnnm*) in different axis orientations. Lattice parameters from Bindi et al. (2014)

et al. 2014). In this structure, Mg^{2+} and Si^{4+} are disordered over the octahedral sites. Distorted Mg and Si octahedra are charge compensated for by H^+ ions, which may occupy one of two positions depicted between the octahedra (Fig. 5.16). When cations with different valence states (i.e., Mg^{2+}, Al^{3+}, and Si^{4+}) become disordered at the octahedral sites, a fluctuation of the hydrogen position occurs to maintain the charge neutrality. Similar to δ-AlOOH, phase Egg and Phase D, the hydrogen bond symmetrization also occurs in Phase H at high pressure, and the hydrogen bond of this phase is already symmetric at the transition from Phase D at 40 GPa (Tsuchiya 2013). Like in Phase D, the hydrogen bond symmetrization probaby helps to stabilize hydrous materials under high-pressure conditions.

Tsuchiya (2013) predicted that pure Phase H had a narrower stability field near 52 GPa at static 0 K, above which it would dissociate into bridgmanite and H_2O (ice VIII). This coincides with experiments performed by Ohtani et al. (2014), in which the stability field of pure Phase H was limited at *P*–*T* conditions up to 60 GPa and ~1600 K, above which Phase H decomposes into bridgmanite and H_2O.

Considering that Mg- and Si-containing δ-AlOOH has the same as Phase H *Pnnm* structure (Komatsu et al. 2011), one may suggest that Phase H ($MgSiO_4H_2$) and δ-AlOOH ($AlAlO_4H_2$) could form a complete solid solution at high pressures (Bindi et al. 2014). Experiments have shown that intermediate Al-rich compositions can be synthesized together with perovskite and post-perovskite under slab geotherm conditions (Ohira et al. 2014). In association with Phase D, Phase H contains little Al, which strongly partitions into Phase D. At pressure above ~50 GPa, where Phase D is no longer stable, Phase H has the ability to incorporate a significant amount of Al, present in hydrous compositions in the deep lower mantle (Bindi et al. 2015). In this case, the effect of the Al component in Phase H is expected to extend its stability field, which is similar to that in Phase D. Phase H may be considered as an end-member of hydrous-phase solid solution AlOOH–$MgSiO_4H_2$. Experiments by Ohira et al. (2014) in a high-pressure DAC demonstrated that at 68 GPa and 2010 K, hydrous δ-phase contains at least 44 mol% of $MgSiO_4H_2$ component; at 128 GPa and 2190 K, it contains 23 mol% of Phase H component. In other series of the DAC experiments in system MgO–Al_2O_3–SiO_2–H_2O, Phase H was stable between ~45 and 115 GPa in all bulk compositions studied (Walter et al. 2015). Most recent experiments in the $MgSiO_4H_2$–AlOOH–SiO_2 system within the 20-60 GPa pressure range demonstrated that phase H and δ-AlOOH form an ideal mixture and that hydrogen in phase H is bonded in a single well at all pressures, and only the pure phase results in pressure-induced symmetrization, while the solid solution maintains asymmetric bonding to high pressure (Panero and Caracas 2017). Thus, Phase H and δ-AlOOH are likely to be stable hydrous phases in the slabs and transport significant amounts of water into the deep lower mantle.

The presence of hydrogen in the lower mantle may influence other phase transitions. It sharpens the width of the bridgmanite to post-perovskite transition. The unit cell volume of bridgmanite at ambient conditions increases systematically with increasing synthesis pressure above ~55 GPa, possibly representing an increase in Al concentration and potentially hydrogen concentration with depth (Inoue et al. 2010).

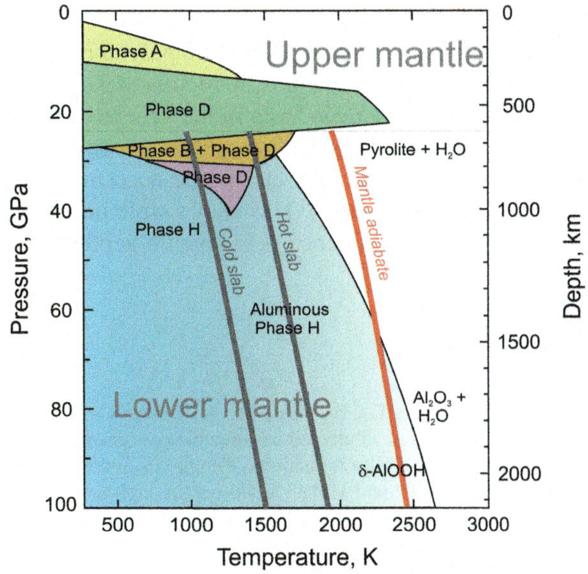

Fig. 5.17 Phase diagram for dense hydrous magnesium silicates in the lower mantle. Temperature profiles in slabs after Kirby et al. (1996). Mantle adiabate after Katsura et al. (2010). Modified after Nishi et al. (2014). Used with a permission of the Nature Geoscience

Bridgmanite in equilibrium with phases D and H has a relatively low Al concentration, and Al partitions preferentially into the hydrous phases (Walter et al. 2015).

The stability of all DHMS phases that are expected to be major inventories of water in subducting slabs are summarized in Fig. 5.17. Phase H in this composition is formed by the reaction of Phase D and periclase, which is assumed to occur at the same pressure as the reaction of Phase D and brucite, although this boundary could be shifted towards higher pressures, defined by the reaction from Phase D to Phase H plus stishovite (Nishi et al. 2014).

High-pressure aluminium-rich phases D and H are more likely to exist in the mafic association because the ultramafic association does not contain enough Al to make these phases stable in the lower-mantle conditions. Phases D and H can survive only in cold subducting slabs because, along an adiabatic geotherm, solid hydrous silicates are not stable. However, to date, none of the DHMS phases were identified in the natural environment.

5.7 Conclusions

The mafic association plays a less important role in the lower mantle compared with the ultramafic one. Firstly, it is much rarer because of the relatively small volume of subducted lithospheric slabs within the lower mantle. Secondly, the structural changes are not so noticeable compared with that in the ultramafic association. For example, a possible NAL to CF phase transition, accompanied by an increase in density, bulk modulus and share wave anisotropy, could be seismically detectable.

However, it is compensated for, in the bulk slab composition, by the faster sound velocity of SiO_2 and its high-pressure polymorphic phase(s). The transitions may be observed only locally within subducting slabs. This means that the mafic association does not play a major role in the general structure of the lower mantle. On the other hand, subducting slabs may cause local seismic anomalies and deliver elements such as Si, Al and water to the deep mantle, forming local water reservoirs within the lower mantle. This may result in local seismic and chemical heterogeneities that can influence deep-mantle geodynamic processes.

References

Andrault, D., Pesce, G., Bouhifd, M. A., Bolfan-Casanova, N., Hénot, J.-M., & Mezouar, M. (2014). Melting of subducted basalt at the core-mantle boundary. *Science, 344*(6186), 892–895.

Akaogi, M. (2007). Phase transitions of minerals in the transition zone and upper part of the lower mantle. In E. Ohtani (Ed.), *Advances in high-pressure mineralogy* (pp. 1–13). Geological Society of America Special Paper 421.

Akaogi, M., Hamada, Y., Suzuki, T., Kobayashi, M., & Okada, M. (1999). High pressure transitions in the system $MgAl_2O_4$–$CaAl_2O_4$: A new hexagonal aluminous phase with implication for the lower mantle. *Physics of the Earth and Planetary Interiors, 115,* 67–77.

Bindi, L., Nishi, M., & Irifune, T. (2015). Partition of Al between phase D and phase H at high pressure: Results from a simultaneous structure refinement of the two phases coexisting in a unique grain. *American Mineralogist, 100,* 1637–1640. doi:10.2138/am-2015-5327

Bindi, L., Nishi, M., Tsuchiya, J., & Irifune, T. (2014). Crystal chemistry of dense hydrous magnesium silicates: The structure of phase H, $MgSiH_2O_4$ synthesized at 45 GPa and 10001000 °C. *American Mineralogist, 99,* 1802–1805. doi:10.2138/am.2014.4994

Boffa Ballaran, T., Frost, D. J., Miyajima, N., & Heidelbach, F. (2010). The structure of a super-aluminous version of the dense hydrous-magnesium silicate phase D. *American Mineralogist, 95,* 1113–1116.

Bolfan-Casanova, N., Andrault, D., Amiguet, E., & Guignot, N. (2009). Equation of state and post-stishovite transformation of Al-bearing silica up to 100 GPa and 3000 K. *Physics of the Earth and Planetary Interiors, 174,* 70–77. doi:10.1016/j.pepi.2008.06.024

Chang, Y. Y., Jacobsen, S. D., Lin, J. F., Bina, C. R., Thomas, S.-M., Wu, J., et al. (2013). Spin transition of Fe^{3+} in Al-bearing dense hydrous magnesium silicate phase D: An alternative explanation for small-scale seismic scatterers in the mid-lower mantle. *Earth and Planetary Science Letters, 382,* 1–9.

Chen, M., Shu, J., Xie, X., & Mao, H.-K. (2003). Natural $CaTi_2O_4$-structured $FeCr_2O_4$ polymorph in the Suizhou meteorite and its significance in mantle mineralogy. *Geochimica et Cosmochimica Acta, 67,* 3937–3942.

Chung, J. I., & Kagi, H. (2002). High concentration of water in stishovite in the MORB system. *Geophysical Research Letters, 29*(21), 2020. doi:10.1029/2002GL015579

Churakov, S. V., & Wunder, B. (2004). Ab-initio calculations of the proton location in topaz-OH, $Al_2SiO_4(OH)_2$. *Physics and Chemistry of Minerals, 31,* 131–141.

Dai, L., Kudo, Y., Hirose, K., Murakami, M., Asahara, Y., Ozawa, H., et al. (2013). Sound velocities of $Na_{0.4}Mg_{0.6}Al_{1.6}Si_{0.4}O_4$ NAL and CF phases to 73 GPa determined by Brillouin scattering method. *Physics and Chemistry of Minerals, 40,* 195–201.

Davies, R. M., Griffin, W. L., O'Reilly, S. Y., & Doyle, B. J. (2004). Mineral inclusions and geochemical characteristics of microdiamonds from the DO27, A154, A21, A418, DO18, DD17 and Ranch Lake kimberlites at Lac de Gras, Slave Craton, Canada. *Lithos, 77*(1–4), 39–55.

Decker, B. F., & Kasper, J. S. (1957). The structure of calcium ferrite. *Acta Crystallographica Sect B Structure Science, 42,* 229–236.

Egglton, R. A., Boland, J. N., & Ringwood, A. E. (1978). High pressure synthesis of a new aluminum silicate: $Al_5Si_5O_{17}(OH)$. *Geochemical Journal, 12,* 191–194.

Finkelstein, G., Dera, P., Jahn, S., Oganov, A. R., Holl, C. M., Meng, Y., et al. (2014). Phase transitions and equation of state of forsterite to 90 GPa from single-crystal X-ray diffraction and molecular modeling. *American Mineralogist, 99*(1), 35–43. doi:10.2138/am.2014.4526

Frost, D. J. (1999). The stability of dense hydrous magnesium silicates in Earth's transition zone and lower mantle. In Y. Fei, C. M. Bertka, & B. O. Mysen (Eds.), *Mantle petrology: Field observations and high pressure experimentation: A tribute to Francis R. (Joe) Boyd* (pp. 283–296). The Geochemical Society Special Publication No. 6.

Frost, D. J., & Fei, Y. (1998). Stability of phase D at high pressure and high temperature. *Journal of Geophysical Research, 103B,* 7463–7474.

Frost, D. J., & Fei, Y. (1999). Static compression of the hydrous magnesium silicate phase D to 30 GPa at room temperature. *Physics and Chemistry of Minerals, 26,* 415–418.

Fukuyama, K., Ohtani, E., Shibazaki, Y., Kagi, H., & Suzuki, A. (2017). Stability field of phase Egg, $AlSi_3OH$ at high pressure and high temperature: possible water reservoir in mantle transition zone. *Journal of Mineralogical and Petrological Sciences, 112,* 31–35. doi:10.2645/jmps.160719e

Funamori, N., Jealoz, R., Miyajima, N., & Fujino, K. (2000). Mineral assemblages of basalt in the lower mantle. *Journal of Geophysical Research, 105*(B11), 26037–26043.

Ganskow, G., & Langenhorst, F. (2014). Stability and crystal chemistry of iron-bearing dense hydrous magnesium silicates. *Chemie der Erde, 74,* 489–496.

Gasparik, T., Tripathi, A., & Parise, J. B. (2000). Structure of a new Al-rich phase, [K, Na]0.9[Mg, Fe]2[Mg, Fe, Al, Si]6O12, synthesized at 24 GPa. *American Mineralogist, 85,* 613–618.

Gautron L., Fitz Gerald J. D., Kesson S. E., Eggleton R. A., & Irifune, T. (1997). Hexagonal Ba-ferrite: A good model for the crystal structure of a new high-pressure phase $CaAl_4Si_2O_{11}$? *Physics of the Earth and Planetary Interiors, 102,* 223–229.

Gautron, L., Kesson, S. E., & Hibberson, W. O. (1996). Phase relations for $CaAI_2Si_2O_8$ (anorthite composition) in the system $CaO–A1_2O_3–SiO_2$ at 14 GPa. *Physics of the Earth and Planetary Interiors, 97,* 71–81.

Ghosh, S., & Schmidt, M. W. (2014). Melting of phase D in the lower mantle and implications for recycling and storage of H_2O in the deep mantle. *Geochimica et Cosmochimica Acta, 145,* 72–88.

Gleason, A. E., Jeanloz, R., & Kunz, M. (2008). Pressure-temperature stability stud-ies of FeOOH using X-ray diffraction. *American Mineralogist, 93,* 1882–1885.

Gleason, A. E., Quiroga, C. E., Suzuki, A., Pentcheva, R., & Mao, W. L. (2013). Symmetrization driven spin transition in ε-FeOOH at high pressure. *Earth and Planetary Science Letters, 379,* 49–55. doi:10.1016/j.epsl.2013.08.012

Guignot, N., & Andrault, D. (2004). Equations of state of Na–K–Al host phases and implications for MORB density in the lower mantle. *Physics of the Earth and Planetary Interiors, 143–144,* 107–128.

Hayman, P. C., Kopylova, M. G., & Kaminsky, F. V. (2005). Lower mantle diamonds from Rio Soriso (Juina, Brazil). *Contributions to Mineralogy and Petrology, 149*(4), 430–445.

Hirose, K., & Fei, Y. (2002). Subsolidus and melting phase relations of basaltic composition in the uppermost lower mantle. *Geochimica et Cosmochimica Acta, 66,* 2099–2108.

Hirose, K., Fei, Y., Ma, Y., & Mao, H.-K. (1999). The fate of subducted basaltic crust in the Earth's lower mantle. *Nature, 397*(6714), 53–56.

Hirose, K., Takafuji, N., Sata, N., & Ohishi, Y. (2005). Phase transition and density of subducted MORB crust in the lower mantle. *Earth and Planetary Science Letters, 237,* 239–251.

Hushur, A., Manghnani, M. H., Smyth, J. R., Williams, Q., Hellebrand, E., Lonappan, D., et al. (2011). Hydrogen bond symmetrization and equation of state of phase D. *Journal Geophysical Research, 116,* B06203. doi:10.1029/2010JB008087

Hutchison, M. T. (1997). *Constitution of the deep transition zone and lower mantle shown by diamonds and their inclusions.* Unpublished PhD Thesis. University of Edinburgh, UK. Vol. 1, 340 pp., Vol. 2 (Tables and Appendices), 306 pp.

Imada, S., Hirose, K., & Ohishi, Y. (2011). Stabilities of NAL and Ca-ferrite-type phases on the join $NaAlSiO_4–MgAl_2O_4$ at high pressure. *Physics and Chemistry of Minerals, 38,* 557–560. doi:10.1007/s00269-011-0427-2

Inoue, T., Wada, T., Sasaki, R., & Yurimoto, H. (2010). Water partitioning in the Earth's mantle. *Physics of the Earth and Planetary Interiors, 183,* 245–251.

Irifune, T., Fujino, K., & Ohtani, E. (1991). A new high-pressure form of $MgAl_2O_4$. *Nature, 349* (6308), 409–411. doi:10.1038/349409a0

Irifune, T., & Ringwood, A. E. (1993). Phase transformations in subducted oceanic crust and buoyancy relationships at depths of 600–800 km in the mantle. *Earth and Planetary Science Letters, 117*(1–2), 101–110.

Irifune, T., Ringwood, A. E., & Hibberson, W. O. (1994). Subduction of continental crust and terrigenous and pelagic sediments: An experimental study. *Earth and Planetary Science Letters, 126,* 351–368.

Kaminsky, F. V., Wirth, R., & Schreiber, A. (2015). A microinclusion of lower-mantle rock and some other lower-mantle inclusions in diamond. *Canadian Mineralogist, 53*(1), 83–104. doi:10.3749/canmin.1400070

Kaminsky, F. V., Zakharchenko, O. D., Davies, R., Griffin, W. L., Khachatryan-Blinova, G. K., & Shiryaev, A. A. (2001). Superdeep diamonds from the Juina area, Mato Grosso State, Brazil. *Contributions to Mineralogy and Petrology, 140*(6), 734–753.

Kanzaki, M. (1989). High pressure phase relations in the system $MgO–SiO_2–H_2O$. *Eos, Transactions American Geophysical Union, 70*(15), 508.

Kanzaki, M. (1991). Stability of hydrous magnesium silicates in the mantle transition zone. *Physics of the Earth and Planetary Interiors, 66,* 307–312.

Kanzaki, M. (2010). Crystal structure of a new high-pressure polymorph of topaz-OH. *American Mineralogist, 95,* 1349–1352. doi:10.2138/am.2010.3555

Kato, C., Hirose, K., Komabayashi, T., Ozawa, H., & Ohishi, Y. (2013). NAL phase in K-rich portions of the lower mantle. *Geophysical Research Letters, 40,* 5085–5088. doi:10.1002/grl.50966

Katsura, T., Yoneda, A., Yamazaki, D., Yoshino, T., & Ito, E. (2010). Adiabatic temperature profile in the mantle. *Physics of the Earth and Planetary Interiors, 183,* 212–218. doi:10.1016/j.pepi.2010.07.001

Kawai, K., & Tsuchiya, T. (2012) Phase stability and elastic properties of the NAL and CF phases in the $NaMg_2Al_5SiO_{12}$ system from first principles. *American Mineralorist, 97,* 305–314. DPI: 10.2138/am.2012.3915

Kesson, S. E., Fitz Gerald, J. D., & Shelley, J. M. (1994). Mineral chemistry and density of subducted basaltic crust at lower mantle pressures. *Nature, 372,* 767–769.

Kirby, S. H., Stein, S., Okal, E. A., & Rubie, D. C. (1996). Metastable mantle phase transformations and deep earthquakes in subducting oceanic lithosphere. *Reviews of Geophysics, 34*(2), 261–306.

Kojitani, H., Hisatomi, R., & Akaogi, M. (2007). High-pressure phase relations and crystal chemistry of calcium ferrite-type solid solutions in the system $MgAl_2O_4–Mg_2SiO_4$. *American Mineralogist, 92,* 1112–1118.

Kojitani, H., Iwabuchi, T., Kobayashi, M., Miura, H., & Akaogi, M. (2011). Structure refinement of high-pressure hexagonal aluminous phases $K_{1.00}Mg_{2.00}Al_{4.80}Si_{1.15}O_{12}$ and $Na_{1.04}Mg_{1.88}Al_{4.64}Si_{1.32}O_{12}$. *American Mineralogist, 96*(9), 1248–1253. doi:10.2138/Am.2011.3638

Komatsu, K., Kuribayashi, T., Sano, A., Ohtani, E., & KudohY. (2006). Redetermination of the high-pressure modification of AlOOH from single-crystal synchrotron data. *Acta Crystallograph Sect E 62*(11), i216–i218, doi:10.1107/S160053680603916X

Komatsu, K., Sano-Furukawa, A., & Kagi, H. (2011). Effects of Mg and Si ions on the symmetry of δ -AlOOH. *Physics and Chemistry of Minerals, 38*(9), 727–733.

Kudoh, Y., Finger, L. W., Hazen, R. M., Prewitt, C. T., Kanzaki, M., & Veblen, D. R. (1993). Phase E: A high pressure hydrous silicate with unique crystal chemistry. *Physics and Chemistry of Minerals, 19,* 357–360.

Kudoh, Y., Kuribayashi, T., Suzuki, A., Ohtani, E., & Kamada, T. (2004). Space group and hydrogen sites of δ-AlOOH and implications for a hypothetical high-pressure form of Mg $(OH)_2$. *Physics and Chemistry of Minerals, 31,* 360–364.

Kudoh, Y., Nagase, T., Mizohata, H., Ohtani, E., Sasaki, S., M. Tanaka, M. (1997) Structure and crystal chemistry of phase G, a new hydrous magnesium silicate synthesized at 22 GPa and 1050 °C. *Geophysical Research Letters, 24*, 1051–1054.

Kudoh, Y., Nagase, T., Sasaki, S., Tanaka, M., & Kanzaki, M. (1995). Phase F, a new hydrous magnesium silicate synthesized at 1000 °C and 17 GPa: Crystal structure and estimated bulk modulus. *Physics and Chemistry of Minerals, 22,* 295–299.

Kurashina, T., Hirose, K., Ono, S., Sata, N., & Ohishi, Y. (2004). Phase transition in Al-bearing $CaSiO_3$ perovskite: Implications for seismic discontinuities in the lower mantle. *Physics of the Earth and Planetary Interiors, 145,* 67–74.

Kuribayashi, T., Sano-Furukawa, A., & Nagase, T. (2013). Observation of pressure-induced phase transition of δ-AlOOH by using single-crystal synchrotron X-ray diffraction method. *Physics and Chemistry Minerals, 41*(4), 303–312. doi:10.1007/s00269-013-0649-6

Lakshtanov, D. L., Sinogeikin, S. V., Konstantin D. Litasov, K. D., Vitali B. Prakapenka, V. B., Hellwig, H., et al. (2007). The post-stishovite phase transition in hydrous Al-bearing SiO_2 in the lower mantle of the Earth. *Proceedings of the National Academy of Sciences of the USA 104*, 13588–13590.

Li, S., Ahuja, R., & Johansson, B. (2006). The elastic and optical properties of the high-pressure hydrous phase δ-AlOOH. *Solid State Communications, 137*(1–2), 101–106. doi:10.1016/j.ssc.2005.08.031

Lin, J.-F., Speciale, S., Mao, Z., & Marquardt, H. (2013). Effects of the electronic spin transitions of iron in lower mantle minerals: implications for deep mantle geophysics and geochemistry. *Reviews of Geophysics, 51*(2), 244–275.

Litasov, K. D., Kagi, H., Shatskiy, A., Ohtani, E., Lakshtanov, D. L., Bass, J. D., et al. (2007a). High hydrogen solubility in Al-rich stishovite and water transport in the lower mantle. *Earth and Planetary Science Letters, 262*(2007), 620–634. doi:10.1016/j.epsl.2007.08.015

Litasov, K. D., & Ohtani, E. (2005). Phase relations in hydrous MORB at 18–28 GPa: Implications for heterogeneity of the lower mantle. *Physics of the Earth and Planetary Interiors, 150,* 239–263. doi:10.1016/j.pepi.2004.10.010

Litasov, K. D., & Ohtani, E. (2007). Effect of water on the phase relations in Earth's mantle and deep water cycle. In E. Ohtani (Ed.), *Advances in High-pressure Mineralogy* (pp. 115–156). Geological Society of America.

Litasov, K. D., Ohtani, E., Nishihara, Y., Suzuki, A., & Funakoshi, K. (2008). Thermal equation of state of Al- and Fe-bearing phase D. *Journal Geophysical Research, 113,* B08205. doi:10.1029/2007JB004937

Litasov, K. D., Ohtani, E., Suzuki, A., & Funakoshi, K. (2007b). The compressibility of Fe- and Al-bearing phase D to 30 GPa. *Physics and Chemistry of Minerals, 34,* 159–167.

Litasov, K., Ohtani, E., Suzuki, A., & Kawazoe, T. (2004). Absence of density crossover between basalt and peridotite in the cold slabs passing through 660 km discontinuity. *Geophysical Research Letters, 31,* L24607. doi:10.1029/2004GL021306

Liu, L. (1977). High pressure $NaAlSiO_4$: The first silicate calcium ferrite isotype. *Geophysical Reseach Letters, 4*(5), 183–186. doi:10.1029/GL004i005p00183

Liu, L.-G. (1986). Phase transformations in serpentine at high pressures and temperatures and implications for subducting lithosphere. *Physics of the Earth and Planetary Interiors, 42,* 255–262.

Liu, L.-G. (1987). Effects of H_2O on the phase behavior of the forsterite-enstatite system at high pressures and temperatures and implications for the Earth. *Physics of the Earth and Planetary Interiors, 49,* 142–167.

Mainprice, D., Le Page, Y., Rodgers, J., & Jouanna, P. (2007). Predicted elastic properties of the hydrous D phase at mantle pressures: Implications for the anisotropy of subducted slabs near 670-km discontinuity and in the lower mantle. *Earth Planetary Science Letters, 259*(3–4), 283–296. doi:10.1016/j.epsl.2007.04.053

Mashino, I., Murakami, M., & Ohtani, E. (2016). Sound velocities of δ-AlOOH up to core-mantle boundary pressures with implications for the seismic anomalies in the deep mantle. *Journal of Geophysical Research: Solid Earth, 121,* 595–609. doi:10.1002/2015JB012477

Miura, H., Hamada, Y., Suzuki, T., Akaogi, M., Miyajima, N., & Fujino, K. (2000). Crystal structure of $CaMg_2Al_6O_{12}$, a new Al-rich high pressure form. *American Mineralogist, 85,* 1799–1803.

Miyajima, N., Fujino, K., Funamori, N., Kondo, T., & Yagi, T. (1999). Garnet–perovskite transformation under conditions of the Earth's lower mantle: An analytical transmission electron microscopy study. *Physics of the Earth and Planetary Interiors, 116,* 117–131.

Miyajima, N., Yagi, T., Hirose, K., Kondo, T., Fujino, K., & Miura, H. (2001). Potential host phase of aluminum and potassium in the Earth's lower mantle. *American Mineralogist, 86,* 740–746.

Mookherjee, M., Karki, B. B., Stixrude, L., Lithgow-Bertelloni, C. (2012). Energetics, equation of state, and elasticity of NAL phase: Potential host for alkali and aluminum in the lower mantle. *Geophysical Research Letters 39*, LI9306.
DOI 10.1029/2012GL053682

Mookherjee, M., Tsuchiya, J., & Hariharan, A. (2016). Crystal structure, equation of state, and elasticity of hydrous aluminosilicate phase, topaz-OH ($Al_2SiO_4(OH)_2$) at high pressures. *Physics of the Earth and Planetary Interiors, 251,* 24–35. doi:10.1016/j.pepi.2015.11.006

Nishi, M., Irifune, T., Tsuchiya, J., Tange, Y., Nishihara, Y., Fujino, K., et al. (2014). Stability of hydrous silicate at high pressures and water transport to the deep lower mantle. *Nature Geoscience, 7*(3), 224–227.

Nishihara, Y., & Matsukage, K. N. (2016). Iron-titanium oxyhydroxides as water carriers in the Earth's deep mantle. *American Mineralogist, 101,* 919–927.

Northrup, P. A., Leinenweber, K., & Parise, J. B. (1994). The location of H in the high-pressure synthetic $Al_2SiO_4(OH)_2$ topaz analogue. *American Mineralogist, 79,* 401–404.

Ohira, I., Ohtania, E., Sakaia, T., Miyahara, M., Hirao, N., Ohishi, Y., et al. (2014). Stability of a hydrous δ-phase, AlOOH–$MgSiO_2(OH)_2$, and a mechanism for water transport into the base of lower mantle. *Earth and Planetary Science Letters, 401,* 12–17. doi:10.1016/j.epsl.2014.05.059

Ohtani, E., Amaike, Y., Kamada, S., Sakamaki, T., & Hirao, N. (2014). Stability of hydrous phase H $MgSiO_4H_2$ under lower mantle conditions. *Geophysical Research Letters, 41*(23), 8283–8287. doi:10.1002/2014GL061690

Ohtani, E., Litasov, K., Hosoya, T., Kubo, T., & Kondo, T. (2004). Water transport into the deep mantle and formation of a hydrous transition zone. *Physics of the Earth and Planetary Interiors, 143,* 255–269.

Ohtani, E., Litasov, K., Suzuki, A., & Kondo, T. (2001a). Stability field of new hydrous phase, δ - AlOOH, with implications for water transport into the deep mantle. *Geophysical Research Letters, 28,* 3991–3993.

Ohtani, E., Mitzobata, H., Kudoh, Y., & Nagase, T. (1997). A new hydrous silicate, a water reservoir, in the upper part of the lower mantle. *Geophysical Research Letters, 24,* 1047–1050.

Ohtani, E., Toma, M., Litasov, K., Kubo, T., & Suzuki, A. (2001b). Stability of dense hydrous magnesium silicate phases and water storage capacity in the transition zone and lower mantle. *Physics of the Earth and Planetary Interiors, 124,* 105–117.

Ono, S. (1999). High temperature stability of phase "Egg", $AlSiO_3(OH)$. *Contributions to Mineralogy and Petrology, 137,* 83–89.

Ono, A., Akaogi, M., Kojitani, H., Yamashita, K., & Kobayashi, M. (2009). High-pressure phase relations and thermodynamic properties of hexagonal aluminous phase and calcium–ferrite phase in the systems $NaAlSiO_4$–$MgAl_2O_4$ and $CaAl_2O_4$–$MgAl_2O_4$. *Physics of the Earth and Planetary Interiors, 174,* 39–49.

Ono, S., Ito, E., & Katsura, T. (2001). Mineralogy of subducted basaltic crust (MORB) from 25 to 37 GPa, and chemical heterogeneity of the lower mantle. *Earth and Planetary Science Letters, 190,* 57–63.

Otte, K., Pentcheva, R., Schmahl, W. W., & Rustad, J. R. (2009). Pressure-induced structural and electronic transitions in FeOOH from first principles. *Physical Review B, 80,* 205116. doi:10.1103/PhysRevB.80.205116

Pamato, M. G., Kurnosov, A., Boffa Ballaran, T., Trots, D.M., Caracas, R., & Frost, D. J. (2014). Hexagonal $Na_{0.41}[Na_{0.125}Mg_{0.79}Al_{0.085}]_2[Al_{0.79}Si_{0.21}]_6O_{12}$ (NAL phase): Crystal structure refinement and elasticity. *American Mineralogist 99*(8–9), 1562–1569.

Pamato, M. G., Myhill, R., Ballaran, T. B., Frost, D. J., Heidelbach, F., & Miyajima, N. (2015). Lower-mantle water reservoir implied by the extreme stability of a hydrous aluminosilicate. *Nature Geoscience, 8*(1), 75–79.

Panero, W.R., & Caracas, R. (2017). Stability of phase H in the $MgSiO_4H_2$–AlOOH–SiO_2 system. *Earth and Planetary Science Letters, 463*, 171–177. doi:10.1016/j.epsl.2017.01.033

Panero, W. R., & Stixrude, L. P. (2004). Hydrogen incorporation in stishovite at high pressure and symmetric bonding in δ-AlOOH. *Earth Planetary Science Letters, 221,* 421–431.

Prewitt, C. T., & Parise, J. B. (2000). Hydrous phases and hydrogen bonding at high pressure. In R. M. Hazen & R. T. Downs (Eds.), *High-temperature and high-pressure crystal chemistry* (Vol. 41, pp. 309–333). Virginia: Reviews in Mineralogy and Geochemistry, Mineralogical Society of America, Chantilly.

Reagan, M. M., Gleason, A. E., Daemen, L., Xiao, Y., & Mao, W. L. (2016). High-pressure behavior of the polymorphs of FeOOH. *American Mineralogist, 101,* 1483–1488. doi:10.2138/am-2016-5449

Reid, A. F., & Ringwood, A. E. (1969). Newly observed high pressure transformations in Mn_3O_4, $CaAl_2O_4$, and $ZrSiO_4$. *Earth and Planetary Science Letters, 6,* 205–208. doi:10.1016/0012-821X(69)90091-0

Ricolleau, A., Perrillat, J.-P., Fiquet, G., Daniel, I., Matas, J., Addad, A., et al. (2010). Phase relations and equation of state of a natural MORB: Implications for the density profile of subducted oceanic crust in the Earth's lower mantle. *Journal Geophysical Research, 115,* B08202. doi:10.1029/2009JB006709

Ringwood, A. E., & Major, A. (1967). High-pressure reconnaissance investigations in the system Mg_2SiO_4–MgO–H_2O. *Earth and Planetary Science Letters, 2,* 130–133.

Rosa, A. D., Mezouar, M., Garbarino, G., Bouvier, P., Ghosh, S., Rohrbach, A., et al. (2013a). Single-crystal equation of state of phase D to lower mantle pressures and the effect of hydration on the buoyancy of deep subducted slabs. *Journal of Geophysical Research: Solid Earth, 118* (12), 6124–6133. doi:10.1002/2013JB010060

Rosa, A. D., Sanchez-Valle, C., & Ghosh, S. (2012). Elasticity of phase D and implication for the degree of hydration of deep subducted slabs. *Geophysical Reseach Letters, 39,* L06304. doi:10.1029/2012GL050927

Rosa, A. D., Sanchez-Valle, C., Nisr, C., Evans, S. R., Debord, R., & Merkel, S. (2013b). Shear wave anisotropy in textured phase D and constraints on deep water recycling in subduction zones. *Earth Planetary Science Letters, 377–378,* 13–22.

Sano, A., Ohtani, E., Kondo, T., Hirao, N., Sakai, T., Sata, N., et al. (2008). Aluminous hydrous mineral δ-AlOOH as a carrier of hydrogen into the core-mantle boundary. *Geophysical Reseach Letters, 35,* L03303.

Sano, A., Ohtani, E., Kubo, T., & Finakoshi, K. (2004). In situ X-ray observation of decomposition of hydrous aluminum silicate $AlSiO_3OH$ and aluminum oxide hydroxide δ-AlOOH at high pressure and temperature. *Journal of Physics and Chemistry of Solids, 65,* 1547–1554.

Sano-Furukawa, A., Komatsu, K., Vanpeteghem, C. B., & Ohtani, E. (2008). Neutron diffraction study of δ-AlOOD at high pressure and its implication for symmetrization of the hydrogen bond. *American Mineralogist, 93*(10), 1558–1567. doi:10.2138/am.2008.2849

Schmidt, M. W. (1995). Lawsonite: Upper pressure stability and formation of higher density hydrous phases. *American Mineralogist, 80,* 1286–1292.

Schmidt, M. W., Finger, L. W., Ross, R. J., & Dinnebier, R. E. (1998). Synthesis, crystal structure, and phase relations of $AlSiO_3OH$, a high-pressure hydrous phase. *American Mineralogist, 83* (7–8), 881–888.

Shieh, S. R., Duffy, T. S., Liu, Z., & Ohtani, E. (2009). Highpressure infrared spectroscopy of the dense hydrous magnesium silicates phase D and phase E. *Physics of the Earth and Planetary Interiors, 175*(3–4), 106–114. doi:10.1016/j.pepi.2009.02.002

Shieh, S. R., Mao, H. K., Hemley, R. J., & Ming, L. C. (1998). Decomposition of phase D in the lower mantle and the fate of dense hydrous silicates in subducting slabs. *Earth and Planetary Science Letters, 159,* 13–23.

Shieh, S. R., Mao, H.-K., Hemley, R. J., & Ming, L. C. (2000). In situ X-ray diffraction studies of dense hydrous magnesium silicates at mantle conditions. *Earth and Planetary Science Letters, 177,* 69–80.

Shinmei, T., Irifune, T., Tsuchiya, J., & Funakoshi, K.-I. (2008). Phase transition and compression behavior of phase D up to 46 GPa using multi-anvil apparatus with sintered diamond anvils. *High Pressure Research, 28*(3), 363–373. doi:10.1080/08957950802246514

Stachel, T., Harris, J. W., Brey, G. P., & Joswig, W. (2000). Kankan diamonds (Guinea) II: lower mantle inclusion parageneses. *Contributions to Mineralogy and Petrology, 140*(1), 16–27.

Suzuki, A., Ohtani, E., & Kamada, T. (2000). A new hydrous phase δ-AlOOH synthesized at 21 GPa and 1000 °C. *Physics and Chemistry Mineral, 27,* 689–693.

Tappert, R., Foden, J., Stachel, T., Muehlenbachs, K., Tappert, M., & Wills, K. (2009). The diamonds of South Australia. *Lithos, 112S,* 806–821.

Thomson, A. R., Kohn, S. C., Bulanova, G. P., Smith, C. B., Araujo, D., EIMF, et al. (2014). Origin of sub lithospheric diamonds from the Juina 5 kimberlite (Brazil): Constraints from carbon isotopes and inclusion Compositions. Contributions to Mineralogy and Petrology 168, 1081, 29 pp.

Tschauner, O., Ma, Ch., Beckett, J. R., Prescher, C., Prakapenka, V. B., & Rossman, G. R. (2014). Discovery of bridgmanite, the most abundant mineral in Earth, in a shocked meteorite. *Science, 346*(6213), 1100–1102. doi:10.1126/science.1259369

Tsuchiya, J. (2013). First principles prediction of a new high-pressure phase of dense hydrous magnesium silicates in the lower mantle. *Geophysical Research Letters, 40,* 4570–4573. doi:10.1002/grl.50875

Tsuchiya, J., & Tsuchiya, T. (2011). First-principles prediction of a high-pressure hydrous phase of AlOOH. *Physics Review B, 83,* 054115. doi:10.1103/PhysRevB.83.054115

Tsuchiya, J., Tsuchiya, T., & Tsuneyuki, S. (2005). First-principles study of hydrogen bond symmetrization of phase D under high pressure. *American Mineralogist, 90,* 44–49.

Tsuchiya, J., Tsuchiya, T., Tsuneyuki, S., & Yamanaka, T. (2002). First principles calculation of a high-pressure hydrous phase, δ-AIOOH. *Geophysical Research Letters, 29*(19), 1909. doi:10.1029/2002GL015417

Umemoto, K., Kawamura, K., Hirose, K., & Wentzcovitch, R. M. (2016). Post-stishovite transition in hydrous aluminous SiO_2. *Physics of the Earth and Planetary Interiors, 255,* 18–26. doi:10.1016/j.pepi.2016.03.008

Vanpeteghem, C., Ohtani, E., Kondo, T., Takemura, K., & Kikegawa, T. (2003). Compressibility of phase Egg $AlSiO_3OH$: Equation of state and role of water at high pressure. *American Mineralogist, 88*(10), 1408–1411.

Vanpeteghem, C. B., Sano, A., Komatsu, K., Ohtani, E., & Suzuki, A. (2007). Neutron diffraction study of aluminous hydroxide δ-AlOOD. *Physics Chemistry Minerals, 34*(9), 657–661. doi:10.1007/s00269-007-0180-8

Walter, M. J., Kohn, S. C., Araujo, D., Bulanova, G. P., Smith, C. B., Gaillou, E., et al. (2011). Deep mantle cycling of oceanic crust: Evidence from diamonds and their mineral inclusions. *Science, 334,* 54–57.

Walter, M. J., Thomson, A. R., Wang, W., Lord, O. T., Ross, J., McMahon, S. C., et al. (2015). The stability of hydrous silicates in Earth's lower mantle: Experimental constraints from the systems $MgO–SiO_2–H_2O$ and $MgO–Al_2O_3–SiO_2–H_2O$. *Chemical Geology, 418,* 16–29.

Wicks, J. K., & Duffy, T. S. (2016). Crystal structures of minerals in the lower mantle. In H. Terasaki & R. A. Fischer (Eds.), *Deep earth: Physics and chemistry of the lower mantle and core. Geophysical Monograph* (Vol. 217, pp. 69–87).

Wirth, R., Vollmer, C., Brenker, F., Matsyuk, S., & Kaminsky, F. (2007). Nanocrystalline hydrous aluminum silicate in superdeep diamonds from Juina (Mato Grosso State, Brazil). *Earth and Planetary Science Letters, 259*(3–4), 384–399.

Wu, X., Wu, Y., Lin, J.-F., Liu, J., Mao, Z., Guo, X., et al. (2016a). Two-stage spin transition of iron in FeAl-bearing phase D at lower mantle. *Journal of Geophysical Research Solid Earth, 121*(9), 6411–6420. doi:10.1002/2016JB013209

Wu, Y., Wu, X., Lin, J.-F., McCammon, C. A., Xiao, Y., Chow, P., et al. (2016b). Spin transition of ferric iron in the NAL phase: Implications for the seismic heterogeneities of subducted slabs in the lower mantle. *Earth and Planetary Science Letters, 434,* 91–100. doi:10.1016/j.epsl.2015.11.011

Wunder, B., Rubie, D. C., Ross, C. R., II, Medenbach, O., Seifert, F., & Schreyer, W. (1993). Synthesis, stability and properties and of $Al_2SiO_4(OH)_2$: A fully hydrated analogue of topaz. *American Mineralogist, 78,* 285–297.

Xue, X., Kanzaki, M., Fukui, H., Ito, E., & Hashimoto, T. (2006). Cation order and hydrogen bonding of high-pressure phases in the Al_2O_3–SiO_2–H_2O system: An NMR and Raman study. *American Mineralogist, 91,* 850–861. doi:10.2138/am.2006.2064

Xue, X., Kanzaki, M., & Shatskiy, A. (2008). Dense hydrous magnesium silicates, phase D, and superhydrous B: New structural constraints from one- and two-dimensional 29Si and 1H NMR. *American Mineralogist, 93,* 1099–1111.

Yamada, H., Matsui, Y., & Eiji, I. (1983). Crystal–chemical characterization of $NaAlSiO_4$ with the $CaFe_2O_4$ structure. *Mineralogical Magazine, 47,* 177–181.

Yamamoto, K., & Akimoto, S. (1977). The system MgO–SiO_2–H_2O at high pressures and temperatures-stability field for hydroxyl-chondrodite, hydroxyl-clinohumite and 10 -phase. *American Journal of Science, 277,* 288–312.

Yamanaka, T., Uchida, A., & Nakamoto, Y. (2008). Structural transition of post-spinel phases $CaMn_2O_4$, $CaFe_2O_4$, and $CaTi_2O_4$ under high pressures up to 80 GPa. *American Mineralogist, 93*, 1874–1881. doi 10.2138/am.2008.2934

Yang, H., Prewitt, C. T., & Frost, D. J. (1997). Crystal structure of the dense hydrous magnesium silicate, phase D. *American Mineralogist, 82,* 651–654.

Zedgenizov, D. A., Kagi, H., Shatsky, V. S., & Ragozin, A. L. (2014). Local variations of carbon isotope composition in diamonds from Sao-Luis (Brazil): Evidence for heterogenous carbon reservoir in sublithospheric mantle. *Chemical Geology, 240*(1–2), 114–124.

Zedgenizov, D. A., Shatsky, V. S., Panin, A. V., Evtushenko, O. V., Ragozin, A. L., & Kagi, H. (2015). Evidence for phase transitions in mineral inclusions in superdeep diamonds of the Sao Luiz deposit, Brazil. *Russian Geology and Geophysics, 56*(1), 296–305.

Chapter 6
Carbonatitic Lower-Mantle Mineral Association

Abstract In addition to ultramafic and mafic associations, a primary natrocarbonatitic association occurs in the lower mantle. To date, it was identified as inclusions in diamonds from the Juina area, Mato Grosso State, Brazil. It comprises almost 50 mineral species: carbonates, halides, fluorides, phosphates, sulphates, oxides, silicates, sulphides and native elements. In addition, volatiles are also present in this association. Among oxides, coexisting periclase and wüstite were identified, pointing to the formation of the natrocarbonatitic association at a depth greater than 2000 km. Some iron-rich (Mg,Fe)O inclusions in diamond are attributed to the lowermost mantle. The initial lower-mantle carbonatitic melt formed as a result of low-fraction partial melting of carbon-containing lower-mantle material, rich in P, F, Cl and other volatile elements at the core–mantle boundary. During ascent to the surface, the initial carbonatitic melt dissociated into two immiscible parts, a carbonate-silicate and a chloride-carbonate melt. The latter melt is parental to the natrocarbonatitic lower-mantle association. Diamonds with carbonatitic inclusions were formed in carbonatitic melts or high-density fluids.

6.1 General

In addition to oxides and other minerals comprising ultramafic and mafic associations in the lower mantle, the third mineral association was unexpectedly identified in deep Earth as being carbonatitic (Kaminsky 2012). Earlier singular grains of calcite and magnesite have been reported from both upper-mantle (Wang et al. 1996; Sobolev et al. 1997) and lower-mantle (Brenker et al. 2007) associations. Multiple finds of saline or hydrous-saline fluid inclusions in fibrous diamond (Smith 2014 and references therein) further implied the presence of carbonatitic media during the formation of diamond in the mantle. In recent years, a series of carbonatitic minerals were identified as inclusions in lower-mantle diamond from Juina in Mato Grosso State, Brazil (Wirth et al. 2009; Kaminsky et al. 2009, 2013, 2015a, 2016a), confirming the presence of carbonatitic mineral associations in deep Earth.

F.V. Kaminsky, *The Earth's Lower Mantle*, Springer Geology,
DOI 10.1007/978-3-319-55684-0_6

On the other hand, experiments under high pressure indicate a significant role of carbonatitic media in the formation of diamond. For example, 'super-deep' diamonds have been experimentally crystallized in melts of the lower mantle, diamond parental carbonate–magnesiowüstite–bridgmanite–carbon system at pressures of up to 60 GPa (Litvin et al. 2014). The following chapter is based on our recent publication (Kaminsky et al. 2016a).

6.2 Lower-Mantle Carbonatitic Association in Diamonds from the Juina Area, Brazil

6.2.1 General Characteristics

To date, almost 50 mineral species of the lower mantle carbonatitic association have been identified as inclusions in diamonds from the Juina, Mato Grosso State, Brazil (Fig. 6.1), which preserved them from various dissociation reactions that probably would have occurred during ascent to the surface. Some of the carbonatitic

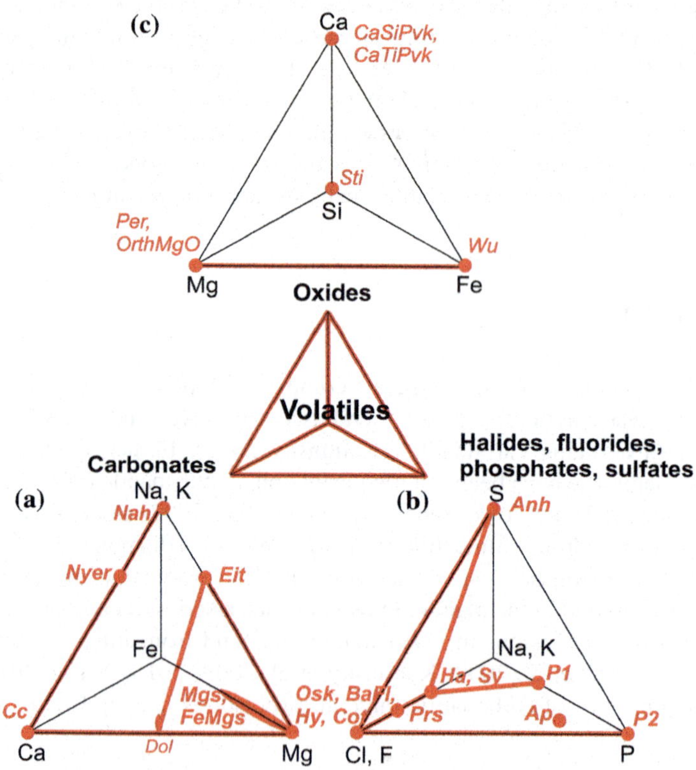

Fig. 6.1 Paragenetic associations of minerals in the primary carbonatitic association. **a** Carbonates. **b** Halides, fluorides, phosphates and sulphates. **c** Oxides. Tie *lines* are shown in *red*. Mineral indices, see Table 6.1. Modified after Kaminsky et al. (2016a). Used with a permission of Springer

Table 6.1 List of minerals of the primary carbonatitic association in the deep Earth (Kaminsky et al. 2016a)

Mineral	Index	Formula	Association	References
Carbonates				
Calcite	Cc	$CaCO_3$	CaSiPvk + CaTiPvk + Ol	Brenker et al. (2007)
			Mont + Wo II + Cusp	Wirth et al. (2009)
			Per + Wu	Kaminsky et al. (2009)
			Sp + Ap	Kaminsky et al. (2009)
			Sp + Phl(?)	
			Nyer + Wo II + CaGa	
			Nah + Phl	
			Dol + Ap + Sp + Sulf + Per + Wu + Phl(?)	Kaminsky et al. (2015a)
			CaSiPvk + $CaTiO_3$ + $CaMgSi_2O_6$	Zedgenizov et al. (2016)
Dolomite	Dol	$CaMg(CO_3)_2$	Ilm	Wirth et al. (2009)
			Sp + Phl	
			Cc + Ap + Sp + Sulf + Per + Wu + Phl(?)	Kaminsky et al. (2015a)
			MMgs + Eit + Ha + Sy + Phl + Pent + Vi + Sp + P1	Kaminsky et al. (2013)
Magnesite	Mgs	$Mg(CO_3)$	Dol + Eit + Ha + Sy + Phl + Pent + Vi + Sp + P1	Kaminsky et al. (2013)
			FeMgs + Fe^0	
Fe-magnesite	FeMgs	$(Mg,Fe)(CO_3)$	Mgs + Fe^0	Kaminsky et al. (2013)
Eitelite	Eit	$Na_2Mg(CO_3)_2$	Dol + Mgs + Ha + Sy + Phl + Pent + Vi + Sp + P1	Kaminsky et al. (2013)
Nyerereite	Nyer	$(Na,K)_2Ca(CO_3)_2$	Cc + Wo II + CaGa + Ap + Wu(?)	Kaminsky et al. (2009)
			Ol + Sp + Ap + Phl(?)	
Nahcolite	Nah	$NaHCO_3$	Cc + Phl	Kaminsky et al. (2009)
Halides				
Halite	Ha	NaCl	Coes + Sy + Hy + Cot + Plat + TiO_2 + Anh	Wirth et al. (2009)
			Dol + Mgs + Eit + Sy + Phl + Pent + Sp + P1	Kaminsky et al. (2015a)
			No	Kaminsky et al. (2015a)

(continued)

Table 6.1 (continued)

Mineral	Index	Formula	Association	References
Sylvite	Sy	KCl	Coes + Ha + Hy + Cot + Plat + TiO_2 + Anh	Wirth et al. (2009)
			Dol + Mgs + Eit + Ha + Phl + Pent + Vi + Sp + P1	Kaminsky et al. (2013)
Hydrophilite	Hy	$CaCl_2$	Coes + Ha + Sy + Cot + Plat + TiO_2 + Anh	Wirth et al. (2009)
Cotunnite	Cot	$PbCl_2$	Coes + Ha + Sy + Hy + Plat + TiO_2 + Anh	Wirth et al. (2009)
Fluorides				
Oskarssonite	Osk	AlF_3	P2 + BaFl	Kaminsky et al. (2013)
			SiO_2 + Fe–O + Mil	
			Hem + SiO_2	
Parascandolaite	Prs	$KMgF_3$	Hc + OrthMgO	Kaminsky et al. (2015a)
New (no name)	BaFl	$(Ba,Sr)AlF_5$	P2 + Osk	Kaminsky et al. (2013)
Sulfates				
Anhydrite	Anh	$CaSO_4$	Coes + Ha + Sy + Hy + Cot + Plat + TiO_2	Wirth et al. (2009)
Phosphates				
Apatite	Ap	$Ca_5(PO_4]_3(F,Cl)$	Nyer + Ol + Sp + Phl(?)	Kaminsky et al. (2009)
			Nyer + Cc + Wo II + Wu(?)	
			Cc + Sp + Wu	
			Cc + Dol + Sp + Sulf + Per + Wu + Phl(?)	Kaminsky et al. (2015a)
New (no name)	P1	$Na_4Mg_3(PO_4)_2(P_2O_7)$	Dol + Mgs + Eit + Ha + Sy + Phl + Pent + Vi + Sp	Kaminsky et al. (2013)
New (no name)	P2	$Fe_2Fe_5(P_2O_7)_4$	Osk + BaFl	Kaminsky et al. (2013)
Oxides				
CaSi-perovskite	CaSiPvk	$CaSiO_3$	Cc + Ol + CaTiPvk	Brenker et al. (2007)
CaTi-perovskite	CaTiPvk	$CaTiO_3$	Cc + Ol + CaSiPvk	Brenker et al. (2007)
Periclase	Per	MgO	Cc + Wu	Kaminsky et al. (2009)
			Dol + Ca + Ap + Sp + Sulf + Wu + Phl(?)	Kaminsky et al. (2015a)

(continued)

Table 6.1 (continued)

Mineral	Index	Formula	Association	References
Orrthorhombic MgO	OrthMgO	MgO	Parasc + Hc	Kaminsky et al. (2015a)
Wüstite	Wu	FeO	Cc + Per	Kaminsky et al. (2009)
			Cc + Sp + Ap	
			Nyer + Cc + Ap + Wo II	
			Dol + Ca + Ap + Sp + Sulf + Per + Phl(?)	Kaminsky et al. (2015a)
Coesite	Coes	SiO_2	Ha + Sy + Hy + Cot + Plat + TiO_2 + Anh	Wirth et al. (2009)
Non-crystalline		SiO_2	Osk + Fe–O + Mil	Kaminsky et al. (2013)
Ilmenite	Ilm	$FeTiO_3$	Dol	Wirth et al. (2009)
			Sp + Phl	
Rutile(?) (α-PbO_2 structure)	Ru(?)	TiO_2	Coes + Ha + Sy + Hy + Cot + Plat + Anh	Wirth et al. (2009)
Ilmenorutile(?)	IlmRu(?)	$(Ti,Nb,Fe)O_2$	Osk + Hem + SiO_2	Kaminsky et al. (2013)
Hematite	Hem	Fe_2O_3	Osk + SiO_2	Kaminsky et al. (2013)
Plattnerite	Plat	PbO_2	Coes + Ha + Sy + Hy + Cot + Ru + Anh	Wirth et al. (2009)
Spinel-hercynite	Hc	$FeAl_2O_4$	Parasc + OrthMgO	Kaminsky et al. (2015a)
Spinel-magnesioferrite	Sp	$(Mg,Fe)(Al,Fe)l_2O_4$	Dol + Phl + Ilm	Wirth et al. (2009)
			Nyer + Ol + Ap + Phl(?)	Kaminsky et al. (2009)
			Cc + Phl(?)	
Spinel-magnetite	Sp	Fe_3O_4	Cc + Ap	Kaminsky et al. (2009)
Spinel non-specified	Sp	Mg–Fe–Al–O	Dol + Ca + Ap + Sulf + Per + Wu + Phl(?)	Kaminsky et al. (2015a)
Silicates				
"Olivine"	Ol	$(Mg,Fe)_2SiO_4$	Cc + $CaSiO_3$ + $CaTiO_3$	Brenker et al. (2007)
			Nyer + Sp + Ap + Phl(?)	Kaminsky et al. (2009)
Wollastonite-II (high)	Wo II	$CaSiO_3$	Cc + Cusp + Mont	Wirth et al. (2009)
			Cc + Nyer + CaGa + Ap + Wu	Kaminsky et al. (2009)

(continued)

Table 6.1 (continued)

Mineral	Index	Formula	Association	References
Monticellite	Mont	$CaMgSiO_4$	Cc + Wo II + Cusp	Wirth et al. (2009)
Cuspidine	Cusp	$Ca_4(Si_2O_7](F,OH)_2$	Cc + Wo II + Mont	Wirth et al. (2009)
Ca–Garnet	CaGa	$Ca_3(Fe,Zr,Ti)_2((Si,Al)O_4]_3$	Cc + Nyer + Wo II + Ap + Wu	Kaminsky et al. (2009)
Phlogopite	Phl	$KMg_3(Si_3AlO_{10}](F,OH)_2$	Cc	Brenker et al. (2007)
			Dol + Ilm + Sp	Wirth et al. (2009)
			Nyer + Ol + Sp + Ap	Kaminsky et al. (2009)
			Cc + Sp	
			Nah + Cc	Kaminsky et al. (2009)
			Dol + Mgs + Eit + Ha + Sy + Pent + Vi + Sp + P1	Kaminsky et al. (2013)
			Dol + Ca + Ap + Sp + Sulf + Per + Wu	Kaminsky et al. (2015a)
Sulphides				
Pentlandite	Pent	$(Fe,Ni)_9S_8$	Dol + Mgs + Eit + Ha + Sy + Phl + Vi + Sp + P1	Kaminsky et al. (2013)
Violarite	Vi	$FeNi_2S_4$	Dol + Mgs + Eit + Ha + Sy + Phl + Pent + Sp + P1	
Millerite	Mil	NiS	Osk + SiO_2 + Fe–O	Kaminsky et al. (2013)
Fe(-Ni) Sulfide	Sulf	Fe–Ni–S	Dol + Ca + Ap + Sp + Per + Wu + Phl(?)	Kaminsky et al. (2015a)
Fe–Ni–Cu Sulfide	Sulf	Fe–Ni–Cu–S	(Cc + Wo II + Cusp + Mont)	Wirth et al. (2009)
Native elements				
Ni–iron	Fe^0	Fe–Ni	Mag + FeMgs	8/102 (4)

association species occur in union with lower-mantle minerals of the ultramafic association, such as ferropericlase (Kaminsky et al. 2009) and CaSi-perovskite (Brenker et al. 2007). This close association is strong evidence of their formation within the lower mantle.

The following mineral classes can be distinguished among these minerals.

1. Carbonates.
2. Halides, fluorides, phosphates and sulphates.
3. Oxides.
4. Silicates.
5. Sulphides and native elements.

In addition, volatiles are present in this association.

The first three groups of minerals are the major species. They form paragenetic associations among themselves (Fig. 6.1). The compositions of individual minerals are variable.

6.2.2 Carbonates

Carbonates (Fig. 6.1a) are represented by two subgroups: (a) calcite–dolomite–magnesite (including ferromagnesite, breunnerite and sideroplesite), and (b) sodium carbonates: eitelite, nyerereite and nahcolite. The minerals form paragenetic assemblages between similar cation species: calcite with nyerereite and nahcolite, and dolomite and magnesite with eitelite.

Magnesite and dolomite, in some samples, comprise the majority of the inclusions. The inclusions are euhedral in shape, which implies that it has the 'negative' morphology that is common to syngenetic mineral inclusions in diamond (Fig. 6.2). Some grains of dolomite are enriched with Fe and Mn, and are possibly ankerite (based on the observed EDX spectra).

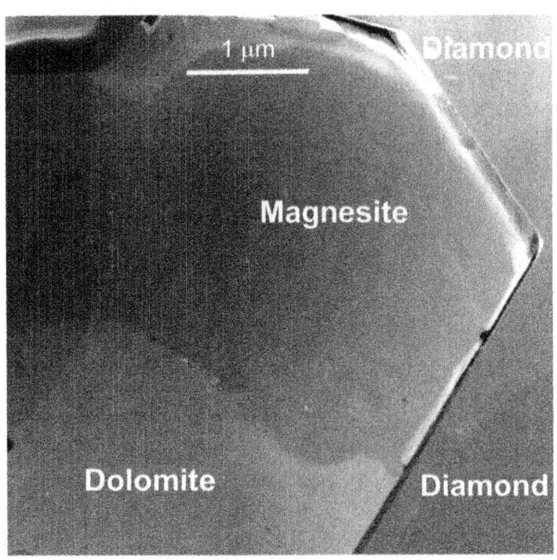

Fig. 6.2 Syngenetic inclusion of dolomite and magnesite in diamond. TEM image. Scale bar 1 µm. From Kaminsky et al. (2016a). Used with a permission of Springer

Recently, dolomite was reported as a mineral inclusion in diamond from the Machado area in Brazil, where it occurs in association with bridgmanite, ferropericlase and majorite (Burnham et al. 2016). One of the grains is dolomite with Ca/Mg close to the stoichiometry (0.967), while the other grain is, by its composition (Ca/Mg = 0.383), closer to magnesite.

Calcite occurs both as single isolated crystals, 10–30 μm in size (Kaminsky et al. 2009) and in association with other minerals (Figs. 6.1 and 6.3), including CaSi-perovskite (Brenker et al. 2007), which supports its formation within the lower mantle. Some analyzed grains contain admixtures of Sr (0.74 at.%) and Ba (trace contents), confirming their primary origin. The structure of calcite in grains associated with CaSi-perovskite was proven by both Raman spectra and X-ray diffraction (Brenker et al. 2007).

Calcite, dolomite and magnesite are primary mineral phases in the lower-mantle association (Figs. 6.2 and 6.3). This is in contrast to the occurrence of these carbonate minerals in the well-studied natrocarbonatitic lavas from the Oldoinyo Lengai, Tanzania, where calcite and dolomite (magnesite is not recorded in those lavas) are considered to have formed through transformation of the primary natrocarbonatitic association minerals (e.g., Zaitsev and Keller 2006).

Sodium carbonates are nyerereite $(Na,K)_2(Ca,Sr,Ba)(CO_3)_2$, nahcolite $NaHCO_3$ and eitelite $Na_2Mg(CO_3)_2$. Nyerereite was discovered in the early 1960s by J.B. Dawson in modern natrocarbonatitic lavas and the ashes of the Oldoinyo Lengai volcano in northern Tanzania (Dawson 1962). Nahcolite is one of the intermediate products of the hydration of nyerereite where it is widespread in altered natrocarbonatites, forming nests, veinlets and thin veins inside carbonatite flows. These minerals are characteristic for Tanzanian natrocarbonatites; in both associations (from the deep Earth and from Tanzanian volcanoes), they are similar to each other.

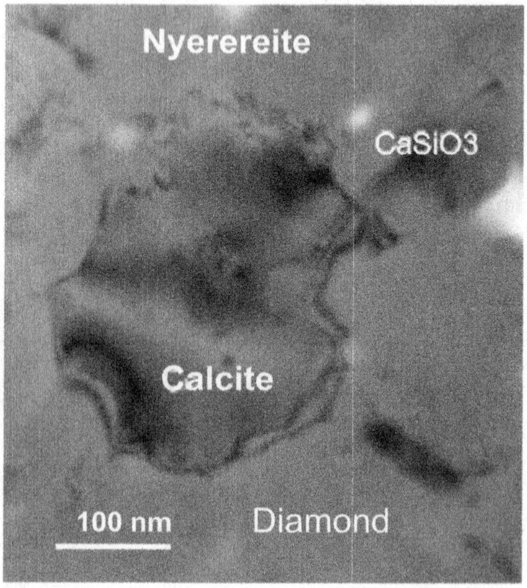

Fig. 6.3 Calcite and nyerereite in carbonatitic inclusion in diamond. TEM image. Scale bar 100 nm. From Kaminsky et al. (2016a). Used with a permission of Springer

There is no doubt that other carbonates may be found in the lower-mantle carbonatitic association. For example, tychite, sodium sulphatocarbonate $Na_6Mg_2(CO_3)_4(SO_3)$ was recently identified in mantle xenoliths from the Udachnaya-East pipe, Siberia (Sharygin et al. 2016), and potassium carbonate K_2CO_3 showed the phase transition with adopting the crystal structure of β-Na_2CO_3 under high pressure (Gavryushkin et al. 2016).

6.2.3 Halides, Fluorides, Phosphates and Sulphates

The second group of minerals is represented by a variety of compounds from volatile elements (Cl, F, P and S): halides, fluorides, phosphates and sulphates (Fig. 6.1b). Despite differences in their compositions, these minerals closely associate with one another and with the aforementioned carbonates.

Halides. In addition to the most common halides of sylvite and hydrophillite, the rare mineral of cotunnite ($PbCl_2$) also occurs. In contrast to halite and sylvite, which are typical for the Oldoinyo Lengai rocks (e.g., Zaitsev and Keller 2006), cotunnite was previously identified only as a post-volcanic (fumarole) product in active volcanoes in association with other halides, or as a secondary mineral in some lead mines. In the lower-mantle, carbonatite cotunnite associates (in addition to other halides) with another rare lead mineral, plattnerite (PbO_2).

Fluorides are another major class within this mineral group. Among them, oskarssonite (AlF_3), parascandolaite ($KMgF_3$) and a new mineral $(Ba,Sr)AlF_5$ were identified (Kaminsky et al. 2013, 2015a). They form assemblages in close association with phosphates.

Phosphates. Among phosphates, in addition to common apatite, two new mineral species were identified: mixed-anion ($Na_4Mg_3(PO_4)_2(P_2O_7)$ and Fe-diphosphate ($Fe_2Fe_5(P_2O_7)_4$ (Kaminsky et al. 2013). The phosphates form irregular grains in dolomite and fluoride.

Anhydrite is the only sulphate in the carbonatitic association and forms assemblages with all halides, but not with the carbonates; whereas phosphates associate with both carbonates and halides, as well as with fluorides.

6.2.4 Oxides

Oxides (Fig. 6.1c), which comprise the major part of the lower mantle, are subordinate in the carbonatitic association. Two major subgroups are distinguished within this class in the studied carbonatitic associations: (a) perovskites and (b) Mg- and Fe-oxides.

Perovskites are represented by CaSi-perovskite and CaTi-perovskite; to date, no bridgmanite has been observed. The **Mg- and Fe-oxides** do not form a compositional spectrum, unlike in the ultramafic association, where Mg-index of ferropericlase varies from 36 to 90. By contrast, Mg- and Fe-oxides from the carbonatitic association have only 7–15 at.% Fe and 2–15 at.% Mg, respectively, and may be considered to be almost pure periclase and wüstite. Such ultimate, immiscible compositions form at high pressures, above 83–86 GPa (Dubrovinsky et al. 2000, 2001); this process is described in detail above (Sect. 4.3.3).

In addition to minerals of these two subgroups, single grains of **accessory oxides**, such as ilmenite, rutile with an α-PbO_2 structure, ilmenorutile, hematite and plattnerite (PbO_2) occur. The presence of plattnerite in association with cotunnite ($PbCl_2$) may not be accidental. Various spinel accessory varieties (hercynite, magnesioferrite and magnetite) also occur; their role is unclear.

Silica. Of particular interest is the presence of silica. In studied specimens, SiO_2 occurs in the form of coesite. However, initially it may have been stishovite or post-stishovite phases, which are unstable under decreasing pressure conditions.

6.2.5 Silicates

Silicates also occur as single grains in the carbonatitic association. Some of them (wollastonite, monticellite and cuspidine) are secondary phases, formed as a result of the decomposition of CaSi-perovskite during the ascent of the host diamond to the surface (Kaminsky et al. 2016b). A mineral phase with a composition of olivine has been identified not only in the carbonatitic association, but within the ultramafic one as well (see Sect. 4.11.1). It has not yet been studied structurally, and its nature remains unclear.

6.2.6 Sulphides and Native Iron

The other minerals, occurring in the carbonatitic association—sulphides (pentlandite, violarite and others) and native iron—belong to the matrix of the deep Earth. Native iron contains 3.5 at.% Ni (Kaminsky et al. 2013) and, in that way, is similar to meteoritic kamacite. These minerals should be considered to be accessory phases.

6.3 Natrocarbonatitic Associations

Four major types of carbonatite are known: calcite carbonatite, dolomite carbonatite, ferrocarbonatite, and natrocarbonatite (International 1989). While the first three types are distributed widely, natrocarbonatites were known only recently and in one location: as recent lavas in the Oldoinyo Lengai volcano in Tanzania ('lengaites'; Kresten 1983), and within the Kerimasi and Tinderet carbonatites

(Mitchell 2005; Zaitsev et al. 2014). Natrocarbonatite rocks are composed mainly of nyerereite $Na_2Ca(CO_3)_2$ and gregoryite $(Na,K,Ca)_2CO_3$, with minor sylvite and fluorite in the groundmass. As such, these demonstrate comparable mineralogical and bulk chemical (Na–Cl–F) features to the carbonatitic association identified in lower-mantle Juina, Brazil, diamonds. In addition, the Juina carbonatitic association is enriched in REE, Sr and Nb, which is common for primary carbonatites. This demonstrates that the two associations, lengaite and Juina carbonatite, may be considered to be representatives of the same, natrocarbonatitic type.

There are, however, some differences in the compositions of the two associations. For example, in addition to Na-carbonates, calcite, dolomite and magnesite also occur in Juina carbonatite entrapped in diamond. Furthermore, calcite and dolomite in the lower-mantle carbonatitic association represent primary crystallizing phases, while in lengaites this mineral association is the result of the transformation of a 'primary' natrocarbonatitic association (Zaitsev and Keller 2006). Fluorite was not identified in Juina samples, while other fluorides observed in Juina are not characteristic for lengaites. Such differences may be the result of different source media compositions, their significantly differing depths of origin, and paragenesis (see Sect. 6.6).

6.4 Depth of Origin of Lower-Mantle Carbonatitic Association and the Stability of Carbonates Under Lower-Mantle Conditions

Minerals comprising the deep mantle natrocarbonatitic association are stable within a wide range of pressure–temperature (*P–T*) conditions. With the exception of secondary minerals such as monticellite, wollastonite, among others, they (or their polymorphs) are stable under conditions found within the lower mantle.

Carbonates are the major rock-forming minerals in the association. Close assemblages of carbonates with lower-mantle oxides as inclusions in diamonds from the Juina area have been identified, and include: calcite with CaSi-perovskite and $CaTiO_3$ (Brenker et al. 2007); and calcite, dolomite and nyerereite with periclase and wüstite (Kaminsky et al. 2009). The relationships between these phases and associations support an origin for the Juina natrocarbonatitic association within the lower mantle, i.e., below the 660 km discontinuity. These data agree with both experimental and theoretical data on the stability of carbonates in the deep Earth.

Magnesite is considered to be the primary carbonate phase within the deep Earth (Brenker et al. 2007; Lin et al. 2012; Scott et al. 2013). It is believed to be the most stable carbonate under conditions prevalent in the lower mantle; however ab initio calculations concluded that calcite is more stable than magnesite in the Earth's lower mantle at pressures above 100 GPa (Pickard and Needs 2015). While it has been suggested that magnesite will decompose into MgO and CO_2 at extreme pressures (Irving and Wyllie 1973) or, in association with SiO_2 into ferropericlase with the carbon-forming diamond (e.g., Seto et al. 2008), both experimental and theoretical studies have demonstrated that it is stable under pressure conditions of up to 113–115 GPa ($\sim$2500 km depth) and T = 2100–2200 K, when magnesite $R\bar{3}c$ structure transforms into a pyroxene-type orthorhombic $MgCO_3$ ('magnesite II', Isshiki et al. 2004; Skorodumova et al. 2005; Oganov et al. 2006, 2008; Panero and Kabbes 2008), and then, at $\sim$200 GPa from the pyroxene structure into the $CaTiO_3$ structure (Skorodumova et al. 2005). Under such extreme pressure conditions, only at very high temperatures (>2700 K) would magnesite decompose, with the formation of MgO and diamond (Solopova et al. 2015). Recently, a reaction of magnesite II with silica ($CaCl_2$-type SiO_2 or seifertite), resulting in the formation of bridgmanite and diamond was observed at pressure conditions above 80 GPa and temperatures above 1800–2000 K (Maeda et al. 2017).

Iron-containing magnesite in deep lower mantle transforms into orthorhombic $(Mg,Fe)CO_3$ phase II (Liu et al. 2015) and may form a monoclinic unusual stoichiometry $Mg_2Fe_2C_4O_{13}$ with tetrahedrally coordinated carbon (Merlini et al. 2015).

$CaCO_3$. The calculated stability of $CaCO_3$ reaches $\sim$80 GPa pressure and T = 3300 K (Ivanov and Deutsch 2002). Within a pressure range of 1–5 GPa and T = 800–2000 K, calcite changes its rhombohedral R-3c structure to that of rhombohedral aragonite (Irving and Wyllie 1973; Ivanov and Deutsch 2002), although some triclinic (P–1) modifications ($CaCO_3$-III and, at higher pressures, $CaCO_3$-VI) are stable at pressures of up to 30 GPa (Merlini et al. 2012a, 2014; Koch-Müller et al. 2016). At pressures near 50 GPa, the aragonite orthorhombic structure undergoes a transition to a trigonal, post-aragonite structure (Santillán and Williams 2004a), and, after reaching 137 GPa (i.e., at the core–mantle boundary), the pyroxene-type $C222_1$ structure of $CaCO_3$ may be expected (Oganov et al. 2006; Ono et al. 2007). More recent ab initio random structure searching (AIRSS) suggests that in the range of 32–48 GPa another, $P2_1/c$-1 structure with sp^2 bonded carbon atoms is more stable than aragonite and post-aragonite phases; and the pyroxene-type structure with four-fold coordinated carbon atoms should be the most stable $CaCO_3$ phase at pressures from 76 GPa to well over 100 GPa (Pickard and Needs 2015).

The grain from natural sample, where calcite associates with CaSi-perovskite, was identified not only by Raman spectrum, but by XRD as well, and is demonstrated to have the calcite structure (Brenker et al. 2007). This indicates that the calcite rhombohedral structure can be stable in the natural environment and under pressure conditions of up to at least 14–24 GPa (Shim et al. 2000). In the other studied natural samples, calcite was identified by analogy on the basis of its composition only; aragonite orthorhombic or trigonal structure of some of those grains may not be excluded.

Dolomite forms a syngenetic assemblage with magnesite (Fig. 6.2; Kaminsky et al. 2013). In another sample, a spherical inclusion of wüstite + periclase in dolomite was identified, pointing to its formation under pressure conditions exceeding 86 GPa (Kaminsky et al. 2015a). This is in agreement with the experimental data of Mao et al. (2011), who demonstrated that $CaMg(CO_3)_2$ is stable under lower-mantle conditions at pressures of up to 83 GPa (and possibly higher) and contrasts with earlier data on the decomposing of dolomite into magnesite + aragonite under lower-mantle *P–T* conditions (Biellmann et al. 1993; Luth 2001; Shirasaka et al. 2002). Recent experiments confirm that high-pressure polymorphism in dolomite could stabilize $CaMg(CO_3)_2$; this composition transforms at ~17 GPa into 'dolomite-II' (with a monoclinic structure according to Santillán et al. 2003, or an orthorhombic structure according to Mao et al. 2011, or a triclinic structure for Fe-dolomite, according to Merlini et al. 2012b) and then, at ~35–41 GPa, into 'dolomite-III' (with monoclinic structure, according to Mao et al. 2011 or a triclinic structure in the case of Fe-dolomite, according to Merlini et al. 2012a, b).

Volume differences between the three major carbonates: $CaCO_3$, $MgCO_3$ and $CaMg(CO_3)_2$ under high pressures are minimal, implying that energetic differences between these phases are small (Santillán et al. 2003) and they all may be present in the lower mantle, depending on the chemical composition of the media or within the same association under variable conditions. Among carbonates, the species with small radii of divalent cations (less than 1 Å; i.e., Mg^{2+} with 0.89 Å and Fe^{2+} with 0.78 Å radii) are more stable, while carbonates with large radii (Ca^{2+} with 1.12 Å radius) are less stable (Santillán and Williams 2004b). However, real mantle composition depends initially on the chemical composition of the media; and $CaCO_3$ and $CaMg(CO_3)_2$ may both be predominant mineral phases if the media is enriched in Ca. Rhombohedral $R\bar{3}c$-structured carbonates are the most stable under high-pressure conditions (Santillán et al. 2005); the boundary of the calcite-I to aragonite phase transition may be up to 44 GPa, as determined from $CaCO_3$ and $CdCO_3$ (Liu and Lin 1997; Santillán and Williams 2004b).

Other minerals from the natrocarbonatitic association are experimentally proven to be stable at ultrahigh-pressure conditions as well: eitelite $\geq$ 21 GPa (Kiseeva et al. 2013), parascandolaite $\geq$ 50 GPa (Aguado et al. 2008).

Of further particular interest are finds in the lower-mantle natrocarbonatitic association of almost pure periclase and wüstite. In one of the samples, two mineral species occur as coexisting, euhedral nano-inclusions in a porous carbonate matrix: periclase with Mg = 0.85–0.92 and wüstite with Mg = 0.02–0.15; no ferropericlase or magnesiowüstite of intermediate composition (which is characteristic for the oxide lower mantle) were observed (Kaminsky et al. 2009). Recently, in another sample, almost pure periclase and wüstite (with admixture of 5 wt% Fe and Mg, respectively) were observed, forming a spherical-shaped inclusion in a porous dolomite-calcite matrix: the core is composed of wüstite, and the rim is composed of periclase, both with spinel-type structures (see Fig. 4.15; Sect. 4.3.3) (Kaminsky et al. 2015a). It has been demonstrated experimentally that ferropericlase dissociates into almost pure wüstite and an Mg-rich phase at 86 GPa and 1000 K (Dubrovinsky et al. 2000, 2001). As such, the natrocarbonatitic association in the deep Earth formed at depth conditions corresponding to or exceeding 86 GPa, and corresponding to a depth in excess of 2000 km. A recent detailed study of another magnesiowüstite of composition ~$(Mg_{0.35}Fe_{0.65})O$ included in another diamond from the Juina area led to the conclusion that its origin was in one of the ultra-low velocity zones at the base of the mantle, with this reaching the surface through diapiric upwelling culminating in kimberlite eruption (Wirth et al. 2014).

A major factor in the stability of carbonate minerals and, hence, the formation of the carbonatite association in the lower mantle is oxygen fugacity. Recent thermodynamic calculations have shown that typical values of oxygen fugacity in zones of diamond formation in the lower mantle lie between the iron–wüstite buffer and six logarithmic units above this level (Ryabchikov and Kaminsky 2013). These calculations are proven by the formation of ferropericlase in the lower-mantle within a wide range of $\Delta \log f_{O_2}$ (IW), from 1.58 to 7.76 (Kaminsky et al. 2015b). This observation makes formation of the carbonatitic association within the lower mantle realistic.

6.5 Origin of the Natrocarbonatitic Association in the Deep Earth

6.5.1 Carbon, Diamond and Hotspots

In numerous works, the presence of carbonate species in the lower mantle has been considered to be a result of the subduction of lithospheric plates and entrained sedimentary carbonates into the lower mantle (e.g., Mao et al. 2011 and references therein). An alternative explanation, however, suggests the origin of carbonate minerals in the lower mantle may instead be juvenile carbon. Deep Earth's interior is a primary reservoir for this element; it has been calculated that more than 90% of Earth's carbon is stored deep within this reservoir (e.g., Javoy 1997; McDonough 2003). The Earth's core, where the carbon content is suggested at ~5 wt%, supplies carbon to the mantle over geological time via the mechanism of grain boundary diffusion that results in the formation of C–H–O volatiles (e.g., Hayden and Watson 2008). As a result, the Earth's mantle is the largest reservoir of carbon (Dasgupta and Hirschmann 2010), with an average C concentration of 100–120 ppm (Zhang and Zindler 1993; McDonough 2003).

It has been supposed, based on a series of petrologic and isotopic evidence, that carbonatite magmas may originate in the deepest parts of the mantle (Dalou et al. 2009; Bell and Simonetti 2010; Collerson et al. 2010) and be linked to the production of large igneous provinces (Ernst and Bell 2010). The data for the existence of carbonatitic inclusions in lower-mantle diamond support these ideas. It may be suggested that the formation of carbonatite and superdeep diamond initiated within the lower-mantle, in regions where hotspot/plumes occur. The observed presence of diamond in modern Hawaiian plume products, where these occur in a low-degree melt glass, rich in C, Cl, and comprising an exsolved, predominant CO_2, gas-fluid phase (Wirth and Rocholl 2003) gives a basis for such a proposition. Moreover, recent finds of carbonatitic and iron carbide-containing inclusions within the same diamond grain (Kaminsky et al. 2015a) confirm the suggestion regarding the formation of carbonatitic liquid in the deepest parts of the mantle.

'Primary', deep hotspot plumes (Courtillot et al. 2003) are thought to originate at the thermal boundary layer at the base of the lower mantle (e.g., Zhong 2006; Boschi et al. 2007). In some areas of the core–mantle boundary (CMB), anomalous seismic properties within thin zones (~ 10 km thick and 50–100 km wide) were identified as Ultra-Low Velocity Zones (ULVZ) due to their drop in seismic velocities greater than 10% relative to the background mantle. They are considered to be zones of partial melting, which may result from vigorous, small-scale convection or instabilities in the thermal boundary layer at the base of the mantle (Wen and Helmberger 1998; McNamara et al. 2010). ULVZ material can become entrained in mantle plumes (McNamara et al. 2010).

6.5.2 *Formation of Carbonatitic Partial Melts in the Lowermost Mantle*

The adiabatic temperature coefficient in the lower mantle was found to be 0.3 K/km (lower than in the upper mantle), and the adiabatic temperature at a depth of 2700 km was estimated at 2730 ± 50 K, if convective heat still dominates in this region (Katsura et al. 2010). However, the lower mantle is not seismically homogeneous (Bunge et al. 2001), and 3D spherical convection is not applicable to the entire deep Earth. There are data, based on the elasticity of silicate perovskite, demonstrating that the real temperature profile in the lower mantle is much steeper than the average mantle adiabatic (AMA), and below 1500 km the actual geotherm is super-adiabatic (da Silva et al. 2000). The computer simulations predict non-adiabatic excess of 100°–300° for the lower mantle because a significant proportion of mantle heat source is internal (Bunge et al. 2001). Calculations based on the measurement of perovskite shear modulus also demonstrate that lower mantle seismic properties may not match an adiabatic geotherm; geotherms may have larger temperature gradients (from 0.5 to 0.9 K/km between 800 and 2700 km), and indeed the temperature at a depth of 2700 km in some areas may reach 3400 K (Matas et al. 2007).

The pyrolite solidus temperature at the CMB was suggested to be near that of the core temperature (~ 3800–4000 K), suggesting possible partial melting of the lower mantle (Zerr et al. 1998). More recent experimental data on the solidus temperature at the CMB (136 GPa pressure) are even higher: 4180 ± 150 K for fertile peridotite (Fiquet et al. 2010) and 4150 ± 150 K for the chondritic composition (Andrault et al. 2011). These data are based on a study of both pyrolitic and chondritic solidi as 'dry' systems composed of only major elements (Si, Al, Fe, Ca and Mg), while the presence of alkalis and particularly C, O, H and other volatiles depress the solidus position drastically. The experimental data performed at 10–15 GPa pressure conditions demonstrated that the presence of C–O–H fluid and/or even the minor admixture of alkalis in peridotite produce carbonatitic melt at temperatures lower than adiabatic ones by as much as 400–500 °C (Litasov et al.

2013a, b; Shatskiy and Litasov 2015). It was demonstrated in experiments at up to 80 GPa, that the melting behaviour of simple carbonate systems is a suitable proxy for many carbonate-bearing lithologies (Thomson et al. 2014).

In this case, even a slight increase of temperature initiates the process of **partial melting**, which is observed in ULVZs at the CMB by P- and S-wave velocity reductions (Lay et al. 2004 and references therein). Such areas of partial melting are likely to be the roots to mantle hotspots. These hotspots are believed to be caused by the presence of chemical heterogeneities with high concentrations of fusible elements (Andrault et al. 2011). It has been calculated theoretically (Wyllie and Ryabchikov 2000), and proven experimentally (Dalton and Presnall 1998), that low-fraction melts generated near the solidus of carbon-containing mantle lherzolite at high enough oxygen fugacity values should have a carbonatite composition, with CO_2 contents of ~45 wt%.

Given that estimates of the average concentrations of carbon and phosphorus in mantle peridotite are closely similar (Palme and O'Neill 2003), the *near-solidus melts should have a carbonate–phosphate composition* (Ryabchikov and Hamilton 1993a, b), exactly as is observed in the diamond-hosted Juina carbonatite association. In experiments, low-fraction volatile-rich carbonatitic melts extract many incompatible elements from parental peridotite (Ryabchikov et al. 1991). The natural carbonatite association is rich in chlorine, fluorine and other elements, which form in addition to carbonates and phosphates, fluorides and chlorides (Kaminsky et al. 2009, 2013).

As a result, carbonatitic melts can be produced from volatile- and/or alkali-rich peridotite (chondrite) at the base of the lower mantle, in the roots to mantle hotspots at the CMB.

6.5.3 Evolution of Carbonatitic Melts in the Lower Mantle

The suggested model for the formation and evolution of carbonatite melts in the deep Earth is illustrated in Fig. 6.4.

The **first, super-adiabatic, stage A** occurs within the lowermost mantle, starting from the CMB (D″ layer). A local increase in the thermal condition, caused most likely by processes in the upper core, triggers low-degree partial melting of lower-mantle material enriched in carbon and other volatiles. Within chambers, approximately 10–20 km thick and 50–100 km wide, carbonatitic melts containing $\geq$45 wt% CO_2 are formed. Such chambers, which are roots to future hotspots, begin to migrate upwards creating local 3D convection. Partial carbonatitic melts comprise 1–2 vol% of the mantle chamber material, and are strongly enriched in volatiles. These partial melts infiltrate mantle lithologies and percolate into regions of the mantle. Carbonate melts have been shown to be highly mobile; they can travel upward into and through the upper mantle at velocities exceeding hundreds to thousands of meters on time scales of 0.1–1 m.y.; i.e., several orders of magnitude

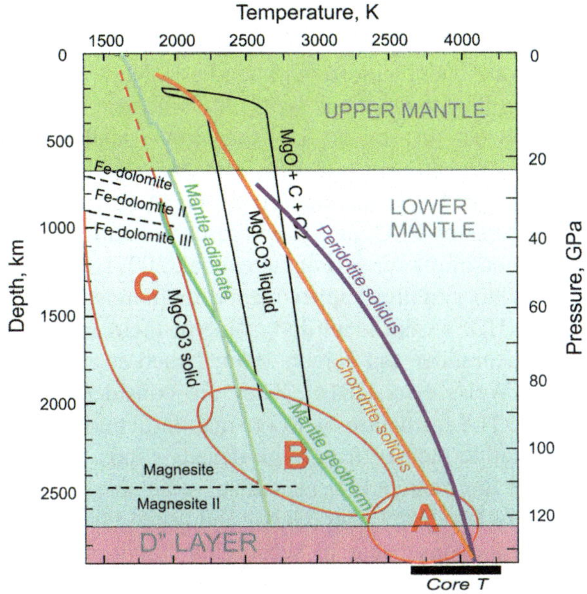

Fig. 6.4 A model for the formation and development of the carbonatitic association in the deep Earth. Modified after Kaminsky et al. (2016a). Used with a permission of Springer

higher than those previously found for basaltic liquids (Hammouda and Laporte 2000; Kono et al. 2014).

During the **second stage B**, partial melts develop at near-adiabatic conditions. During this stage, oxide phases can be crystallized; some of them may form immiscible wüstite + periclase associations under pressure values greater than 83–86 GPa (Dubrovinsky et al. 2000, 2001). The temperature conditions within this stage correspond to a multicomponent, completely miscible carbonate liquid system (Litvin et al. 2014). According to experimental data, as temperature falls, the invariant eutectic assemblage Mg–Fe–magnesite $(Mg,Fe)CO_3$ + Na-carbonate $Na_2(Mg,Fe)(CO_3)_2$ + liquid is formed (Litvin et al. 2014).

At lower *P–T* parameters, the initial carbonatitic system dissolves into two parts. Experiments at 5–16.5 GPa and high temperature demonstrated the presence of two immiscible carbonatitic systems: carbonate-silicate and chloride-carbonate (Litasov and Ohtani 2009; Safonov et al. 2011). The latter one may be considered to be an analogue to the natrocarbonatitic Juina lower-mantle association, while carbonate-silicate association has yet to be observed in mantle conditions. At high temperatures, the two immiscible liquids tend to coexist (Litasov and Ohtani 2009); however, the temperature for this immiscibility is not yet known. One may suggest that it occurs within the *P–T* conditions of the near-adiabatic stage B, where two immiscible carbonatitic melts separate. To date, only the chloride-carbonate (natrocarbonatitic) association is known, however, there is no reason to exclude the existence of the second one in the deep Earth.

Most likely, during the near-adiabatic stage B, formation of the 'ultra-deep' diamond occurs. Carbonatite media is favourable to diamond formation. It has been established experimentally that a concentration barrier of diamond nucleation

(CBDN) arises at a particular ratio of carbonate to silicate components (Litvin 2009). The CBDN position depends on the chemical composition of the system. In multicomponent silicate-carbonatite solvent, the CBDN is within the range of carbonatite compositions (<50 wt% silicates). Diamonds have been experimentally crystallized in melts of the lower-mantle diamond parental carbonate–magnesiowüstite–bridgmanite–carbon system, at conditions found within the upper part of the lower-mantle (20–30 GPa) (Litvin et al. 2014). It is apparent that admixtures of P, Cl and F within carbonatitic melt are of great importance in the formation of diamond (Palyanov et al. 2007; Sonin et al. 2008). Within both fibrous and monocrystalline upper-mantle diamonds, carbonatitic and saline high-density fluids (HDF) were identified, some of them are purely carbonatitic in composition (e.g., Schrauder and Navon 1994; Izraeli et al. 2001; Klein-BenDavid et al. 2006, 2009; Weiss et al. 2014). They are considered to be relics of diamond-forming media. The formation of lower-mantle diamonds may have happened either from carbonatitic melt or from high-density fluids.

The third stage C occurs at sub-adiabatic conditions. Various carbonate phases ($CaCO_3$, $MgCO_3$, $CaMg(CO_3)_2$ and others) form within this stage and major phase transformations occur in subsolidus conditions with the participation of remnant fluid phases. In experiments, the assemblage Mg–Fe–magnesite (Mg,Fe) CO_3 + Na-carbonate $Na_2(Mg,Fe)(CO_3)_2$, with Na_2CO_3 solidification at the lowermost isobaric temperature 1300 K (Litvin et al. 2014) was observed.

At or near to the Earth's surface, diamonds with carbonatitic inclusions are delivered, as xenocrysts, by kimberlitic volcanism. The kimberlite magmas form at shallower depth as a result of partial melting of peridotite within the upper mantle and/or transition zone. The diamonds entrained by kimberlitic magma en route to the surface are formed both in the lower mantle and upper mantle/transition zone (dominating) and, as such, contain inclusion suites of both carbonatitic and ultramafic mineral associations.

6.6 Conclusions

In addition to ultramafic and mafic associations, a primary carbonatitic association occurs in the lower mantle. It is predominantly carbonate (natrocarbonate)–halide–fluoride in composition with a minor (accessory) admixture of oxides, silicates and other minerals. It is close, by mineral composition, to the natrocarbonatitic association.

The initial lower-mantle carbonatitic melt formed as a result of low-fraction partial melting of lower-mantle material at the CMB. It was rich in P, F, Cl and other volatile elements. During ascent to the surface, the initial carbonatitic melt dissociated into two immiscible parts, a carbonate-silicate and a chloride-carbonate melt. The latter melt is parental to the observed natrocarbonatitic lower-mantle association. Diamonds with carbonatitic inclusions were formed in carbonatitic melts or high-density natrocarbonatitic fluids.

References

Aguado, F., Rodriguez, F., Hirai, S., Walsh, J. N., Lennie, A., & Redfern, S. A. T. (2008). High-pressure behaviour of $KMgF_3$ perovskites. *High Pressure Research, 28,* 539–544.

Andrault, D., Bolfan-Casanova, N., Lo Nigro, G., Bouhifd, M. A., Garbarino, G., & Mezoua, M. (2011). Solidus and liquidus profiles of chondritic mantle: Implication for melting of the Earth across its history. *Earth and Planetary Science Letters, 24,* 251–259.

Bell, K., & Simonetti, A. (2010). Source of parental melts to carbonatites—Critical isotopic constraints. *Mineralogy and Petrology, 98,* 77–89.

Biellmann, C., Gillet, P., Guyot, F., Peyronneau, J., & Reynard, B. (1993). Experimental evidence for carbonate stability in the Earth's lower mantle. *Earth and Planetary Science Letters, 118,* 31–41.

Boschi, L., Becker, T. W., & Steinberger, B. (2007). Mantle plumes: Dynamic models and seismic images. *Geochemistry, Geophysics, Geosystems, 8*(10), Q10006. doi:10.1029/2007GC001733

Brenker, F. E., Vollmer, C., Vincze, L., Vekemans, B., Szymanski, A., Janssens, K., et al. (2007). Carbonates from the lower part of transition zone or even the lower mantle. *Earth and Planetary Science Letters, 260,* 1–9.

Bunge, H.-P., Ricard, Y., & Matas, J. (2001). Non-adiabaticity in mantle convection. *Geophysical Research Letters, 28*(5), 879–882.

Burnham, A. D., Bulanova, G. P., Smith, C. B., Whitehead, S. C., Kohn, S. C., Gobbo, L., et al. (2016). Diamonds from the Machado River alluvial deposit, Rondônia, Brazil, derived from both lithospheric and sublithospheric mantle. *Lithos, 265,* 199–213. doi:10.1016/j.lithos.2016.05.022

Collerson, K. D., Williams, Q., Ewart, A. E., & Murphy, D. T. (2010). Origin of HIMU and EM-1 domains sampled by ocean island basalts, kimberlites and carbonatites: The role of CO_2-fluxed lower mantle melting in thermochemical upwellings. *Physics of the Earth and Planetary Interiors, 181,* 112–131.

Courtillot, V., Davaille, A., Besse, J., & Stock, J. (2003). Three distinct types of hotspots in the Earth's mantle. *Earth and Planetary Science Letters, 205,* 295–308.

Da Silva, C. R. S., Wentzcovitch, R. M., Patel, A., Price, G. D., & Karato, S. I. (2000). The composition and geotherm of the lower mantle: Constraints from the elasticity of silicate perovskite. *Physics of the Earth and Planetary Interiors, 118,* 103–109.

Dalou, C., Koga, K. T., Hammouda, T., & Poitrasson, F. (2009). Trace element partitioning between carbonatitic melts and mantle transition zone minerals: Implications for the source of carbonatites. *Geochimica et Cosmochimica Acta, 73,* 239–255.

Dalton, J. A., & Presnall, D. C. (1998). Carbonatitic melts along the solidus of model lherzolite in the system CaO-MgO-Al_2O_3-SiO_2-CO_2 from 3 to 7 GPa. *Contributions to Mineralogy and Petrology, 131,* 123–135.

Dasgupta, R., & Hirschmann, M. M. (2010). The deep carbon cycle and melting in Earth's interior. *Earth and Planetary Science Letters, 198*(1–2), 1–13. doi:10.1016/j.epsl.2010.06.039

Dawson, J. B. (1962). The geology of Ol Doinyo Lengai. *Bulletin of Volcanologique, 24,* 348–387.

Dubrovinsky, L. S., Dubrovinskaia, N. A., Annersten, H., Halenius, E., & Harryson, H. (2001). Stability of $(Mg_{0.5}Fe_{0.5})O$ and $(Mg_{0.8}\ Fe_{0.2})O$ magnesiowüstites in the lower mantle. *European Journal of Mineralogy, 13*(5), 857–861.

Dubrovinsky, L. S., Dubrovinskaia, N. A., Saxena, S. K., Annersten, H., Halenius, E., Harryson, H., et al. (2000). Stability of ferropericlase in the Lower Mantle. *Science, 289*(5478), 430–432.

Ernst, R. E., & Bell, K. (2010). Large igneous provinces (LIPs) and carbonatites. *Mineralogy and Petrology, 98,* 55–76.

Fiquet, G., Auzende, A. L., Siebert, J., Corgne, A., Bureau, H., Ozawa, H., et al. (2010). Melting of peridotite to 140 Gigapascals. *Science, 329,* 1516–1518. doi:10.1126/science.1192448

Gavryushkin, P.N., Rashenko, S.V., Shatskiy, A.F., Litasov, K.D., & Ancharov, A.I. (2016). Compressibility and phase transitions of potassium carbonate at pressures below 30 kbar. *Journal of Structural Chemistry, 57*(7), 1485–1488. doi:10.1134/S0022476616070258

Hammouda, T., & Laporte, D. (2000). Ultrafast mantle impregnation by carbonatite melts. *Geology, 28,* 283–285.

Hayden, L. A., & Watson, E. B. (2008). Grain boundary mobility of carbon in Earth's mantle: A possible carbon flux from the core. *Proceedings of the National Academy of Sciences of the USA, 105*(25), 8537–8541.

International Union of Geological Sciences, & Le Maitre, R. W. (1989). *A classification of igneous rocks and glossary of terms: Recommendations of the International Union of Geological Sciences Subcommission on the Systematics of Igneous Rocks.* Oxford: Blackwell.

Irving, A. J., & Wyllie, P. J. (1973). Melting relationships in CaO–CO_2 and MgO–CO_2 to 36 kbar with comments on CO_2 in the mantle. *Earth and Planetary Science Letters, 20,* 220–225.

Isshiki, M., Irifune, T., Hirose, K., Ono, S., Ohishi, Y., Watanuki, T., et al. (2004). Stability of magnesite and its high-pressure form in the lowermost mantle. *Nature, 427*(6969), 60–63.

Ivanov, A. B., & Deutsch, A. (2002). The phase diagram of $CaCO_3$ in relation to shock compression and decomposition. *Physics of the Earth and Planetary Interiors, 129,* 131–143.

Izraeli, E. S., Harris, J. W., & Navon, O. (2001). Brine inclusions in diamonds: A new upper mantle fluid. *Earth and Planetary Science Letters, 187,* 323–332.

Javoy, M. (1997). The major volatile elements of the Earth: Their origin, behavior, and fate. *Geophysical Research Letters, 24,* 177–180.

Kaminsky, F. V. (2012). Mineralogy of the lower mantle: A review of 'super-deep' mineral inclusions in diamond. *Earth-Science Reviews, 110*(1–4), 127–147. doi:10.1016/j.earscirev.2011.10.005

Kaminsky, F., Matzel, J., Jacobsen, B., Hutcheon, I., & Wirth, R. (2016a). Isotopic fractionation of oxygen and carbon in decomposed lower-mantle inclusions in diamonds. *Mineralogy and Petrology, 110*(2–3), 379–385. doi:10.1007/s00710-015-0401-7

Kaminsky, F. V., Ryabchikov, I. D., McCammon, C., Longo, M., Abakumov, A. M., Turner, S., et al. (2015a). Oxidation potential in the Earth's lower mantle as recorded from ferropericlase inclusions in diamond. *Earth and Planetary Science Letters, 417,* 49–56. doi:10.1016/j.epsl.2015.02.029

Kaminsky, F. V., Ryabchikov, I. D., & Wirth, R. (2016b). A primary natrocarbonatitic association in the Deep Earth. *Mineralogy and Petrology, 110*(2–3), 387–398. doi:10.1007/s00710-015-0368-4

Kaminsky, F., Wirth, R., Matsyuk, S., Schreiber, A., & Thomas, R. (2009). Nyerereite and nahcolite inclusions in diamond: Evidence for lower-mantle carbonatitic magmas. *Mineralogical Magazine, 73*(5), 797–816. doi:10.1180/minmag.2009.073.5.797

Kaminsky, F. V., Wirth, R., & Schreiber, A. (2013). Carbonatitic inclusions in Deep Mantle diamond from Juina, Brazil: New minerals in the carbonate-halide association. *The Canadian Mineralogist, 51,* 669–688. doi:10.3749/canmin.51.5.669

Kaminsky, F. V., Wirth, R., & Schreiber, A. (2015b). A microinclusion of lower-mantle rock and some other lower-mantle inclusions in diamond. *Canadian Mineralogist, 53*(1), 83–104. doi:10.3749/canmin.1400070

Katsura, T., Yoneda, A., Yamazaki, D., Yoshino, T., & Ito, E. (2010). Adiabatic temperature profile in the mantle. *Physics of the Earth and Planetary Interiors, 183,* 212–218.

Kiseeva, E. S., Litasov, K. D., Yaxley, G. M., Ohtani, E., & Kamenetsky, V. S. (2013). Melting and phase relations of carbonated eclogite at 9–21 GPa and the petrogenesis of alkali rich melts in the deep mantle. *Journal of Petrology, 54*(8), 1555–1583.

Klein-BenDavid, O., Logvinova, A. M., Schrauder, M., Spetius, Z. V., Weiss, Y., Hauri, E., et al. (2009). High-Mg carbonatitic microinclusions in some Yakutian diamonds: A new type of diamond-forming fluid. *Lithos, 112S,* 648–659.

Klein-BenDavid, O., Wirth, R., & Navon, O. (2006). TEM imaging and analysis of microinclusions in diamonds: A close look at diamond-growing fluids. *American Mineralogist, 91*(2–3), 353–365.

Koch-Müller, M., Jahn, S., Birkholz, N., Ritter, E., & Schade, U. (2016). Phase transitions in the system $CaCO_3$ at high P and T determined by in situ vibrational spectroscopy in diamond anvil cells and first-principles simulations. *Physics and Chemistry of Minerals, 43,* 545. doi:10.1007/s00269-016-0815-8

Kono, Y., Kenney-Benson, C., Hummer, D., Ohfuji, H., Park, C., Shen, G., et al. (2014). Ultralow viscosity of carbonate melts at high pressures. *Nature Communications, 5,* 5091. doi:10.1038/ncomms6091

Kresten, P. (1983). Carbonatite nomenclature. *Geologische Rundschau, 72,* 389–395.

Lay, T., Garnero, E. J., & Williams, Q. (2004). Partial melting in a thermo-chemical boundary layer at the base of the mantle. *Physics of the Earth and Planetary Interiors, 146,* 441–467.

Lin, J.-F., Liu, J., Jacobs, C., & Prakapenka, V. B. (2012). Vibrational and elastic properties of ferromagnesite across the electronic spin-pairing transition of iron. *American Mineralogist, 97,* 583–591.

Litasov, K. D., & Ohtani, E. (2009). Phase relations in the peridotite–carbonate–chloride system at 7.0–16.5 GPa and the role of chlorides in the origin of kimberlite and diamond. *Chemical Geology, 262,* 29–41.

Litasov, K. D., Shatskiy, A., & Ohtani, E. (2013a). Earth's mantle melting in the presence of C-O-H-bearing fluid. In S. Karato (Ed.), *Physics and chemistry of the deep Earth* (pp. 38–65). New York: Wiley-Blackwell.

Litasov, K. D., Shatskiy, A., Ohtani, E., & Yaxley, G. M. (2013b). Solidus of alkaline carbonatite in the deep mantle. *Geology, 41,* 79–82.

Litvin, Y. A. (2009). The physicochemical conditions of diamond formation in the mantle matter: Experimental studies. *Russian Geology and Geophysics, 50,* 1188–1200.

Litvin, Y., Spivak, A., Solopova, N., & Dubrovinsky, L. (2014). On origin of lower-mantle diamonds and their primary inclusions. *Physics of the Earth and Planetary Interiors, 228,* 176–185.

Liu, J., Lin, J.-F., & Prakapenka, V. B. (2015). High-pressure orthorhombic ferromagnesite as a potential deep-mantle carbon carrier. *Scientific Reports, 5,* 7640.

Liu, L., & Lin, C. (1997). A calcite-aragonite-type phase transition in $CdCO_3$. *American Mineralogist, 82,* 643–646.

Luth, R. W. (2001). Experimental determination of the reaction aragonite + magnesite = dolomite at 5 to 9 GPa. *Contributions to Mineralogy and Petrology, 141,* 222–232.

Maeda, F., Ohtani, E., Kamada, S., Sakamaki, T., Hirao, N., & Ohishi, Y. (2017). Diamond formation in the deep lower mantle: A high-pressure reaction of $MgCO_3$ and SiO_2. *Scientific Reports, 7,* 40602. doi:10.1038/srep40602

Mao, Z., Armentrout, M., Rainey, E., Manning, C. E., Dera, P., Prakapenka, V. B., et al. (2011). Dolomite III: A new candidate lower mantle carbonate. *Geophysical Research Letters, 38,* L22303. doi:10.1029/2011GL049519

Matas, J., Bass, J. D., Ricard, Y., Mattern, E., & Bukowinsky, M. S. (2007). On the bulk composition of the lower mantle: Predictions and limitations from generalized inversion of radial seismic profiles. *Geophysical Journal International, 170,* 764–780.

McDonough, W. F. (2003). Compositional model for the Earth's core. In H. D. Holland & K. K. Turekian (Eds.), *Treatise on geochemistry* (pp. 547–568). Oxford: Pergamon.

McNamara, A. K., Garnero, E. J., & Rost, S. (2010). Tracking deep mantle reservoirs with ultra-low velocity zones. *Earth and Planetary Science Letters, 299,* 1–9.

Merlini, M., Crichton, W. A., Chantel, J., Guignard, J., & Poli, S. (2014). Evidence of interspersed co-existing $CaCO_3$-III and $CaCO_3$-IIIb structures in polycrystalline $CaCO_3$ at high pressure. *Mineralogical Magazine, 78*(2), 225–233.

Merlini, M., Crichton, W., Hanfland, M., Gemmi, M., Müller, H., Kupenko, I., et al. (2012b). Structures of dolomite at ultrahigh pressure and their influence on the deep carbon cycle. *Proceedings of the National Academy of Sciences of the U.S.A., 109,* 13509–13514.

Merlini, M., Hanfland, M., & Crichton, W. (2012a). $CaCO_3$-III and $CaCO_3$-VI, high-pressure polymorphs of calcite: Possible host structures for carbon in the Earth's mantle. *Earth and Planetary Science Letters, 333,* 265–271.

Merlini, M. Hanfland, M., Salamat, A., Petitgirard, S., & Müller, H. (2015). The crystal structures of $Mg_2Fe_2C_4O_{13}$, with tetrahedrally coordinated carbon, and $Fe_{13}O_{19}$, synthesized at deep mantle conditions. *American Mineralogist, 100*(8–9), 2001–2004.

Mitchell, R. H. (2005). Carbonatites and carbonatites and carbonatites. *Canadian Mineralogist, 43,* 2049–2068.

Oganov, A. R., Glass, C. W., & Ono, S. (2006). High-pressure phases of $CaCO_3$: Crystal structure prediction and experiment. *Earth and Planetary Science Letters, 241*(1/2), 95–103.

Oganov, A. R., Ono, S., Ma, Y., Glass, C. W., & Garcia, A. (2008). Novel high-pressure structures of $MgCO_3$, $CaCO_3$ and CO_2 and their role in Earth's lower mantle. *Earth and Planetary Science Letters, 273*(1–2), 38–47.

Ono, S., Kikegawa, T., & Ohishi, Y. (2007). High-pressure transition of $CaCO_3$. *American Mineralogist, 92,* 1246–1249.

Pickard, C. J., & Needs, R. J. (2015). Structures and stability of calcium and magnesium carbonates at mantle pressures. *Physical Review B, 91,*104101. doi:10.1103/PhysRevB.91.104101

Palme, H., & O'Neill, H. S. C. (2003). Cosmochemical estimates of mantle composition. In R. W. Carlson (Ed.), *Treatise on geochemistry (the mantle and core)* (Vol. 2, pp. 1–38). Amsterdam: Elsevier.

Palyanov, Yu N, Shatsky, V. S., Sobolev, N. V., & Sokol, A. G. (2007). The role of mantle ultrapotassic fluids in diamond formation. *Proceedings of National Academy of Sciences USA, 104*(22), 9122–9127.

Panero, W. R., & Kabbes, J. E. (2008). Mantle-wide sequestration of carbon in silicates and the structure of magnesite II. *Geophysical Research Letters, 35,* L14307. doi:10.1029/2008gl034442

Ryabchikov, I. D., & Hamilton, D. L. (1993a). Interaction of carbonate-phosphate melts with mantle peridotites at 20-35 kbar. *South African Journal of Geology, 96,* 143–148.

Ryabchikov, I. D., & Hamilton, D. L. (1993b). Near-solidus carbonate–phosphate melts in mantle peridotites. *Geokhimiya, 12,* 1151–1160.

Ryabchikov, I. D., & Kaminsky, F. V. (2013). Redox potential of diamond formation processes in the lower mantle. *Geology of Ore Deposits, 55*(1), 1–12.

Ryabchikov, I. D., Orlova, G. P., Senin, V. G., & Trubkin, N. V. (1991). Interphase distribution of rare earth elements during partial melting in the system peridotite-carbonate-phosphate. *Geologhiya Rudnykh Mestorozhdeniy, 3,* 78–86 (in Russian).

Safonov, O. G., Kamenetsky, V. S., & Perchuk, L. L. (2011). Links between carbonatite and kimberlite melts in chloride-carbonate-silicate systems: Experiments and application to natural assemblages. *Journal of Petrology, 52,* 1307–1331.

Santillán, J., Katalli, K., & Williams, Q. (2005). An infrared study of carbon-oxygen bonding in magnesite to 60 GPa. *American Mineralogist, 90,* 1669–1673.

Santillán, J., & Williams, Q. (2004a). A high pressure X-ray diffraction study of aragonite and the post-aragonite phase transition in $CaCO_3$. *American Mineralogist, 89,* 1348–1352.

Santillán, J., & Williams, Q. (2004b). A high-pressure infrared and X-ray study of $FeCO_3$ and $MnCO_3$: Comparison with $CaMg(CO_3)_2$-dolomite. *Physics of the Earth and Planetary Interiors, 143–144,* 291–304.

Santillán, J., Williams, Q., & Knittle, E. (2003). Dolomite-II: A high-pressure polymorph of $CaMg(CO_3)_2$. *Geophysical Research Letters, 30*(2), 1054. doi:10.1029/2002GL016018

Schrauder, M., & Navon, O. (1994). Hydrous and carbonatitic mantle fluids in fibrous diamonds from Jwaneng, Botswana. *Geochimica et Cosmochimica Acta, 58,* 761–771.

Scott, H. P., Doczy, V. M., Frank, M. R., Hasan, M., Lin, J.-F., & Yang, J. (2013). Magnesite formation from MgO and CO_2 at the pressures and temperatures of Earth's mantle. *American Mineralogist, 98*(7), 1211–1218.

Seto, Y., Hamane, D., Nagai, T., & Fujino, K. (2008). Fate of carbonates within oceanic plates subducted to the lower mantle, and a possible mechanism of diamond formation. *Physics and Chemistry of Minerals, 35*(4), 223–229.

Sharygin, I. S., Golovin, A. V., Korsakov, A. V., & Pokhilenko, N. P. (2016). Tychite in mantle xenoliths from kimberlites: The first find and a new genetic type. *Doklady Earth Sciences, 467* (1), 270–274. doi:10.1134/S1028334X16030065

Shatskiy, A. F., & Litasov, K. D. (2015). *Formation of carbonates and a mechanism for the migration of carbonate melts through the Earth's mantle* (246 pp). Novosibirsk: Publishing House of the Siberian Branch of the Russian Academy of Sciences. (in Russian).

Shim, S. H., Duffy, T., & Shen, G. (2000). The stability and P–V–T equation of state of $CaSiO_3$ perovskite in the Earth's lower mantle. *Journal of Geophysical Research, 105*(B11), 25955–25968.

Shirasaka, M., Takahashi, E., Nishihara, Y., Matsukage, K., & Kikegawa, T. (2002). In situ X-ray observation of the reaction dolomite = aragonite + magnesite at 900–1300 K. *American Mineralogist, 87*(7), 922–930.

Skorodumova, N. V., Belonoshko, A. B., Huang, L., Anuja, R., & Johansson, B. (2005). Stability of the $MgCO_3$ structures under lower mantle conditions. *American Mineralogist, 90,* 1008–1011.

Smith, E. (2014). *Fluid inclusions in fibrous and octahedrally-grown diamonds* (Ph.D. thesis). University of British Columbia, 195 p.

Sobolev, N. V., Kaminsky, F. V., Griffin, W. L., Yefimova, E. S., Win, T. T., Ryan, C. G., et al. (1997). Mineral inclusions in diamonds from the Sputnik kimberlite pipe, Yakutia. *Lithos, 39,* 135–157.

Solopova, N. A., Dubrovinsky, L., Spivak, A. V., Litvin, Yu A, & Dubrovinskaia, N. (2015). Melting and decomposition of $MgCO_3$ at pressures up to 84 GPa. *Physics and Chemistry of Minerals, 42*(1), 73–81. doi:10.1007/s00269-014-0701-1

Sonin, V. M., Zhimulev, E. I., Chepurov, A. I., & Fedorov, I. I. (2008). Diamond stability in NaCl and NaF melts at high pressure. *Doklady Earth Sciences, 420*(4), 641–643.

Thomson, A. R., Walter, M. J., Lord, O. T., & Kohn, S. C. (2014). Experimental determination of melting in the systems enstatite-magnesite and magnesite-calcite from 15 to 80 GPa. *American Mineralogist, 99,* 1544–1554.

Wang, A., Pasteris, J. D., Meyer, H. O. A., & Dele-Duboi, M. L. (1996). Magnesite-bearing inclusion assemblage in natural diamond. *Earth and Planetary Science Letters, 141*(1–4), 293–306.

Weiss, Y., Kiflawi, I., Davies, N., & Navon, O. (2014). High density fluids and the growth of monocrystalline diamonds. *Geochimica et Cosmochimica Acta, 141,* 145–159.

Wen, L., & Helmberger, D. V. (1998). Ultra-low velocity zones near the core-mantle boundary from broadband PKP precursors. *Science, 279,* 1701–1703.

Wirth, R., Dobrzhinetskaya, L., Harte, B., Schreiber, A., & Green, H. W. (2014). High-Fe (Mg, Fe)O inclusion in diamond apparently from the lowermost mantle. *Earth and Planetary Science Letters, 404,* 365–376.

Wirth, R., Kaminsky, F., Matsyuk, S., & Schreiber, A. (2009). Unusual micro- and nano-inclusions in diamonds from the Juina Area, Brazil. *Earth and Planetary Science Letters, 286*(1–2), 292–303.

Wirth, R., & Rocholl, A. (2003). Nano-crystalline diamond from the Earth mantle underneath Hawaii. *Earth and Planetary Science Letters, 211*(3–4), 357–369.

Wyllie, P. J., & Ryabchikov, I. D. (2000). Volatile components, magmas, and critical fluids in upwelling mantle. *Journal of Petrology, 41,* 1195–1206.

Zaitsev, A. N., & Keller, J. (2006). Mineralogical and chemical transformation of Oldoinyo Lengai natrocarbonatites, Tanzania. *Lithos, 91,* 191–207.

Zaitsev, A. N., Wenzel, T., Vennemann, T., & Markl, G. (2014). Tinderet volcano, Kenya: An altered natrocarbonatite locality? *Mineralogical Magazine, 77*(3), 213–226.

Zedgenizov, D. A., Ragozin, A. L., Kalininaa, V. V., & Kagi, H. (2016). The mineralogy of Ca-rich inclusions in sublithospheric diamonds. *Geochemistry International, 54*(10), 890–900. doi:10.1134/S0016702916100116

Zerr, A., Diegeler, A., & Boehler, R. (1998). Solidus of the Earth's deep mantle. *Science, 281* (5374), 243–246.

Zhang, Y., & Zindler, A. (1993). Distribution and evolution of carbon and nitrogen in Earth. *Earth and Planetary Science Letters, 117,* 331–345.

Zhong, S. (2006). Constraints on thermochemical convection of the mantle from plume heat flux, plume excess temperature, and upper mantle temperature. *Journal of Geophysical Research, 111,* B04409. doi:10.1029/2005JB003972

Chapter 7
Diamond in the Lower Mantle

Abstract Diamond contains mineral inclusions of all three lower-mantle associations, juvenile ultramafic, mafic and carbonatitic; it is also an accessory mineral in all these associations. While the first two associations coexist with diamond, the carbonatitic association is a parental medium for the lower-mantle diamond. Physical and chemical characteristics of lower-mantle diamond differ from ones of lithospheric origin. Most of the lower-mantle diamonds are 'nitrogen-free' Type II variety. The others are usually low-nitrogen stones with the average nitrogen aggregation rate of 94%. The high proportion of nitrogen-aggregated diamonds suggests that they had a prolonged residence in the lower mantle under high-T conditions, which resulted in an almost complete transformation of single-atomic and paired nitrogen centers into polyatomic complexes. In contrast to lithospheric diamonds, almost all analyzed lower-mantle ones (70–89%) have noticeable levels of hydrogen centers (up to 4–6 cm^{-1}). The isotopic compositions of lower-mantle diamonds are located within a narrow range: from −5.45 to −1.26‰ $\delta^{13}C$ VPDB, with an average value of −4.36‰ ± 2.28‰ (2σ). It may be considered as the juvenile lower-mantle carbon isotopic composition. The isotopic composition of nitrogen for lower-mantle diamonds is located within a close range, from −5.2 to −1.0‰ $\delta^{15}N_{atm}$, with an average value of $\delta^{15}N_{atm}$ = −3.00‰ ± 2.37‰ $\delta^{15}N_{atm}$. Lower-mantle diamond was formed in carbonate-oxide parental melts and fluids, which experienced fractional crystallization with the decrease of temperature and changes in the melt composition. The most important role in this process belongs to the carbonate component in the parental melt.

7.1 General

Diamond exhibits a uniquely wide stability field of pressure and temperature with an approximate melting temperature above 5000 K and a stable pressure of up to twice the pressure of the Earth's center (Correa et al. 2006; Oganov et al. 2013). It is expected to be a planetary 'survivor of catastrophe', such as the Moon-forming collision on Earth. Only at ~1000 GPa (1 TPa) does the cubic diamond structure ($Fd\bar{3}m$) transform into the tetrahedral BC8 structure (Oganov et al. 2013 and

F.V. Kaminsky, *The Earth's Lower Mantle*, Springer Geology,
DOI 10.1007/978-3-319-55684-0_7

references therein). Owing to its wide stability and metastability, diamond is a host of lower-mantle mineral inclusions, encapsulating and delivering them to the Earth's surface. The tensions in a diamond lattice, surrounding some lower-mantle inclusions, owe their simultaneous origin in deep Earth. Diamond contains mineral inclusions of all three lower-mantle associations, juvenile ultramafic, mafic and carbonatitic; and it is an accessory mineral in all of these associations. Physical and chemical characteristics of lower-mantle diamond differ from those of lithospheric origin.

7.2 Morphology and Internal Structure

Similar to diamonds from the upper mantle, the morphology of the lower-mantle diamonds include octahedra, dodecahedroids, combination-type crystals, and their twins and aggregates (Fig. 7.1). Singular tetrahedra were also identified as well (Kaminsky et al. 2001, 2009). In addition to these morphological forms, there are diamonds with an uncertain habit, either irregular aggregates or fragments of crystals, bounded by spallation surfaces on all sides. No cubic forms of the diamond crystals (which are usually considered to be formed under lower temperatures) or even cubic faces were found among the lower-mantle diamonds studied to date. In all of the kimberlitic pipes, dodecahedroids and tetrahexahedroids are most common, comprising 28–70% of the stones from different kimberlites and placer deposits in the Juina and Machado areas (Brazil), independently of the diamond size fraction within a size range of 1–16 mm; octahedra comprise from 0 to 13%. The average (Dod + Tetr)/Oct ratio in different deposits varies from 2.9 to 13.1. Considering the formation of the dodecahedral and tetrahexahedral rounded forms of diamond as the result of the resorption of octahedra, this exhibits a long history of diamond within the lower mantle. Deformation laminations (DeVries 1975) are visible on at least 20% of the diamonds.

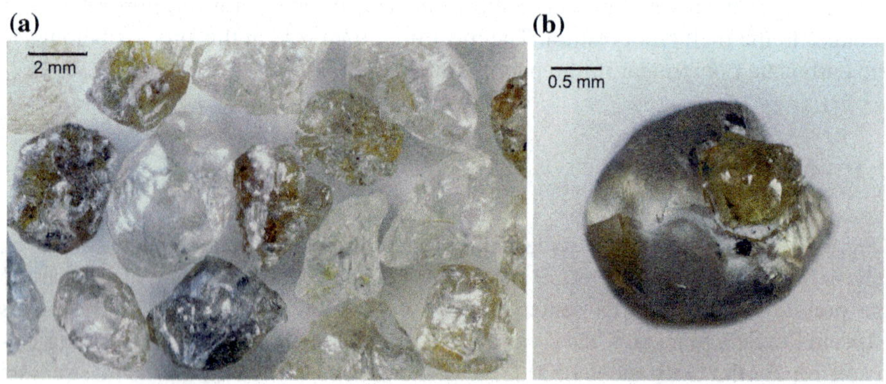

Fig. 7.1 Diamonds with lower-mantle inclusions from the Juina area, Brazil. **a** General view. **b** Dodecahedroid with an ingrowth of a smaller crystal and an inclusion of ferropericlase (*black*)

Dodecahedroids and tetrahexahedroids in the Juina area show a very limited number of varieties compared with such crystals from other localities worldwide. They have different degrees of distortion; most of them are elongated. The surfaces of dodecahedroidal crystals show evidence of resorption: mat or shagreen micro-relief (Fig. 7.2a). The most common shagreen surfaces usually consist of overlapping, very fine hummocks of different sizes and irregular shapes (Fig. 7.2b). Such crystals are commonly semi-transparent with a silky lustre. No other surface features or accessory forms were identified in dodecahedral and tetrahexahedral diamond crystals from the Juina area. Such a characteristic feature of the Juina diamonds is in contrast to the wide variety of surface features generally present on dodecahedral forms of the diamond crystals from other regions (e.g., Kukharenko 1955; Tolansky 1955; Bartoshinsky 1962; Robinson 1980; Orlov 1987).

Octahedra among diamonds from the Juina area are, like dodecahedroids and tetrahexahedroids, quite uniform in shape. They frequently show polycentric face growth (Fig. 7.3). Combination-type diamonds comprise 9–24% of all stones, i.e. they are more frequent than pure octahedra.

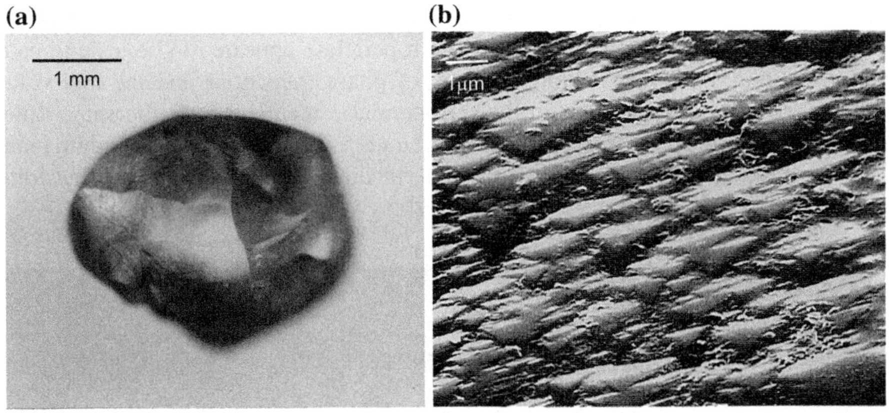

Fig. 7.2 Dodecahedroid with shagreen texture structure. **a** General appearance. **b** Surface detail

Fig. 7.3 Octahedron with polycentric development of crystal faces and a small inclusion of ferropericlase on a ridge

Coloured stones predominate among lower-mantle diamonds; brown and brownish stones comprise 55–80% indicating plastic deformations at the final stages of their crystallization (Robinson et al. 1989).

The internal structure in diamond, revealed through examination of cut diamonds from Rio Soriso with cathodoluminescence (CL), demonstrated concentric structures with varying nitrogen concentrations (Fig. 7.4). Episodic resorption and growth are visible in some crystals (Fig. 7.4a) and most diamonds have experienced a final episode of resorption before exhumation, which is reflected in a great number of dodecahedroids and tetrahexahedroids among Juina diamonds (Hayman et al. 2005).

CL images of lower-mantle diamonds from the Collier-4 pipe usually demonstrate internal cracks infilled with later diamond generation, these are broken and resorbed crystals or display very complex growth patterns (Bulanova et al. 2010). Good examples of complicated internal structure correlated with the distribution of carbon isotopic composition was demonstrated for the São Luis diamonds by Zedgenizov et al. (2014).

Cayzer et al. (2008) and Zedgenizov et al. (2015) studied the internal structure of lower-mantle diamonds from Brazil with the use of micro-Raman and the electron backscatter diffraction (EBSD) techniques. In the EBSD studies, misorientation of the diamond lattice around inclusions of ferropericlase appeared to be insignificant (not exceeding 1°–2°), indicating the lack of phase transitions and the low volumetric thermal expansion around the ferropericlase inclusions. At the same time, well-defined plastic deformations around inclusions of bridgmanite, CaSi-perovskite, CaTi-perovskite and SiO_2 in diamond were observed, reaching

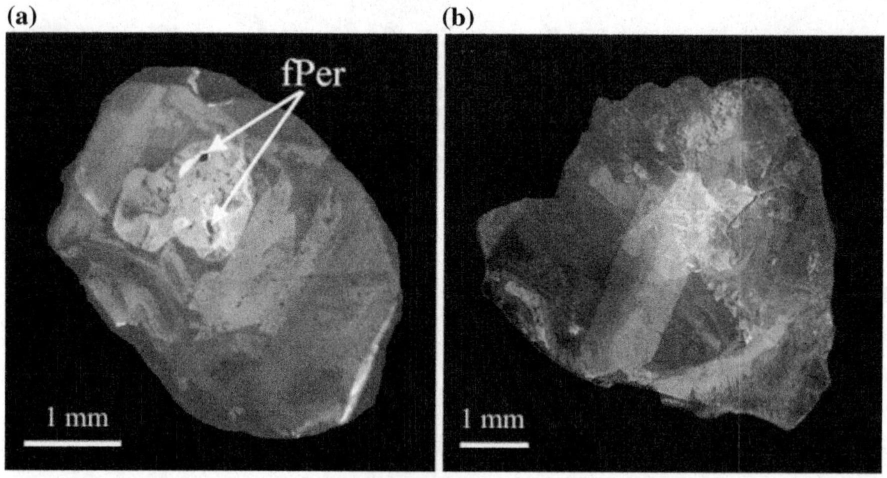

Fig. 7.4 Cathodoluminescence images of lower-mantle diamonds from Rio Soriso, Brazil. **a** Internal structure of diamond with inclusions of bridgmanite, ferropericlase and 'olivine'. Note the strongly resorbed bright CL core with inclusions of ferropericlase, episodic octahedral growth and dissolution events and the general weakening of CL from core to rim. **b** Internal structure of diamond with an octahedral core. The variation in CL intensity is a function of nitrogen concentration, which in turn is controlled by the growth face developing. From Hayman et al. (2005). Used with a permission of Springer

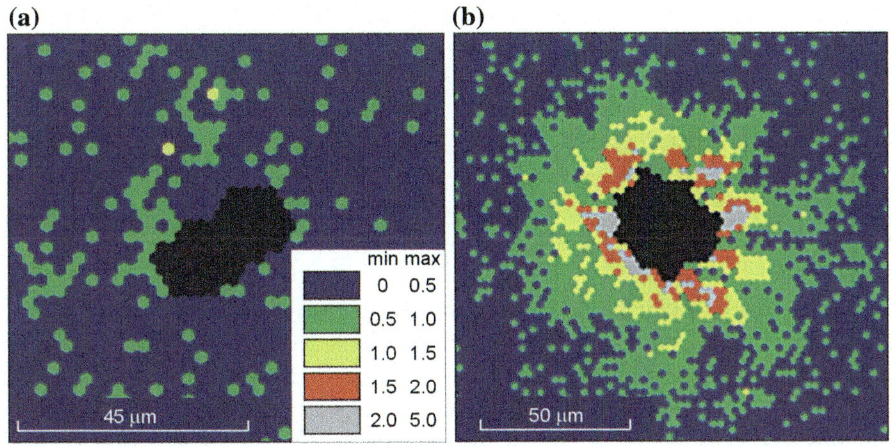

Fig. 7.5 EBSD maps of diamond sites around inclusions of ferropericlase (**a**) and bridgmanite (**b**) in diamond. The grade of the misorientation of the diamond lattice is shown on a scale in corresponding colours. From Zedgenizov et al. (2015). © Elsevier

4°–7° of misorientation of the diamond lattice (Fig. 7.5). In the Raman studies, the width of the inclusion influence on diamond was found to be greater for perovskites than for ferropericlase (Cayzer et al. 2008). Plastic deformations around these minerals indicate considerable stress around the inclusions, which can be interpreted as a consequence of the phase transitions of these minerals, accompanied by a significant volumetric effect, which caused considerable plastic deformations of the surrounding diamond lattice.

7.3 Luminescence of Diamond

7.3.1 Photoluminescence (PL)

In different areas, approximately 14–26% of the lower-mantle diamonds do not fluoresce; in some areas (such as Machado placer deposit) most diamonds have nil or very weak photoluminescence (Bulanova et al. 2008). Among the fluorescent stones, the ones with blue colour predominate (40–80%); the others have green to turquoise (9–15%), pink (4–15%), and yellow to brown (2–8%) colours of PL, varying in quantities in different deposits. Green and turquoise PL colours are commonly associated with brown and yellow stones, thus indicating that the defects responsible for brown and yellow body colour also have control on PL colours.

Spectral PL characteristics of the Juina diamonds measured using various excitation wavelengths (313, 473, and 532 nm) are quite distinct from other diamonds (Yuryeva et al. 2015). The overall presence of the 490.7 nm, H3, and H4 centers in the luminescence spectra attests to strong plastic deformations in these diamonds. The 558.5 nm PL center is one of the most common defects in 'nitrogen-free' samples. It is accompanied by the EPR center with a *g*-factor of

2.00285. The 536 and 576 nm systems totally dominated the PL spectra of superdeep diamonds, while none of lithospheric diamonds (e.g., from the Mir pipe, Yakutia) with similar nitrogen characteristics showed these PL centers, as well as the 558.5 nm center. The neutral vacancy known as the GR1 center has probably occurred in a number of crystals due to radiation damage in the post-growth period.

7.3.2 Cathodoluminescence (CL)

Cathodoluminescence (CL) in lower-mantle diamonds is also weak and, in general, homogeneous (Hayman et al. 2005; Bulanova et al. 2008). Among stones studied from the Rio Soriso area, blue and turquoise (38% each), green (13%) and purple luminescence (11%) colours were apparent. CL spectroscopy indicates that green and blue CL colours have two broad peaks between 435–445 nm and 483–512 nm, with a shift to higher numbers for green CL. Absorption at 440 nm is found in many different diamonds worldwide with a variety of nitrogen concentrations and aggregation states (Pratesi et al. 2003). The stronger peak observed for diamonds with green CL (~510 nm) may be related to H3(N_2 + V) center strain (Zaitsev 2001).

7.3.3 Laser-Induced Luminescence (LIL)

The laser-induced luminescence spectra, luminescence decay kinetics and distribution of radiation in diamonds from Juina, Brazil, were investigated in detail by Martynovich et al. (2013) and Mironov et al. (2015) with pulsed semiconductor lasers with a wavelength of 375, 405, 470, 532 and 640 nm. Like in other types of luminescence, the distribution of the luminescence centers in the Juina diamonds were found to be sufficiently homogeneous, which can be affiliated with the diamond growth at constant high pressure–temperature (*P–T*) conditions and their post-origin conditions of conservation, significantly exceeding the *P–T* values for lithospheric diamonds. The octahedral zoning observed in some samples implies the tangential growth of diamond layers in stable conditions (Fig. 7.6).

Juina diamonds have a multi-component spectra of luminescence, which are more complex than those in lithospheric diamonds. Impurity structural nitrogen centers of various degrees of aggregation (with vacancies) were detected in diamonds: H4 (N_4 + 2V), N3 (N_3 + V), H3 (N_2 + V) and 575 nm (N + V) containing four, three, two and one nitrogen atom(s), respectively, as well as the red band with a maximum at 690–700 nm and the lines of zero-phonon transitions at 491, 523 and 536 nm. The visible non-uniformities in the distribution of luminescence over the area of the diamonds are determined by relatively small fluctuations in the intensities of individual bands in the spectra.

Luminescence kinetics in the temperature range of 80–480 K and the red region of the spectrum for the Juina diamonds were studied by Stepanov et al. (2016). Spectral components with decay time constants of 23 and 83 ns were observed at

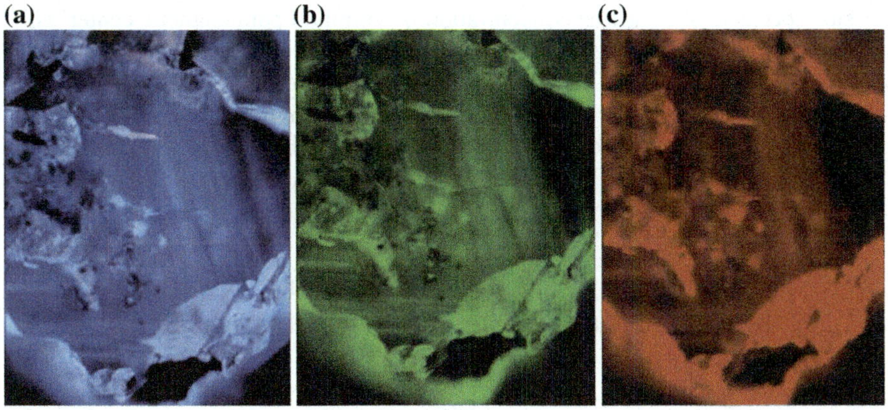

Fig. 7.6 Octahedral zoning in one of the Juina diamonds, revealed from luminescence excited by the use of lasers with wavelength of 375 nm (**a**), 470 nm (**b**) and 532 nm (**c**). From Mironov et al. 2015. © Elsevier

room temperature after being excited by laser radiation with wavelengths of 375 and 532 nm; these spectral characteristics differ considerably from the data published earlier for the luminescence kinetics of NV°- and NV^{-}-centers for lithospheric diamonds (Davies 1977; Beha et al. 2012; Liaugaudas et al. 2012; Inam et al. 2013).

A distinctive feature of the Juina lower-mantle diamonds is a broad smooth band with a maximum at ~700 nm in the luminescence spectra, i.e., almost at the place where the band of the NV^{-}-centers are located. In this band, however, the luminescence decay time (32 ns at 80 K, 23 ns at 300 K, and 3 ns at 480 K) differs considerably from those from NV^{-}-centers that were found earlier for natural lithospheric diamonds from different locations (from 8 to 13 ns) (Collins et al. 1983; Hanzawa et al. 1997; Vins and Pestryakov 2006). Another important feature of the Juina diamonds is an unexpectedly long decay time of NV°-center luminescence in the zero-phonon line at 575 nm and its electronic-vibrational wing: 83 ns, virtually independent of the temperature in the range of 80–480 K. By contrast, values of 19–32 ns were recorded earlier for natural upper-mantle and man-made diamonds (Davies 1977; Beha et al. 2012; Liaugaudas et al. 2012; Inam et al. 2013). It has been suggested that the aforementioned spectral features of the Juina diamonds are caused by their asthenospheric origin (Stepanov et al. 2016).

7.4 Electron Paramagnetic Resonance (EPR) Characteristics of the Juina Diamonds

The EPR spectra have been recorded for 30 diamond crystals from the Juina area (Yuryeva et al. 2015). The most frequently recognized paramagnetic centers (in particular, P2 (N_3 + V) and P1) in natural diamond crystals were not identified, possibly because of their low concentration. A common paramagnetic center with a high concentration as a single line (g-value of 2.00285) was first detected in crystals with a low nitrogen content. Electron irradiation and annealing of the crystals at 500–900 °C show that this center was resistant to annealing in this temperature range and did not interact with radiation-induced defects. Comparison of the EPR and PL data demonstrated that the 2.00285 center and the 558.5 nm system can be interrelated.

7.5 Nitrogen and Hydrogen Concentrations of the Lower-Mantle Diamonds by Infrared Spectroscopy

7.5.1 Infrared (IR) Spectra of the Lower-Mantle Diamond

Infrared spectroscopy is a major tool for the study of structural impurities and defects (mainly nitrogen) in diamond. Using gas chromatography, Kaiser and Bond (1959) established a correlation between the **concentration of nitrogen** in diamond and its spectral characteristics in infrared (IR) (8–16 μm, or 1300–500 cm^{-1}) and ultraviolet (UV) (200–600 nm) wavelengths as the following:

$$N = 5.8 \times 10^{18} \times \alpha_{7.8}(\text{for IR spectra}) = 11.6 \times 10^{18} \times \alpha_{306.5}(\text{for UV spectra});$$

where N is the concentration of nitrogen in the form of paired atoms (A-form) in atoms per 1 cm^3, and α is the absorption coefficient. Since then, additional studies have been performed in this field, and now three major forms of nitrogen are distinguished in diamond: A, B and C (N).

In C (N)-center, one nitrogen atom replaces one carbon atom in a number of carbon sites. This center produces a characteristic infrared absorption spectrum with a sharp peak at 1344 cm^{-1} and a broader feature at 1130 cm^{-1} (Fig. 7.7). Absorption at those peaks is routinely used to measure the concentration of the single nitrogen (Kiflawi et al. 1994).

A-center is formed by a pair of nitrogen atoms jointly substituting for carbon atoms (Davies 1976; Sobolev 1978) and shows up in the IR absorption spectrum with a peak at 1282 cm^{-1} (Fig. 7.7), whereas B-center, forming a sharp peak at 1332 cm^{-1} and a broader feature at 1280 cm^{-1}, is an aggregate of four nitrogen atoms tetrahedrally arranged around a vacancy (Bursill and Glaisher 1985; Jones and Goss 2002; Liu et al. 2016). Experimental works demonstrated that there is a transformation of nitrogen A to B centers (called 'aggregation'), which is a function of time and temperature of the diamond residence in the mantle. This process may be considered as the result of diffusion, in which the diffusion coefficient of nitrogen in diamond strongly depends on temperature (Harte et al. 2009). Within the upper mantle, at relatively low temperatures (1000–1500 °C), this process takes billions of years (Evans 1992; Mendelssohn and Milledge 1995; Taylor and Milledge 1995; Taylor et al. 1996). The diffusion process is much faster in the lower mantle with an

Fig. 7.7 Typical IR spectrum of diamond with major absorption peaks, indicating different nitrogen and hydrogen impurity centres

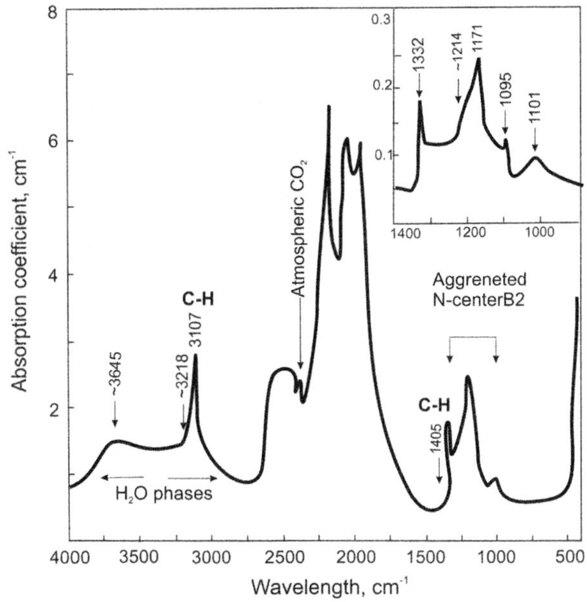

adiabatic temperature profile of 1700–2500 °C. The analysis of the concentration and aggregation rate of nitrogen provides valuable data on the formation and post-genetic history of diamond in the Earth's interior, such as the temperature history of diamonds from the primary sources with known age (Kohn et al. 2016).

Historically, on the basis of the total nitrogen content in diamond and on the presence of different nitrogen centers, diamonds are grouped into two types, type I ('nitrogen-containing') and type II ('nitrogen-free') (Robertson et al. 1934). Type I-diamond contains nitrogen in a number of different forms, whereas the type II-diamond is considered as a 'nitrogen-free' variety (i.e., a diamond containing less than 10–20 at.ppm nitrogen that cannot usually be detected by IR spectroscopy). Type I-diamond was further subdivided into type Ia and type Ib depending on the form that the nitrogen takes within the lattice of the individual crystal. The nitrogen in type Ia-diamond is present in various aggregated forms, while in type Ib-diamond, nitrogen atoms are dispersed in isolated substitutional sites as C (N) center. Most natural diamonds are Type Ia, containing C-centre nitrogen in the amounts not exceeding 50 at.ppm, and synthetic diamonds produced at low temperatures is commonly that of Type Ib. On the basis of the predominance of A- or B-centers, Type Ia is categorized into two subdivisions, type IaA and type IaB, respectively. The majority of natural diamond crystals belong to the transitional IaAB type. Type IaB is very rare in lithospheric diamonds, comprising only ~0.2 % of the total stones (Dobrinets et al. 2013), while *most nitrogen-containing lower-mantle diamonds belong to type IaB with strongly aggregated nitrogen.*

IR spectroscopy provides a means of estimating the concentrations of nitrogen (A and B) impurity centers. The increasingly widespread use of IR spectroscopy in the investigation of natural diamond samples was promoted largely by the refinement of the analytical procedures; now most calculations of the concentration of A- and B-centers in diamond are based on empiric formulas:

$$N_A = K_A \times \mu_{1282(A)};$$
$$N_B = K_B \times \mu_{1282(B)} = K'_B \times \mu_{1175(B)};$$

where N_A and N_B are the concentrations of A- and B-centers in atomic ppm, respectively; $K_A = 16.5 \pm 1$ at.ppm (Boyd et al. 1994), $K_B = 79.4 \pm 8$ at.ppm (Boyd et al. 1995); μ represents the absorption coefficients measured for the corresponding bands in IR spectra (Fig. 7.7).

In addition to A- and B-centers, the 'platelets' P-centers, which form aggregations up to several micrometres in size, exist in most natural diamonds as planar defects along ⟨100⟩ lattice planes, visible via electron microscopy (Evans and Phaal 1962) and in cathodoluminescence (Mendelssohn 1971). For a long time, they were interpreted as large nitrogen complexes forming as a result of ultimate nitrogen aggregation at high temperatures. However, direct measurements of nitrogen in the platelets revealed very little nitrogen (Kiflawi et al. 1998), and the currently accepted model for the origin of the platelets is a large regular array of carbon interstitials (Goss et al. 2003). These platelets produce sharp absorption peaks at 1359–1375 and 330 cm^{-1} in the IR absorption spectra; the relative proportions of the 'platelets' structural impurity in diamond is usually estimated in arbitrary units in absorption coefficient values measured at 1365 cm^{-1} (Fig. 7.7).

In addition to the absorption bands corresponding to the main nitrogen impurity centers, IR absorption lines at 3310, 3237, 3150, 3107, 1405 cm^{-1} reflect the presence of the **structural impurity of hydrogen** in diamond. The sharp absorption peaks at 3107 and 1405 cm^{-1} with accompanying lines (Fig. 7.7) were attributed to C–H bond vibrations (Chrenko et al. 1967; Woods and Collins 1983); later, these peaks were interpreted to be bending and stretching vibrations of C–H bonds, probably located on the interfacial surfaces (Clark et al. 1992). It has also been suggested that the H-related IR absorptions at 1405 and 3107 cm^{-1} in diamond are not simply due to C-H vibrations, but rather to stretching modes of complexes involving hydrogen and nitrogen (Fritsch et al. 2007). Recently, a new atomic model of this center has been proposed with a complex of the N3 defect and hydrogen atom VN_3H (Goss et al. 2014). Peaks at 2785, 3050, 3154 and 3236–3237 cm^{-1} may be attributed to hydrogen centers as well, but they are only 0.1–18% of the 3107 cm^{-1} peak intensity and may be caused by different hydrogen centers (Rondeau et al. 2004). Usually, only one line 3107 cm^{-1} is measured for the evaluation of hydrogen concentration in diamond (Kaminsky and Khachatryan 2001; Zaitsev 2001).

7.5.2 *Nitrogen in the Lower-Mantle Diamonds*

The systematic analyses of nitrogen impurity in diamond from different localities revealed great variations of both total nitrogen content and its aggregation level in various localities (Kaminsky and Khachatryan 2001). Lower-mantle diamonds from all localities are substantially different in their nitrogen impurity characteristics from most other kimberlite-related diamonds. They have a low total nitrogen content (most of them are Type II) and a very high nitrogen aggregation level (Hutchison et al. 1999; Harte et al. 1999; Kaminsky et al. 2001, 2009; Davies et al. 2004; Bulanova et al. 2008; Arajuo et al. 2013; Zedgenizov et al. 2014; Burnham et al. 2016).

In the majority of the studied lower-mantle diamonds, IR absorption in the single-phonon spectral range (400–1700 cm^{-1}) is extremely weak to undetectable. In total, 12–90% of the stones are 'nitrogen-free' type II diamonds with <10–20 atomic ppm nitrogen concentrations. In the IR spectra of the nitrogen-containing stones, the most common peak is at 1175 cm^{-1}, which is related to the nitrogen B-center (Fig. 7.8); its concentration is 21–305 atomic ppm. These type IaB-diamonds comprise 40–90% of the nitrogen-containing diamonds. Some diamonds, mainly from the pipes, have a spectral absorption maximum at 1282 cm^{-1}, which is related to the nitrogen A-center (forming type IaA and transitional type IaAB diamonds), and a narrow band at 1360–1375 cm^{-1}, which is attributed to 'platelets'. The average aggregation rate in the Juina diamonds is 94% (Fig. 7.8). Some diamonds demonstrate zonal structure with a strong increase (from the core to the rim) of the total nitrogen content and aggregation rate (Bulanova et al. 2010; Zedgenizov et al. 2014), consistent with previous data on the increase of the aggregation rate as a function of the total nitrogen concentration in lithospheric diamond (Taylor et al. 1990) and can be explained in terms of constant temperature during growth.

Such extremely high levels of nitrogen aggregation in the lower-mantle diamonds is caused by their long residence within the deep mantle under higher temperatures by ~100–150 °C, compared with lithospheric diamond temperatures. For the continental lithosphere, at temperatures of ~1250 °C, ~3 Ga is required to reach such an aggregation level (Fig. 7.8). The residence time of ~1 Ga is needed

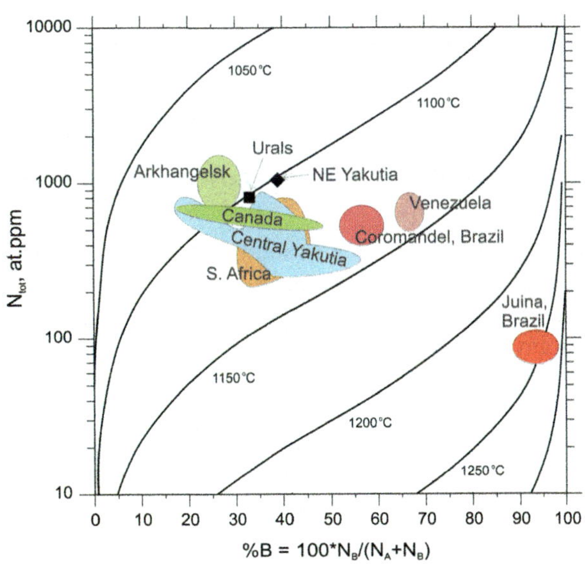

Fig. 7.8 Average nitrogen concentrations and aggregation in the Juina diamonds in comparison with diamonds from different areas. Isotherm curves for 3 Ga after Taylor and Milledge (1995). Modified after Kaminsky and Khachatryan (2001)

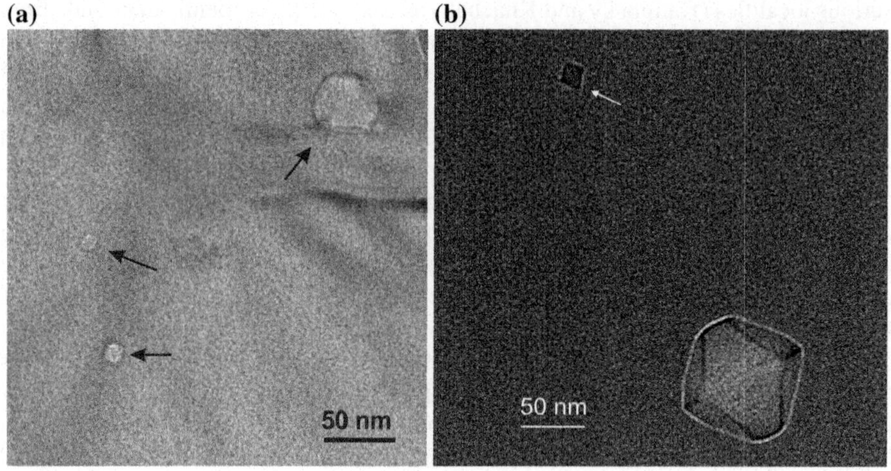

Fig. 7.9 TEM bright-field images of nitrogen inclusions in lower-mantle diamonds from the Juina area, Brazil. **a** Group of inclusions of nitrogen in diamond. **b** Detailed image of a octahedral inclusions of nitrogen. Note the same orientation of the inclusions in the host diamond. Photo by R. Wirth. From Kaminsky et al. (2013, 2015)

for the lower-mantle conditions to have ambient temperatures at ~2000 °C at a depth of 1000 km, according to a representative adiabatic geotherm of the lower mantle (see Fig. 2.3).

Inclusions of free nitrogen, 20–100 nm in size were repeatedly identified in lower mantle diamonds (Kaminsky et al. 2013, 2015). The inclusions have 'negative' shapes and the same crystallographic orientation parallel to the diamond (111) faces (Fig. 7.9). The nitrogen composition of the inclusions was confirmed by electron energy loss spectroscopy (EELS) and by the presence of a strong N–K edge in the spectrum. Raman spectroscopy of such inclusions indicated their solid cubic nature (Navon et al. 2017).

7.5.3 Hydrogen in the Lower-Mantle Diamonds

Hydrogen is the second (after nitrogen) most important impurity in natural diamond, forming a C–H defect recognisable at the 3107 cm^{-1} in IR spectra. The exact concentration of hydrogen in diamond is still unknown because of the incapability of any existing technique to measure hydrogen directly (Fritsch et al. 2007). Early estimates, based on chemical analyses of burned diamond samples, show hydrogen concentrations from insignificant to 0.5 wt% (Orlov 1987); however, they are not very reliable. To date, the only way to establish the relative quantity of hydrogen in diamond is to measure line 3107 cm^{-1} in its IR spectrum and to consider it as a conventional value of the hydrogen concentration in diamond (Fig. 7.7). So far, no correlation is known to exist between the intensity of this peak and the absolute concentration of hydrogen.

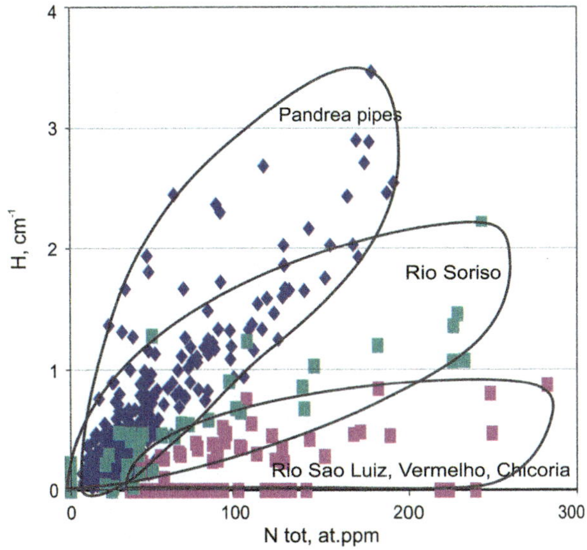

Fig. 7.10 Distribution of hydrogen against total nitrogen content in diamonds from the Juina area, Brazil. Modified after Kaminsky et al. (2009)

The hydrogen center identified in lithospheric diamonds is rather rare; usually the intensity of the 3107 cm^{-1} peak does not exceed 0.1–0.3 cm^{-1}; in high concentrations (up to 10 cm^{-1}), it was met only in a coat of some coated diamonds (Yelisseyev et al. 2004). In contrast to most diamond deposits worldwide, almost all diamonds from the Juina area (70–89%) have noticeable (up to 4–6 cm^{-1}) levels of hydrogen centers (Fig. 7.10). A positive correlation between concentrations of N and H was established for the Juina diamonds both from kimberlite pipes and placer deposits; each deposit shows its own specific correlation (Fig. 7.10). Rondeau et al. (2004) reported a correlation between N and H in 'asteroidal' (section-growth) diamonds from Brazil and India as well, and suggested that this correlation may be due to the trapping of hydrogen at a C–N bond. This data has been confirmed by experiments, according to which high temperature and aggregation progress of A-centers during diamond growth play important roles in the formation of the 3107 cm^{-1} center (Liu et al. 2016).

7.6 Isotopic Compositions of Carbon and Nitrogen in the Lower-Mantle Diamonds

7.6.1 Carbon

Isotopic composition of carbon in upper-mantle diamonds from kimberlites and lamproites varies widely from −39 to +5‰ $\delta^{13}C$ VPDB (e.g., Galimov 1985; Cartigny 2005) (Fig. 7.11).

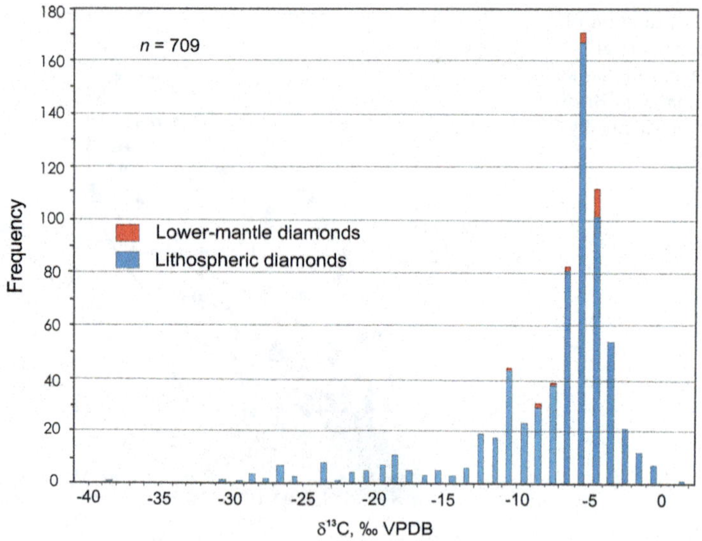

Fig. 7.11 Isotopic composition of carbon in diamonds from kimberlites, lamproites and placer deposits related to them. Shown in the figure is data only from diamonds where both carbon and nitrogen isotopes were analysed. Data from Javoy et al. (1984), Boyd et al. (1987, 1992), Cartigny et al. (1997, 1998, 1999, 2001, 2003, 2004a, b, 2009), Boyd and Pillinger (1994), van Heerden et al. (1995), Stachel and Harris (1997), Hutchison et al. (1999), Griffin et al. (2001), Palot et al. (2009, 2012), Thomassot et al. (2009), Klein-BenDavid et al. (2010)

It was established that 'eclogitic' diamonds occupy the entire range, including both isotopically 'light' and 'heavy' diamonds, while juvenile 'peridotitic' diamonds have a much closer range of values, between −10 and −1‰ δ^{13}C (Sobolev et al. 1979). However, among 'peridotitic' diamonds 'light' samples are also known (Deines et al. 1984, 1991, 1993; Galimov et al. 1994; Cartigny et al. 1999). *Only diamonds with lower-mantle oxide inclusions have suggested juvenile isotopic carbon compositions from −10 to −1‰ δ^{13}C.* Lower-mantle diamonds with carbonate inclusions have δ^{13}C values within the same range, from −8.85 to −2.31‰ (Pinti et al. 2016). The analysis of the distribution of the δ^{13}C demonstrates that the lower-mantle diamonds occupy an even closer range: from −5.60 to −0.78‰ (Hutchison et al. 1999; Kaminsky et al. 2001, 2009; Stachel et al. 2002; Palot et al. 2012; Arajuo et al. 2013; Zedgenizov et al. 2014; Burnham et al. 2016) (Figs. 7.11 and 7.12). There are singular analyses with δ^{13}C values of −6.40, −7.59, −8.31, −9.60‰, and even −11.6 and −0.49‰. The latter two contain inclusions of CaSi-perovskite and may belong to the lowermost part of the transition zone. Excluding these two analyses, the average value of δ^{13}C in lower-mantle diamonds is −4.32‰ ± 2.45‰ (2σ). It may be considered as the juvenile lower-mantle carbon isotopic composition (Fig. 7.12).

Lower-mantle diamonds are isotopically homogeneous: they have core-to-rim variations similar to the average error value, 2.6–3.1‰ (Zedgenizov et al. 2014; Burnham et al. 2016). In some samples, the rim is heavier than the core; in the

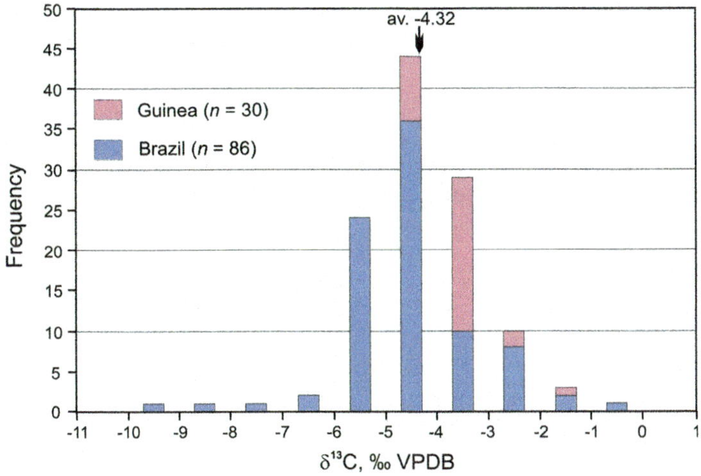

Fig. 7.12 Isotopic composition of carbon in lower-mantle diamonds from different areas. Data from Hutchison et al. (1999), Kaminsky et al. (2001, 2009), Stachel et al. (2002), Palot et al. (2012), Zedgenizov et al. (2014), Burnham et al. (2016), Pinti et al. (2016)

others, the reverse is true. Guinean diamonds are isotopically lighter than the Brazilian diamonds (−3.71 ± 1.30‰ and −4.67 ± 2.37‰, respectively), demonstrating regional (or stratigraphic) carbon isotopic heterogeneity in the lower mantle.

7.6.2 Nitrogen

Variations of the isotopic composition of nitrogen in natural diamonds of different genesis vary within the range of −25‰ and +18‰ $\delta^{15}N_{atm}$ (Fig. 7.13).

The $\delta^{15}N_{atm}$ values for the lower-mantle diamonds are located within a closer range from −5.2 to −1.0‰ (Hutchison et al. 1999; Palot et al. 2012) (Fig. 7.14). Because of a small number of analyses, only a preliminary estimate of the average value of the isotopic composition of nitrogen in the lower mantle can be made: $\delta^{15}N_{atm} = -3.00‰ \pm 2.37‰$ $\delta^{15}N_{atm}$ (2σ). Further analytical works are needed to better refine this estimate.

7.7 Structure of Diamond

Two major high-pressure polymorphs of carbon are known: diamond and lonsdaleite (Bundy and Kasper 1967; Frondel and Marvin 1967; Lonsdale 1971). In contrast to the sp^2 bonding environment of carbon in graphite, each carbon atom in

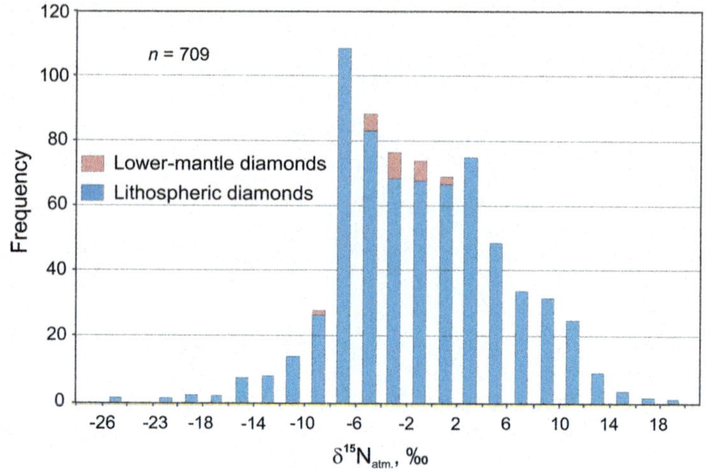

Fig. 7.13 Isotopic composition of nitrogen in diamonds of different genesis. Data from Javoy et al. (1984), Boyd et al. (1987, 1992), Cartigny et al. (1997, 1998, 1999, 2001, 2003, 2004a, b, 2009), Boyd and Pillinger (1994), van Heerden et al. (1995), Stachel and Harris (1997), Hutchison et al. (1999), Griffin et al. (2001), Palot et al. (2009, 2012), Thomassot et al. (2009), Klein-BenDavid et al. (2010)

Fig. 7.14 Isotopic composition of nitrogen in diamonds with lower-mantle inclusions from the Juina area in Brazil. Data from Hutchison et al. (1999), Palot et al. (2012)

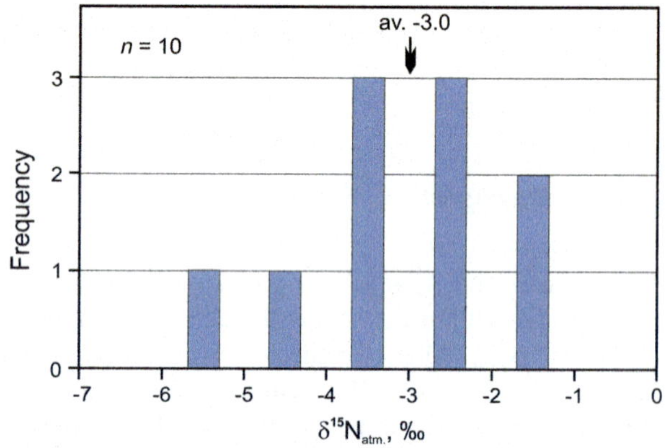

both diamond and lonsdaleite can bind to four adjacent carbon atoms in tetrahedral co-ordination (Hazen et al. 2013). This tetrahedral bonding configuration reflects the hybridization of one 2*s* and three 2*p* orbitals from each carbon atom to form four sp^3 orbitals. The structures of cubic diamond and hexagonal lonsdaleite are similar: both forms of carbon feature tetrahedral co-ordination of C in a 3D framework (Fig. 7.15). However, given a flat layer of linked carbon tetrahedra with all vertices pointed in the same direction (layer A), there exists two ways to stack subsequent layers, in orientations described as B or C. The diamond structure represents a three-layer stacking sequence of [...ABCABC...] along the

(111) cubic direction, whereas lonsdaleite stacking is two-layer [...ABAB...] along the (001) hexagonal direction (Hazen et al. 2013). It should be noted that an alternative model of the lonsdaleite structure as twinning and stacking faults was offered (Nemeth et al. 2014), and Jones et al. (2016) consider the stacking disorder as a reason of the existence of transitional artificial diamond-hexagonal phase varieties, leaving the name 'lonsdaleite' to a natural mineral.

Both diamond and lonsdaleite are metastable phases at ambient conditions. The large cohesive energy of diamond (717 kJ/mol), together with significant energy barriers, gives rise to its high degree of metastability and extremely high melting temperature (5000 K). Carbon in general has a remarkable feature of existence in a broad range of metastable phases that can be formed near ambient conditions and their wide fields of kinetic stability (Oganov et al. 2013). Diamond survives at both low (room conditions) and very high pressures. Similarly, lonsdaleite is known in both meteoritic (Ross et al. 2011; Nakamuta and Toh 2013), terrestrial impact (Massaitis et al. 1972; Valter et al. 1992; Kaminsky 1994; Massaitis 2013; Ohfuji et al. 2015) and metamorphic (Dubinchuk et al. 2010; Shumilova et al. 2011; Godard et al. 2012) rocks; yet its occurrence cannot be interpreted in terms of origin, since fundamental understanding of its metastability remains unresolved (Godard et al. 2012; Kurdyumov et al. 2012).

The graphite → lonsdaleite transitions have been intensely investigated (e.g., Hazen et al. 2013). The mechanism of diamond → lonsdaleite transition is also possible; it was described as a deformation of a homogeneous sphere packing with three contacts per sphere (type 3/10/01) in the common subgroup *Pnna* of *Fd*3*m* and $P6_3/mmc$ (Sowa and Koch 2001); however, the parameters of this transition are, as yet, unknown. Oganov et al. (2013) presented a phase diagram of carbon, in which lonsdaleite from graphite is quenchable and recoverable above ~1070 K at 8–10 GPa; whereas, below ~1070 K, lonsdaleite is unquenchable. This may possibly explain the absence of lonsdaleite even as relics in most observed terrestrial samples except impact ones, in which the quenching took place

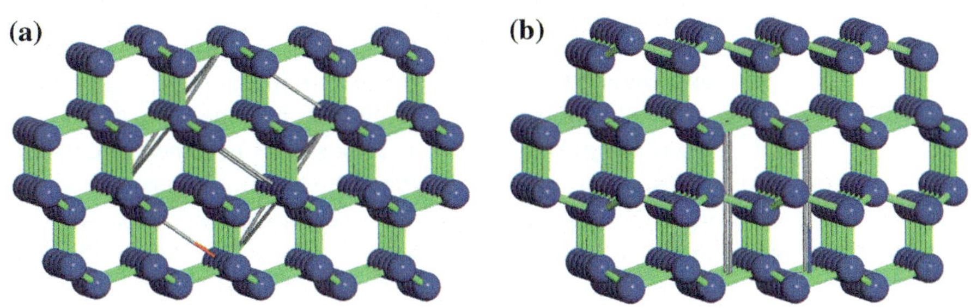

Fig. 7.15 Crystal structures of diamond (**a**) and lonsdaleite (**b**). The diamond crystal structure is cubic (space group *Fd*3*m*; a = 3.560 Å; Z = 8); it features a framework of tetrahedrally co-ordinated carbon atoms. The lonsdaleite crystal structure is hexagonal (space group $P6_3/mmc$; a = 2.52 Å; c = 4.12 Å; Z = 4). From Hazen et al. (2013). Used with permission of Mineralogical Society of America

at high temperatures. In addition, in natural impact events, kinetics and grain size can influence the survival of lonsdaleite. The extremely high energy released during natural impacts and the long duration of a large natural shock pulse ($\sim$1 s) may be many orders of magnitude greater than those that can be replicated in the laboratory (DeCarli et al. 2002).

Recently, Jones et al. (2016) studied the natural diamond shocked under pressure of 60 GPa. They established the presence of the lonsdaleite phase in that diamond and concluded that 'hexagonality' can be achieved by shocking diamond and that intermediate stacking sequences between pure diamond (fully cubic) and lonsdaleite (fully hexagonal) phases may occur, like in natural impact diamonds. One may suggest that the diamond → lonsdaleite transformation may occur not only as a result of shock impact, but under static high *P–T* compression (Bundy and Kasper 1967) and the existence of lonsdaleite in the lower mantle. Gorshkov et al. (1997) reported an intergrowth of hexagonal form and cubic diamond from the Lianing deposit, China; however, this finding has not yet been confirmed and, as such, further studies in this field are necessary.

7.8 Origin of Diamond in the Lower Mantle

7.8.1 Diamond-Parental Medium

Based on mineral inclusions and fluids identified in diamond (as well as on experimental data), it is accepted now that diamond in the mantle was formed within a carbonatitic or oxide-carbonatitic media; the presence of halides stimulates the growth kinetics (e.g., Litvin et al. 2012, 2014; Spivak 2015; Bureau et al. 2016; Palyanov et al. 2016). The parental carbonatitic melt was formed as a result of partial melting of the juvenile lower-mantle material, starting at the CMB (see Sect. 6.5.3). The ascending of the melt to colder domains of the lower mantle with lower f_{O_2} stimulates crystallization of diamond through the reduction of carbonate components (Ryabchikov and Kogarko 2016).

Exsolved carbon exists in the parental diamond-forming medium. The formation of diamond and various syngenetic minerals (inclusions in diamond) takes place simultaneously during crystallization of the juvenile parental medium under temperature and fugitivity decreasing conditions. The relic strains in mineral inclusions in diamond prove their formation and the growth of host diamond crystals simultaneously, i.e., their syngenetic origin during the course of lowering temperatures in a diamond-forming chamber within the lower mantle. Ca-, Mg-, Fe- and Na-carbonates, bridgmanite, CaSi-perovskite, periclase-wüstite solid solution, silica and carbon were major components of the crystallized system.

Genetic classification of the constituents of a diamond-parental medium and, correspondingly, primary major and secondary inclusions in lower-mantle diamonds provide a possibility for the development of a generalized composition diagram for the lower-mantle parental medium (Fig. 7.16). In this diagram, the main tetrahedron, which symbolizes the entire parental diamond-forming medium, is accompanied by smaller satellite tetrahedrons, representing ultramafic, mafic, carbonatitic and minor soluble components. Symbol ‘C’ of carbon as a common component is moved beyond the diagram for simplicity. A satellite triangle of the minor insoluble components is separated from the main one by a symbolic border (hatched-and-dotted line) of complete liquid immiscibility. Thus, the classification indicators of all primary inclusions in lower-mantle diamonds are properly accounted for in the generalized diagram of multicomponent heterogeneous and chemically changeable diamond-parental media compositions.

Compositions of the parental medium are shown as the shaded area at the bottom of the main tetrahedron and are limited by a line of a presumed concentration barrier of diamond nucleation (CBDN), within the compositional triangle of ultramafic–mafic–carbonatite composition. This barrier outlines the upper limit of the oxide constituent content, which results in the termination of diamond nucleation. According to experimental data in carbonate-oxide–silicate systems under upper-mantle *P–T* conditions, the diamond-parental compositions are enriched with

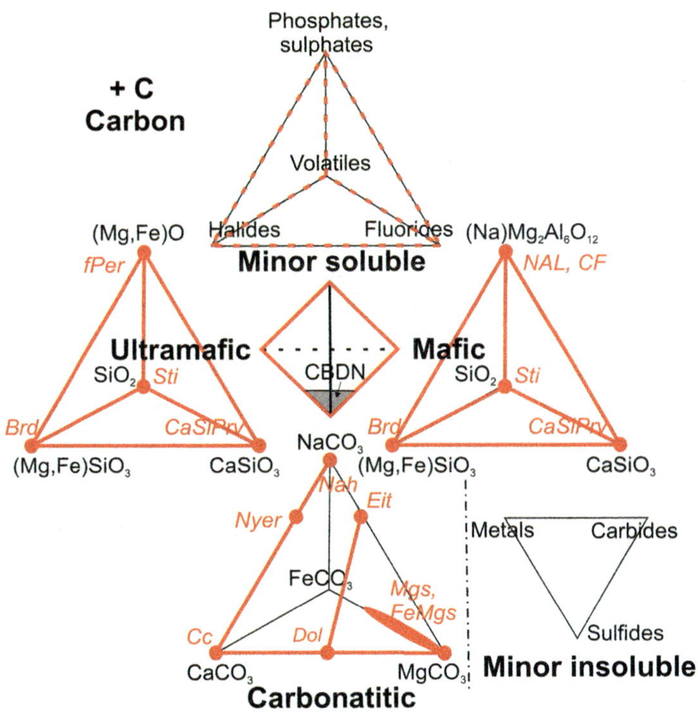

Fig. 7.16 Generalised compositional diagram for the lower-mantle diamond-forming medium for lower-mantle diamond. Symbols—see Fig. 6.1. *CBDN* concentration barrier of diamond nucleation. Modified after Litvin et al. (2014)

carbonate constituents of 65–70 wt% (Litvin et al. 2012); we suggest the same position of the CBDN at least in the carbonate-oxide system of the upper part of the lower mantle.

The generalized composition diagram for parental medium of the lower-mantle diamonds is schematically characteristic for physicochemical conditions of syngenesis for the diamonds and their primary inclusions at parental magmas with changeable contents of ultramafic, mafic and carbonatitic components.

7.8.2 The Mechanism for the Formation of the Lower-Mantle Diamonds

Crystallization of diamond under the 'melt-solution' mechanism was experimentally realized using the LHDAC cell in melts of $MgCO_3$–C, $CaCO_3$–C, Na_2CO_3–C and Mg–Fe–Na-carbonate–magnesiowüstite–bridgmanite–carbon systems within the 20–30 GPa pressure range, i.e., in an ultramafic-carbonatitic system under relevant *P–T* conditions of the transition zone and the lower mantle (Litvin et al. 2014, 2016; Spivak 2015). Both binary $CaCO_3$, $MgCO_3$, Na_2CO_3 and multicomponent $MgCO_3$–$FeCO_3$–Na_2CO_3 carbonates melt congruently, and the melts formed are stable over sufficiently extended *P–T* phase fields in these regions. The wide stability of the carbonate melts makes it thermodynamically possible to consider them as key components of the lower-mantle diamond-parental medium. The parental melts are bound to be capable: (1) to dissolve carbon and form oversaturated carbonate-carbon melt-solutions effective for the crystallization of diamond; (2) to dissolve the lower mantle minerals together with carbon and form the completely miscible carbonate-oxide–carbon diamond-parental melts of multicomponent composition and; (3) to be formed as the diamond-parental medium by the low-temperature boundary of congruent (eutectic) carbonate melting at temperatures compatible with geothermal values for the lower mantle. Testing experiments on diamond formation in representative melt-solutions of carbonate–magnesiowüstite–bridgmanite–carbon systems at expected *P–T* conditions of the transition zone and the lower mantle demonstrated their high efficiency for nucleation and mass crystallization of 'super-deep' diamonds. The kinetic peculiarities provide indirect evidence of high diamond solubility in carbonate and carbonate-oxide melts of the lower mantle. Thus, the 'melt-solution' physicochemical mechanism of 'super-deep' diamond and paragenetic inclusions formation in completely miscible carbonate-oxide melts with dissolved carbon is the most credible.

Another mechanism for lower-mantle diamond formation has been derived based on decomposition of carbonate melts into oxides, oxygen and diamond at high-temperature boundaries of their one-phase fields (e.g., Ivanov and Deutsch 2002) with a significant volume increase on the Hugoniot curve at 3300 K. The effect was explained by the appearance of a supercritical fluid phase of CO_2 as a result of the decomposition of $CaCO_3$ into CaO and CO_2. Then, the diamond-forming reaction could consist of two steps: (1) decomposition of carbonate melt to oxide and CO_2 fluid; and (2) decomposition of CO_2 fluid to diamond and oxygen phase. This two-step reaction of diamond formation at a high-temperature boundary of the carbonate melt stability (where CO_2 comes into being) is in good agreement with the LHDAC experimental data by Tschauner et al. (2001), which showed diamond formation at a decomposition of a strongly compressed CO_2 fluid. However, these results are not applicable to the origin of the lower-mantle diamonds for the following reasons: (1) the decomposition temperatures are too high to be compatible with geothermal values within the lower mantle; and (2) the mechanism of diamond formation by decomposition of carbonates does not comply with the criterion of diamond and inclusion syngenesis and is of no significance for the formation of heterogeneous syngenetic inclusions in association with diamonds (Litvin et al. 2014, 2016).

Reservoirs of parental partially molten medium for lower-mantle diamonds and its inclusions were probably formed within the ultramafic lower-mantle material under chemical interactions with metasomatic agents that originated under processes of the mantle dynamics. Major and minor components and phases of the lower-mantle ultramafic rocks could be involved in the formation of a primary diamond-parent carbonate-oxide medium. Melts of the primary partially molten parental medium with ultramafic oxide components could be subjected to fractional crystallization, resulting in the in situ origin of stishovite-bearing mafic parageneses. Physicochemical mechanisms of lower-mantle diamond formation in the parental melts reservoir is conditioned by the formation of carbon-saturated (in respect to diamond) carbonate-oxide–carbon (carbonatitic) melts. Under the regime of the carbonatite-carbon melt cooling, carbon-saturated carbonatitic melts can turn oversaturated that offers the nucleation and crystallization of diamonds. Accompanied formations of paragenetic minerals in the parental medium has opened opportunities for capturing their fragments by growing diamonds along with xenogenetic minerals penetrated into the parental melts outside (Litvin et al. 2014, 2016). Based on experiments at 26 GPa pressure conditions and variable temperature values, Spivak (2015) compiled a model of fractional crystallization of the diamond-parental melt of the ultramafic-mafic composition. During the course of fractional crystallization with the decrease of temperature, the composition of the melt changes from ultramafic to mafic, with the final formation of the association magnesiowüstite + stishovite + CaSi-perovskite + carbonate + diamond. Bridgmanite disappears as a result of the reaction Fe-bridgmanite + L → magnesiowüstite + stishovite in a peritectic point L + periclase-wüstite solid solution + Fe-bridgmanite + stishovite + CaSi-perovskite + carbonate + diamond. During the process of cooling in a chamber, labile and metastable melt-solution phases appear, which are oversaturated towards diamond, and diamond crystallization occurs.

7.9 Conclusions

Lower-mantle diamond has a series of features, different from those in lithospheric diamond. The morphological and structural peculiarities of lower-mantle diamond indicate its origin and long-time residence under high-*T* conditions. Among morphological forms of diamond, dodecahedroids and tetrahexahedroids (formed as a result of high-*T* octahedral from resorption), while low-*T* cubic forms are absent. All types of luminescence in lower-mantle diamonds are, in general, homogeneous, witnessing the growth of diamond in stable *P–T* conditions.

Lower-mantle diamonds, after their origin, had a long-term residence in deep Earth. This resulted not only in the resorption of crystals but also in low total nitrogen content, a great proportion of 'nitrogen-free' stones, and almost 100% aggregation level in nitrogen-containing stones as a result of the high-temperature diffusion of nitrogen.

Diamond crystals in the lower mantle experienced intense plastic deformation, resulting in the domination of 490.7 nm, H3, and H4s in the PL spectra, which are not characteristic for lithospheric diamonds. The latter center is accompanied by the EPR center with a *g*-factor of 2.00285, which was first detected in crystals with a low nitrogen content. Cathodoluminescence kinetics of the lower-mantle diamonds differ considerably from the lithospheric diamonds for both NV^{o}- and NV^{-}-centers in their much longer decay times. Characteristic brown colouration of numerous lower-mantle diamonds is also a result of their high-*T* plastic deformation.

Most lower-mantle diamonds have a high concentration of hydrogen impurity, which is a specific, non-characteristic for asthenospheric diamond. This feature still needs further study, and first of all a technique for the direct analysis of hydrogen in diamond must be created.

Isotopic compositions of carbon and nitrogen in lower-mantle diamonds exist in narrow ranges; they allow establishment of average isotopic characteristics for juvenile lower mantle: $\delta^{13}C_{VPDB} = -4.32‰ \pm 2.45‰$ (2σ) and $\delta^{15}N_{atm} = -3.00‰ \pm 2.37‰$ (2σ).

Lower-mantle diamond was formed in carbonate-oxide parental magmas, which experienced fractional crystallization with the decrease of temperature and changes in the melt composition. Among the results of this process was the appearance of labile and metastable melt-solution phases, which were oversaturated towards diamond, and diamond crystallization occurred. The most important role in this process belongs to the carbonate component in the parental melt.

References

Arajuo, D. P., Gaspar, J. C., & Bulanova, G. P. (2013). Juina diamonds from kimberlites and alluvials: A comparison of morphology, spectral characteristics and carbon isotope composition. In *Proceedings of the 10th International Kimberlite Conference* (Vol. 1) (Special Issue of the Journal of the Geological Society of India, Vol. 1, pp. 255–269).

Bartoshinsky, Z. V. (1962). *Crystallomorphology of Yakutian diamonds*. Ph.D. thesis, Lvov University. Lvov, 17 pp. (in Russian).

Beha, K., Batalov, A., Manson, N. B., Bratschitsch, R., & Leitenstorfer, A. (2012). Optimum photoluminescence excitation and recharging cycle of single nitrogen-vacancy centers in ultrapure diamond. *Physical Review Letters, 109,* 097404. doi:10.1103/PhysRevLett.109.097404

Boyd, S. R., Kiflawi, I., & Woods, G. S. (1994). The relationship between infrared absorption and A-defect concentration in diamond. *Philosophical Magazine B, 69*(6), 1149–1153.

Boyd, S. R., Kiflawi, I., & Woods, G. S. (1995). Infrared absorption by the B nitrogen aggregate in diamond. *Philosophical Magazine B, 72*(3), 351–361.

Boyd, S. R., Mattey, D. P., Pillinger, C. T., Milledge, H. J., Mendelssohn, M., & Seal, M. (1987). Multiple growth events during diamond genesis: An integrated study of carbon and nitrogen isotopes and nitrogen aggregation state in coated stones. *Earth and Planetary Science Letters, 86*(2–4), 341–353.

Boyd, S. R., & Pillinger, C. T. (1994). A preliminary study of $^{15}N/^{14}N$ in octahedral growth form diamonds. *Chemical Geology, 116,* 43–59.

Boyd, S. R., Pillinger, C. T., Milledge, H. J., Mendelssohn, M. J., & Seal, S. M. (1992). C and N isotopic composition and the infrared absorption spectra of coated diamonds: Evidence for the regional uniformity of CO_2–H_2O rich fluids in lithospheric mantle. *Earth and Planetary Science Letters, 109,* 633–644.

Bulanova, G. P., Smith, C. B., Kohn, S. C., Walter, M. J., Gobbo, L., & Kearns, S. (2008). Machado River, Brazil—A newly recognised ultradeep diamond occurrence. In: *9th International Kimberlite Conference Extended Abstract No. 9IKC-A-00233.*

Bulanova, G. P., Walter, M. J., Smith, C. B., Kohn, S. C., Armstrong, L. S., Blundy, J., et al. (2010). Mineral inclusions in sublithospheric diamonds from Collier 4 kimberlite pipe, Juina, Brazil: Subducted protoliths, carbonated melts and primary kimberlite magmatism. *Contributions to Mineralogy and Petrology, 159*(4), 489–510. doi:10.1007/s00410-010-0490-6

Bundy, F. P., & Kasper, J. S. (1967). Hexagonal diamond—A new form of carbon. *The Journal of Chemical Physics, 46,* 3437.

Bureau, H., Frost, D. J., Bolfan-Casanova, N., Leroy, C., Esteve, I., & Cordier, P. (2016). Diamond growth in mantle fluids. *Lithos, 265,* 4–15. doi:10.1016/j.lithos.2016.10.004

Burnham, A. D., Bulanova, G. P., Smith, C. B., Whitehead, S. C., Kohn, S. C., Gobbo, L., et al. (2016). Diamonds from the Machado River alluvial deposit, Rondônia, Brazil, derived from both lithospheric and sublithospheric mantle. *Lithos, 265,* 199–213. doi:10.1016/j.lithos.2016.05.022

Bursill, L. A., & Glaisher, R. W. (1985). Aggregation and dissolution of small and extended defect structures in type Ia diamond. *American Mineralogist, 70,* 608–618.

Cartigny, P. (2005). Stable isotopes and the origin of diamond. *Elements, 1*(2), 79–84.

Cartigny, P., Boyd, S. R., Harris, J. W., & Javoy, M. (1997). Nitrogen isotopes in peridotite diamonds from Fuxian, China: The mantle signature. *Terra Nova, 9,* 175–179.

Cartigny, P., Chinn, I., Viljoen, K. S., & Robinson, D. (2004a). Early Proterozoic (>1.8 Ga) ultrahigh pressure metamorphism: Evidence from Akluilak microdiamonds. *Science, 304,* 853–855.

Cartigny, P., De Corte, K., Shatsky, V. S., Ader, M., De Paepe, P., Sobolev, N. V., et al. (2001). The origin and formation of metamorphic microdiamonds from the Kokchetav massif, Kazakhstan: A nitrogen and carbon isotopic study. *Chemical Geology, 176*(1–4), 265–281.

Cartigny, P., Farquar, J., Thomassot, E., Harris, J. W., Wing, B., Masterson, A., et al. (2009). A mantle origin for Paleoarchean peridotite diamonds from the Panda kimberlite, Slave Province: Evidence from ^{13}C, ^{15}N and $^{33-34}S$ stable isotope systematics. *Lithos, 112S,* 852–864.

Cartigny, P., Harris, J. W., & Javoy, M. (1998). Eclogitic diamond formation of Jwaneng: No room for a recycled component. *Science, 280*(5368), 1421–1424.

Cartigny, P., Harris, J. W., & Javoy, M. (1999). Eclogitic, peridotitic and metamorphic diamonds and the problems of carbon recycling—The case of Orapa (Botswana). In *Proceedings of the VIIth International Kimberlite Conference* (Vol. 1, pp. 117–124). Cape Town: Red Roof Design.

Cartigny, P., Harris, J. V., Taylor, A., Davies, R., & Javoy, M. (2003). On the possibility of a kinetic fractionation of nitrogen stable isotopes during natural diamond growth. *Geochimica et Cosmochimica Acta, 67*(6), 1571–1576.

Cartigny, P., Stachel, T., Harris, J. W., & Javoy, M. (2004b). Constraining diamond metasomatic growth using C- and N-stable isotopes: Examples from Namibia. *Lithos, 77,* 359–373.

Cayzer, N. J., Odake, S., Harte, B., & Kagi, H. (2008). Plastic deformation of lower mantle diamonds by inclusion phase transformations. *European Journal of Mineralogy, 20*(3), 333–339.

Chrenko, R. M., McDonald, R. S., & Darrow, K. A. (1967). Infra-red spectra of diamond coat. *Nature, 213*(5075), 474–476.

Clark, C. D., Collins, A. T., & Woods, G. S. (1992). Absorption and luminescence spectroscopy. In J. E. Field (Ed.), *The properties of natural and synthetic diamonds* (pp. 35–80). London: Academic Press.

Collins, A. T., Thomaz, M. F., & Jorge, M. I. B. (1983). Luminescence decay time of the 1.945 eV centre in type Ib diamond. *Journal of Physics C: Solid State Physics, 16,* 2177.

Correa, A. A., Bonev, S. A., & Galli, G. (2006). Carbon under extreme conditions: Phase boundaries and electronic properties from first-principles theory. In *Proceedings of the National Academy of the U.S.A.* (Vol. 103(5), pp. 1204–1208). doi:10.1073/pnas.0510489103

Davies, G. (1976). The A nitrogen aggregate in diamond: Its symmetry and possible structure. *Journal of Physics C: Solid State Physics, C9,* L537–L542.

Davies, G. (1977). The optical properties of diamond. In P. L. Walker & P. A. Thrower (Eds.), *Chemistry and physics of carbon* (Vol. 13, pp. 2–143). New York: Marcel Dekker.

Davies, R. M., Griffin, W. L., O'Reilly, S. Y., & Doyle, B. J. (2004). Mineral inclusions and geochemical characteristics of microdiamonds from the DO27, A154, A21, A418, DO18, DD17 and Ranch Lake kimberlites at Lac de Gras, Slave Craton, Canada. *Lithos, 77*(1–4), 39–55.

DeCarli, P. S., Bowden, E., Jones, A. P., & Price, G. D. (2002). Laboratory impact experiments versus natural impact events. *GSA Special Paper, 356,* 595–605.

Deines, P., Gurney, J. J., & Harris, J. W. (1984). Accociated chemical and carbon isotopic composition variations in diamonds from Finsch and Premier kimberlite, South Africa. *Geochimica et Cosmochimica Acta, 48*(2), 325–342.

Deines, P., Harris, J. W., & Gurney, J. J. (1991). The carbon isotopic composition and nitrogen content of lithospheric and astenospheric diamonds from the Jagersfontein and Koffifontein kimberlite, South Africa. *Geochimica et Cosmochimica Acta, 55*(5), 2615–2625.

Deines, P., Harris, J. W., & Gurney, J. J. (1993). Depth-related carbon isotope and nitrogen concentration variability in the mantle below the Orapa kimberlite, Botswana, Africa. *Geochimica et Cosmochimica Acta, 57,* 2781–2796.

DeVries, R. C. (1975). Plastic deformation and "work-hardening" of diamond. *Materials Research Bulletin, 10,* 1193–1200.

Dobrinets, A., Vins, V.G., & Zaitsev, A.M. (2013). HPHT-Treated Diamonds. Berlin: Springer, Heidelberg, p. 257.

Dubinchuk, V. T., Simakov, S. K., & Pechnikov, V. A. (2010). Lonsdaleite in diamond bearing metamorphic rocks of the Kokchetav massif. *Doklady Earth Sciences, 430*(1), 40–42.

Evans, T. (1992). Aggregation of nitrogen in diamond. In J. Field (Ed.), *The properties of natural and synthetic diamond* (pp. 259–290). London: Academic Press.

Evans, T., & Phaal, C. (1962). Imperfections in Type I and Type II diamonds. In *Proceedings of Royal Society A270* (pp. 538–552).

Fritsch, E., Hainschwang, T., Massi, L., & Rondeau, B. (2007). Hydrogen-related optical centers in natural diamond: An update. *New Diamond and Frontier Carbon Technology, 17(2),* 63–89.

Frondel, C., & Marvin, U. B. (1967). Lonsdaleite, a hexagonal polymorph of diamond. *Nature, 214,* 587–589.

Galimov, E. M. (1985). The relation between formation conditions and variations in isotope composition of diamonds. *Geochemistry International, 22*(1), 118–142.

Galimov, E. M., Zakharchenko, O. D., Maltsev, K. A., Makhin, A. I., & Pavlenko, T. A. (1994). The isotopic composition of carbon in diamonds from kimberlitic pipes of the Arkhangelsk district. *Geochemistry, 1,* 67–73.

Godard, G., Frezzotti, M. L., Palmeri, R., & Smith, D. C. (2012). Origin of high-pressure disordered metastable phases (lonsdaleite and incipiently amorphized quartz) in metamorphic rocks: Geodynamic shock or crystal-scale overpressure? In L. F. Dobrzhinetskaya, S. W. Faryad, S. Wallis, & S. Cuthbert (Eds.), *Ultrahigh pressure metamorphism: 25 years after the discovery of coesite and diamond* (pp. 125–148). London: Elsevier.

Gorshkov, A. I., Yanan, B., Bershov, L. V., Ryabchikov, I. D., Sivtsov, A. V., & Lapina, M. I. (1997). Inclusions in diamond from the Liaoning deposit (China) and their genetic meaning. *Geochemistry International, 35*(1), 58–65.

Goss, J. P., Briddon, P. R., Hill, V., Jones, R., & Rayson, M. J. (2014). Identification of the structure of the 3107 cm^{-1} H-related defect in diamond. *Journal of Physics. Condensed Matter, 26*(14), 145801.

Goss, J. P., Coomer, B. J., Jones, R., Fall, C. J., Briddon, P. R., & Öberg, S. (2003). Extended defects in diamond: The interstitial platelet. *Physical Review B, 67,* 165208.

Griffin, W. L., Win, T. T., Davies, R., Wathanakul, P., Andrew, A., Metcalfe, I., et al. (2001). Diamonds from Myanmar and Thailand: Characteristics and possible origins. *Economic Geology, 96*(1), 159–170.

Hanzawa, H., Nisida, Y., & Kato, T. (1997). Measurement of decay time for the NV centre in Ib diamond with a picosecond laser pulse. *Diamond and Related Materials, 6*(11), 1595–1598.

Harte, B., Harris, J. W., Hutchison, M. T., Watt, G. R., & Wilding, M. C. (1999). Lower mantle mineral associations in diamonds from Sao Luiz, Brazil. In Y. Fei, C. M. Bertka, & B. O. Mysen (Eds.), *Mantle petrology: Field observations and high pressure experimentation: A tribute to Francis R. (Joe) Boyd* (No. 6, pp. 125–153). Geochemical Society Special Publication.

Harte, B., Taniguchi, T., & Chakraborty, S. (2009). Diffusion in diamond. II. High-pressure-temperature experiments. *Mineralogical Magazine, 73*(2), 201–204. doi:10.1180/minmag.2009.073.2.201

Hayman, P. C., Kopylova, M. G., & Kaminsky, F. V. (2005). Lower mantle diamonds from Rio Soriso (Juina, Brazil). *Contributions to Mineralogy and Petrology, 149*(4), 430–445.

Hazen, R. M., Downs, R. T., Jones, A. P., & Kah, L. (2013). Carbon mineralogy and crystal chemistry. *Rev Mineral Geochem, 75,* 7–46.

Hutchison, M. T., Cartigny, P., & Harris, J. W. (1999). Carbon and nitrogen compositions and physical characteristics of transition zone and lower mantle diamonds from São Luiz, Brazil. In J. J. Gurney, J. L. Gurney, M. D. Pascoe, & S. H. Richardson (Eds.), *Proceedings of the VIIth International Kimberlite Conference* (Vol. 1, pp. 372–382). Cape Town: Red Roof Design.

Inam, F. A., Grogan, M. D. W., Rollings, M., Gaebel, T., Say, J. M., Bradac, C., et al. (2013). Emission and nonradiative decay of nanodiamond NV centers in a low refractive index environment. *ACS Nano, 7*(5), 3833–3843.

Ivanov, B. A., & Deutsch, A. (2002). The phase diagram of $CaCO_3$ in relation to shock compression and decompression. *Physics of the Earth and Planetary Interiors, 129,* 131–143.

Javoy, M., Pineau, F., & Demaiffe, D. (1984). Nitrogen and carbon isotopic composition in the diamonds of Mbuji Mayi (Zaire). *Earth and Planetary Science Letters, 68*(3), 399–412.

Jones, J., & Goss, J. P. (2002). Theory of aggregation of nitrogen in diamond. *EMIS Datareviews Series, 26,* 127–129.

Jones, J. P., McMillan, P. F., Salzmann, C. G., Alvaro, M., Nestola, F., Prencipe, M., et al. (2016). Structural characterization of natural diamond shocked to 60 GPa; implications for earth and planetary systems. *Lithos, 265,* 214–221. doi:10.1016/j.lithos.2016.09.023

Kaiser, W., & Bond, W. L. (1959). Nitrogen, a major impurity in common type I diamond. *Physical Review, 115*(4), 857–863.

Kaminsky, F. V. (1994). Carbonado and Yakutite: Properties and possible genesis. In H. O. A. Meyer & O. H. Leonardos (Eds.), *Proceedings of the Fifth International Kimberlite Conference: Vol. 2. Diamonds: Characterization, Genesis and Exploration* (pp. 136–143). Brazil.

Kaminsky, F. V., & Khachatryan, G. K. (2001). Characteristics of nitrogen and other impurities in diamond, as revealed by infrared absorption data. *Canadian Mineralogist, 39*(6), 1733–1745.

Kaminsky, F. V., Khachatryan, G. K., Andreazza, P., Araujo, D., & Griffin, W. L. (2009). Super-deep diamonds from kimberlites in the Juina area, Mato Grosso State, Brazil. *Lithos, 112S*(2), 833–842.

Kaminsky, F. V., Wirth, R., & Schreiber, A. (2013). Carbonatitic inclusions in Deep Mantle diamond from Juina, Brazil: New minerals in the carbonate-halide association. *Canadian Mineralogist, 51*(5), 669–688. doi:10.3749/canmin.51.5.669

Kaminsky, F. V., Wirth, R., & Schreiber, A. (2015). A microinclusion of lower-mantle rock and some other lower-mantle inclusions in diamond. *Canadian Mineralogist, 53*(1), 83–104. doi:10.3749/canmin.1400070

Kaminsky, F. V., Zakharchenko, O. D., Davies, R., Griffin, W. L., Khachatryan-Blinova, G. K., & Shiryaev, A. A. (2001). Superdeep diamonds from the Juina area, Mato Grosso State, Brazil. *Contributions to Mineralogy and Petrology, 140*(6), 734–753.

Kiflawi, I., Bruley, J., Luyten, W., & Van Tendeloo, G. (1998). 'Natural' and 'man-made' platelets in type-Ia diamonds. *Philosophical Magazine Part B, 78*(3), 299–314. doi:10.1080/014186398258104

Kiflawi, I., Mayer, A. E., Spear, P. M., Van Wyk, J. A., & Woods, G. S. (1994). Infrared-absorption by the single nitrogen and A defect centres in diamond. *Philosophical Magazine Part B, 69,* 1141–1147.

Klein-BenDavid, O., Pearson, D. G., Nowell, G. M., Ottley, C., McNeill, J. C. R., & Cartigny, P. (2010). Mixed fluid sources involved in diamond growth constrained by Sr–Nd–Pb–C–N isotopes and trace elements. *Earth and Planetary Science Letters, 289*(1/2), 123–133.

Kohn, S. C., Speich, L., Smith, C. B., & Bulanova, G. P. (2016). FTIR thermochronometry of natural diamonds: A closer look. *Lithos, 265,* 148–158. doi:10.1016/j.lithos.2016.09.021

Kukharenko, A. A. (1955). *Diamonds of the Urals* (515 pp.). Moscow: Geoltekhizdat Publishing House. (in Russian).

Kurdyumov, A. V., Britun, V. F., Yarosh, V. V., & Danlienko, A. I. (2012). The influence of shock compression on the graphite transformations into lonsdaleite and diamond. *Journal of Superhard Materials, 34,* 19–27.

Liaugaudas, G., Davies, G., Suhling, K., Khan, R. U., & Evans, D. J. (2012). Luminescence lifetimes of neutral nitrogen-vacancy centres in synthetic diamond containing nitrogen. *Journal of Physics: Condensed Matter, 24*(43), 435503–435507.

Litvin, Y. A., Spivak, A. V., & Kuzyura, A. V. (2016). Fundamentals of the mantle carbonatite concept of diamond genesis. *Geochemistry International, 54*(10), 839–857. doi:10.1134/S0016702916100086

Litvin, Y., Spivak, A., Solopova, N., & Dubrovinsky, L. (2014). On origin of lower-mantle diamonds and their primary inclusions. *Physics of the Earth and Planetary Interiors, 228,* 176–185. doi:10.1016/j.pepi.2013.12.007

Litvin, Y. A., Vasiliev, P. G., Bobrov, A. V., Okoemova, V Yu., & Kuzyura, A. V. (2012). Parental media of natural diamonds and primary mineral inclusions in them: Evidence from physicochemical experiment. *Geochemistry International, 50*(9), 726–759.

Liu, X., Jia, X., Fang, C., & Ma, H.-A. (2016). Diamond crystallization and growth in N–H enriched environment under HPHT conditions. *CrystEngComm, 18,* 8506. doi:10.1039/c6ce02034h

Lonsdale, K. (1971). Formation of lonsdaleite from single-crystal graphite. *American Mineralogist, 56,* 333–336.

Martynovich, E. F., Mironov, V. P., Rakevich, A. L., Stepanov, F. A., Emel'yanova, A. S., Shatskii, V. S., et al. (2013). Spectral-temporal luminescence microscopy of superdeep diamonds of the Juina province. Izv. Vuzov. *Fizika, 56*(2/2), 227–232 (in Russian, English abstract).

Massaitis, V. L. (2013). Impact diamonds of the Popigai astrobleme: Main properties and practical use. *Geology of Ore Deposits, 55*(8), 607–612.

Massaitis, V. L., Futergendler, S. I., & Gnevushev, M. A. (1972). Diamonds in Popigai meteorite crater impactites. *Zapisky Vsesoyuznogo Mineralogicheskogo Obshchestva, 102*(1), 108–112. (in Russian).

Mendelssohn, M. (1971). *The etching of diamond and of the associated minerals garnet and olivine*. Ph.D. thesis, University of London.

Mendelssohn, M. J., & Milledge, H. J. (1995). Geologically significant information from routine analysis of the mid-infrared spectra of diamonds. *International Geological Review, 37,* 95–110.

Mironov, V. P., Rakevich, A. L., Stepanov, F. A., Emel'yanova, A. S., Zedgenizov, D. A., Shatsky, V. S., et al. (2015). Luminescence in diamonds of the São Luiz placer (Brazil). *Russian Geology and Geophysics, 56,* 729–736.

Nakamuta, Y., & Toh, S. (2013). Transformation of graphite to lonsdaleite and diamond in the Goalpara ureilite directly observed by TEM. *American Mineralogist, 98,* 574–581.

Navon, O., Wirth, R., Schmidt, C., Jablon, B. M., Schreiber, A., & Emmanuel, S. (2017). Solid molecular nitrogen (δ-N_2) inclusions in Juina diamonds: Exsolution at the base of the transition zone. *Earth and Planetary Science Letters, 464,* 237–247. doi:10.1016/j.epsl.2017.01.035.

Nemeth, P., Garvie, L. A. J., Aoki, T., Dubrovinskaia, N., Dubrovinsky, L., & Buseck, P. R. (2014). Lonsdaleite is faulted and twinned cubic diamond and does not exist as a discrete material. *Nature Communications, 5,* 5447. doi:10.1038/ncomms6447

Oganov, A. R., Hemley, R. J., Hazen, R. M., & Jones, A. P. (2013). Structure, bonding, and mineralogy of carbon at extreme conditions. *Reviews in Mineralogy and Geochemistry, 75,* 47–77. doi:10.2138/rmg.2013.75.3

Ohfuji, H., Irifune, T., Litasov, K. D., Yamashita, T., Isobe, F., Afanasiev, V. P., et al. (2015). Natural occurrence of pure nanopolycrystalline diamond from impact crater. *Scientific Reports, 5,* 14702. doi:10.1038/srep14702

Orlov, Yu L. (1987). *Mineralogy of diamonds* (p. 235). NY: Wiley.

Palot, M. P., Cartigny, P., Harris, J. W., Kaminsky, F. V., & Stachel, T. (2012). Evidence for deep mantle convection and primordial heterogeneity from N and C stable isotopes in diamond. *Earth and Planetary Science Letters, 357–358,* 179–193.

Palot, M., Cartigny, P., & Viljoen, F. (2009). Diamond origin and genesis: A C and N stable isotope study on diamonds from a single eclogitic xenolith (Kaalvallei, South Africa). *Lithos, 112S,* 758–766.

Palyanov, Y. N., Kupriyanov, I. N., Sokol, A. G., Borzdov, Y. M., & Khokhryakov, A. F. (2016). Effect of CO_2 on crystallization and properties of diamond from ultra-alkaline carbonate melt. *Lithos, 265,* 339–350. doi:10.1016/j.lithos.2016.05.021

Pinti, D. L., Ishida, A., Takahata, N., Sano, Y., Bureau, H., & Cartigny, P. (2016). Micron-scale $\delta^{13}C$ determination by NanoSIMS in a Juina diamond with a carbonate inclusion. *Geochemical Journal, 50*(4), e7–e12. doi:10.2343/geochemj.2.0427

Pratesi, G., Lo Giudice, A., Vishnevsky, S., Manfredotti, C., & Cipriani, C. (2003). Cathodoluminoscence investigation of the Popigai, Ries, and Lappajarvi impact diamonds. *American Mineralogist, 88,* 1778–1787.

Robertson, R., Fox, J. J., & Martin, A. E. (1934). Two types of diamond. *Philosophical Transactions of the Royal Society (London), A263,* 463–535.

Robinson, D. N. (1980). *Surface textures and other features of diamonds*. Ph.D. thesis (unpublished). University of Cape Town, Rondebosch, South Africa, 221 pp.

Robinson, D. N., Scott, J. A., Van Nierkerk, A., & Anderson, V. G. (1989). The sequence of events reflected in the diamonds of some southern African kimberlites. In J. Ross (Ed.), *Kimberlites and Related Rocks*: *Vol. 2. Their mantle/crust setting* (Vol. 14, pp. 990–1000). Special Publication Geological Society of Australia.

Rondeau, B., Fritsch, E., Guiraud, M., Chalain, J. P., & Notari, F. (2004). The historical 'asteriated' hydrogen-rich diamonds: Growth history and sector-dependence impurity incorporation. *Diamond and Related Materials, 13*(9), 1658–1673. doi:10.1016/j.diamond.2004.02.002

Ross, A., Steele, A., Fries, M. D., Kater, L., Downes, H., Jones, A. P., et al. (2011). MicroRaman spectroscopy of diamond and graphite in Almahata Sitta and comparison with other ureilites. *Meteoritics & Planetary Science, 46,* 364–378.

Ryabchikov, I. D., & Kogarko, L. N. (2016). Physicochemical parameters of deep-seated mantle plumes. *Russian Geology and Geophysics, 57*(5), 687–697. doi:10.1016/j.rgg.2015.09.013

Shumilova, T. G., Mayer, E., & Isaenko, S. I. (2011). Natural monocrystalline lonsdaelite. *Doklady Earth Sciences, 441,* 1552–1554.

Sobolev, E. V. (1978). Nitrogen centers and crystal growth of natural diamond. In V. S. Sobolev (Ed.), *Problems of lithosphere and upper mantle petrology* (pp. 245–255). Novosibirsk: Nauka Press (in Russian).

Sobolev, N. V., Galimov, E. M., Ivanovskaya, I. N., & Yefimova, E. S. (1979). Carbon isotopic composition of diamonds containing crystalline inclusions. *Doklady Akademii Nauk SSSR, 249* (5), 1217–1220. (in Russian).

Sowa, H., & Koch, E. (2001). A proposal for a transition mechanism from the diamond to lonsdaleite type. *Acta Crystallographica, A57,* 406–413.

Spivak, A. V. (2015). *Genesis of super-deep diamond and its primary inclusions in the Earth's lower mantle (experimental study).* D.Sc. thesis, Lomonosov Moscow State University, Moscow, 216 pp. (in Russian).

Stachel, T., & Harris, J. W. (1997). Syngenetic inclusions in diamonds from the Birim field (Ghana)—A deep peridotitic profile with a history of depletion and re-enrichment. *Contributions to Mineralogy and Petrology, 127,* 336–352.

Stachel, T., Harris, J. W., Aulbach, S., & Deines, P. (2002). Kankan diamonds (Guinea) III: $\delta^{13}C$ and nitrogen characteristics of deep diamonds. *Contributions to Mineralogy and Petrology, 142*(4), 465–475.

Stepanov, F. A., Mironov, V. P., Rakevich, A. L., Shatsky, V. S., Zedgenizov, D. A., & Martynovicha, E. F. (2016). Red luminescence decay kinetics in Brazilian diamonds. *Bulletin of the Russian Academy of Sciences. Physics, 80*(1), 74–77.

Taylor, W. R., Canil, D., & Milledge, H. J. (1996). Kinetics of Ib to IaA nitrogen aggregation in diamonds. *Geochimica et Comsmochimica Acta, 60*(23), 4725–4733.

Taylor, W. R., Jaques, A. L., & Ridd, M. (1990). Nitorgen-defect aggregation characteristics of some Australasian diamonds: Time-temperature constraints on the source regions of pipe and alluvial diamonds. *American Mineralogist, 75,* 1290–1310.

Taylor, W. R., & Milledge, H. J. (1995). Nitrogen aggregation character, thermal history and stable isotope composition of some xenolith-derived diamonds from Roberts Victor and Finch. In *Sixth International Kimberlite Conference on Extended Abstract, Novosibirsk,* August 1995 (pp. 620–622).

Thomassot, E., Cartigny, P., Harris, J. W., Lorand, J. P., Rollion-Bard, C., & Chaussidon, M. (2009). Metasomatic diamond growth: A multi-isotope study (^{13}C, ^{15}N, ^{33}S, ^{34}S) of sulphide inclusions and their host diamonds from Jwaneng (Botswana). *Earth and Planetary Science Letters, 282,* 79–90.

Tolansky, S. (1955). *The microstructures of diamond surfaces* (p. 67). London: N.A.G. Press.

Tschauner, O., Mao, H., & Hemley, R. J. (2001). New transformations of CO_2 at high pressures and temperatures. *Physical Review Letters, 87*(7), 075701–075704.

Valter, A. A., Eryomenko, G. K., Kvasnitsa, V. N., & Polkanov, Y. A. (1992). *Shock-metamorphic minerals of carbon* (172 p). Kiev: Naukova Dumka (in Russian).

Van Heerden, L. A., Boyd, S. R., Milledge, H. J., & Pillinger, C. T. (1995). The carbon and nitrogen-isotope characteristics of the Argyle and Ellendale diamonds, Western Australia. *International Geology Review, 37,* 39–50.

Vins, V. V., & Pestryakov, E. V. (2006). Color centers in diamond crystals: Their potential use in tunable and femtosecond lasers. *Diamond and Related Materials, 15*(4–8), 569–571. doi:10.1016/j.diamond.2005.11.038

Woods, G. S., & Collins, A. T. (1983). Infrared absorption spectra of hydrogen complexes in type I diamonds. *Journal of Physics and Chemistry of Solids, 44*(5), 471–475. doi:10.1016/0022-3697(83)90078-1.

Yelisseyev, A. P., Pokhilenko, N. P., Steeds, J. W., Zedgenizov, D. A., & Afanasiev, V. P. (2004). Features of coated diamonds from the Snap Lake/King Lake kimberlite dyke, Slave craton, Canada, as revealed by optical topography. *Lithos, 77,* 83–97.

Yuryeva, O. P., Rakhmanova, M. I., Nadolinny, V. A., Zedgenizov, D. A., Shatsky, V. S., Kagi, H., et al. (2015). The characteristic photoluminescence and EPR features of superdeep diamonds (São-Luis, Brazil). *Physics and Chemistry of Minerals, 42,* 707–722. doi:10.1007/s00269-015-0756-7

Zaitsev, A. (2001). *Optical properties of diamonds* (p. 502). Berlin: Springer.

Zedgenizov, D. A., Kagi, H., Shatsky, V. S., & Ragozin, A. L. (2014). Local variations of carbon isotope composition in diamonds from Sao-Luis (Brazil): Evidence for heterogenous carbon reservoir in sublithospheric mantle. *Chemical Geology, 240*(1–2), 114–124.

Zedgenizov, D. A., Shatsky, V. S., Panin, A. V., Evtushenko, O. V., Ragozin, A. L., & Kagi, H. (2015). Evidence for phase transitions in mineral inclusions in superdeep diamonds of the Sao Luiz deposit, Brazil. *Russian Geology and Geophysics, 56*(1), 296–305.

Chapter 8
Role of Spin Crossover and Other Physicochemical Transformations in the Lower Mantle

Abstract Extremely high *P–T* conditions in the lower mantle affect some basic physicochemical properties of elements in minerals. One of the most important is a spin transition, which significantly changes the properties of iron-containing minerals in lower-mantle associations. The iron high-spin to low-spin transition in ferropericlase occurs at depths of 1000–1500 km. It is accompanied by the reduction of the unit cell volume and corresponding seismic velocity variations. However, the spin crossover in ferropericlase is a seismologically transparent transition owing to its gradual nature. Incorporation of iron in bridgmanite is more complex than in ferropericlase. Fe^{2+} in the A site remains in the HS state at all mantle conditions. In contrast, Fe^{3+} undergoes a spin transition in the entire range of lower-mantle conditions. The iron spin transition in bridgmanite does not change the existing seismological model down to the D″ layer. Under high-pressure conditions, chemical elements can obtain dramatic new properties in the lower mantle, including the formation of unexpected crystal structures and completely new counter-intuitive compounds. Some of these compounds are confirmed experimentally. Most of these transformations may occur within the lower mantle in specific compositions, which may produce only accessory mineralization. However, they may play a significant role in the Earth's balance of light elements, in the formation of the primordial carbonatitic association, and influence some major lower-mantle phases, such as periclase with the formation of magnesium peroxide MgO_2.

8.1 General

Extremely high pressure–temperature (*P–T*) conditions in the lower mantle affect not only the properties of the constituent minerals, but also some basic physicochemical properties of the elements in these minerals. One of the most important is the change in electron configuration in iron under high pressure (so-called 'spin transition' or 'spin crossover'), which significantly changes the properties of iron-containing minerals in lower-mantle associations. It was discovered in natural minerals in the 2000s (Badro et al. 2003) and, since then, has been studied in detail

F.V. Kaminsky, *The Earth's Lower Mantle*, Springer Geology,
DOI 10.1007/978-3-319-55684-0_8

both experimentally and theoretically. Spin transition has been demonstrated to dramatically change physical, chemical, rheological and transport properties of lower-mantle minerals, which can carry significant consequences for our understanding of the deep Earth.

Another effect is the change, under high pressure, of electronic configurations, size, electronegativity and chemical hardness of various chemical elements, which leads to the appearance of uncommon chemical compounds that are not characteristic for ambient *P–T* conditions. This effect was discovered recently and theoretical study on this has begun; experimental confirmations of this effect are scarce.

8.2 Iron Spin Crossover

8.2.1 General

A transition between a high-spin (HS) and low-spin (LS) state is known for metal ions in different transition metal compounds. Such a transition may be induced by variations of temperature, pressure or other factors (e.g., Bolvin and Kahn 1995). Iron is the most abundant transition metal in the mantle and spin transition in iron is the most important factor in this respect. For natural mantle iron-containing minerals, the spin transition of the iron *d*-electrons under high pressure was predicted more than 55 years ago (Fyfe 1960), but was observed experimentally only in the 2000s, initially for ferropericlase (Badro et al. 2003, 2004). The spin transition was subsequently studied at *P–T* conditions corresponding to the middle and lower parts of the lower mantle with the use of different techniques for major lower mantle iron-containing minerals, first of all for ferropericlase and bridgmanite (e.g., Li et al. 2004; Speziale et al. 2005; Lin et al. 2005, 2007; Tsuchiya et al. 2006; Li 2013; Dorfman et al. 2015).

This effect, called 'spin transition' or 'spin crossover', is based on the fact that when iron reaches approximately 50 GPa pressure, all six electrons from *3d*-orbital Fe^{2+} became paired, while in the HS state four *3d*-electrons remain unpaired (Fig. 8.1).

According to the crystal field theory (Burns 1993), the Fe^{2+} ion contains five *d*-orbitals, and six electrons are distributed over them, although each orbital is capable of accommodating two electrons with opposite spins (Fig. 8.1). The energy difference between the opposite spins on each orbital is called spin-pairing energy. Orbitals are split into two levels: two e_g orbitals with higher energy, and three t_{2g} orbitals with lower energy. The energy difference between the two levels is called crystal field splitting energy. At the HS state, Fe^{2+} has the largest number of unpaired electrons: only one electron is paired. Upon compression, crystal field splitting energy increases with increasing density; certain electrons occupy lower-energy levels and become paired. This state is called LS and has no unpaired electrons. Quantum number of total spin S in HS configuration is $S = 2$. In the LS configuration, in which all spins are paired, $S = 0$. In some cases, the intermediate

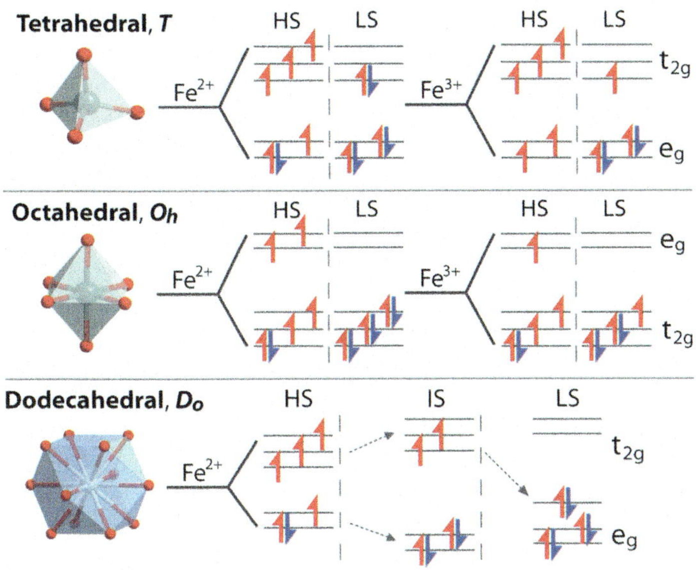

Fig. 8.1 Crystal field splitting diagrams for iron in tetrahedral, octahedral and dodecahedral sites in lower mantle minerals. Iron is shown as a Fe^{2+} or Fe^{3+} cation in high-spin (HS), intermediate spin (only in dodecahedral site; IS), and low-spin (LS) electronic configurations. The crystal field splitting energy (CFSE) can be altered by pressure, temperature, and/or composition. The energy of the e_g orbitals (those which are oriented towards the ligands) is heightened by increased repulsion due to shortened e_g orbital-ligand distance during pressure-induced unit cell distortion, resulting in an overall larger CFSE. When the CFSE surpasses the spin-pairing energy, the spin-pairing transition of 3*d* electrons is more favourable than jumping the energy gap to achieve aligned spins. For example, the LS state with all six 3*d* electrons paired ($S = 0$) in Fe^{2+} occurs at high pressures in lower mantle ferropericlase. From Lin and Weat (2012). Used with a permission of Springer

spin (IS) Fe^{2+} with two unpaired electrons ($S = 1$) was reported (Lin et al. 2008b). However, this has been neither fully confirmed nor excluded in other works; theoretical calculations have found the IS state unstable at lower-mantle pressures (Hsu and Wentzcovitch 2014). The pictures for a non-octahedral environment and for Fe^{3+} ion are similar (Fig. 8.1).

Various experimental techniques have been used to study the spin-state crossover, including X-ray emission spectroscopy (XES) (Badro et al. 2004; Li et al. 2004), X-ray absorption near-edge spectroscopy (XANES) (Narigina et al. 2009), and Mössbauer spectroscopy (MS) (Jackson et al. 2005; Li et al. 2006; Lin et al. 2008; McCammon et al. 2008). Subsequently, density functional theory (DFT) calculations (Umemoto et al. 2008; Bengtson et al. 2009; Hsu et al. 2010, 2011; Hsu and Yu 2012) have appeared to be important to the interpretation of experiments on iron spin transitions. Unlike experiments, DFT methods are not limited in accessible conditions and allow full control of parameters such as the valence state of iron.

During the spin-state crossover, there are strong density changes in iron-containing minerals (by a few percent) with additional changes in elasticity, transport and optical properties. More importantly, the spin crossover significantly modifies the chemical bonding character of Fe^{2+} and Fe^{3+} ions (Burns 1993). The LS iron in the deep mantle behaves like a different element from the HS iron at more shallow depths. There may be phase separation between Fe-rich and Mg-rich phases. The melting temperature of the Fe-rich end-member is probably higher than the Mg-rich end-members. These factors could lead to enrichment of iron in the deep mantle as a result of partitioning between the deep and shallow mantle (Li et al. 2004).

Sometimes there is debate in the literature regarding whether spin crossover in ferropericlase should be considered as a phase transition or not (Speziale et al. 2005). Different molecules or ions within the same crystal exist in different spin states while the crystal is still a single phase, occupying two sublattices (Bolvin and Kahn 1995). When both spin states of iron coexist in the same single crystal, they cannot be considered as two distinct phases. According to the existing experimental studies, there are no structural changes in ferropericlase during the spin transition (Dubrovinsky et al. 2010); thus the process cannot be considered to be a phase transition.

The discovery of the high-to-low (HL) spin transition in Fe^{2+} led to substantial progress in the understanding of the composition and properties of minerals comprising the lower mantle. In the last decade, the discovery of a HL spin crossover of iron in the lower-mantle ferropericlase and bridgmanite has been shown to influence their physical, chemical, rheological and transport properties, which can carry significant geophysical and geochemical consequences for our understanding of the deep Earth (e.g., Badro et al. 2003; Cammarano et al. 2010; Lin et al. 2013; Wu and Wentzcovitch 2014). The spin transition has great importance for the interpretation of geophysical observations within the lower mantle because this transition increases density, incompressibility, mantle seismic velocities, shear modulus, elastic anisotropy, electrical conductivity and iron partitioning into (Mg, Fe)O; and decreases thermal conductivity and bulk sound speed (Lin et al. 2013). First of all, HL spin transition significantly influences the density of minerals because a radius of the Fe^{2+} with a LS is less than a radius of the Fe^{2+} with a HS (0.61 Å compared with 0.78 Å, respectively; i.e., 22% less) (Fei et al. 2007; Lin et al. 2013).

8.2.2 Iron Spin Crossover in Ferropericlase

The iron spin transition in ferropericlase (Mg, Fe)O with cubic B1 structure is a broad and smooth crossover caused by gradual formation of spin-like molecules, which tend to assemble themselves in clusters (Bolvin and Kahn 1995). At room temperature, spin transition changes gradually from HS to LS within pressure intervals from 50–55 to 75–105 GPa, depending on the iron content in the mineral (Tsuchiya et al. 2006; Dubrovinsky et al. 2010). Within the lower mantle, at ambient temperature conditions, the spin crossover in ferropericlase increases; it

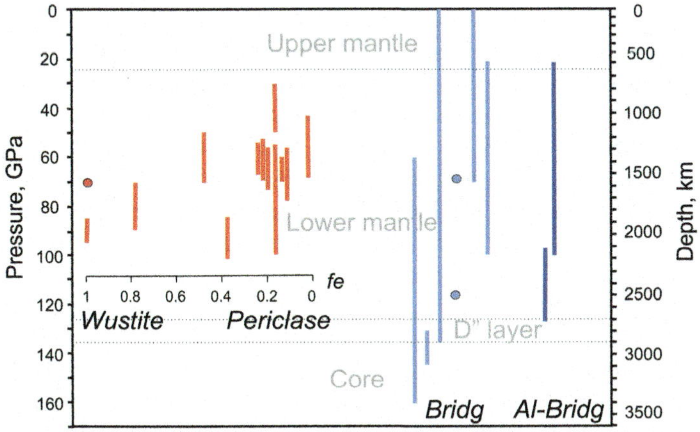

Fig. 8.2 Experimental and theoretical results on spin crossover in ferropericlase and bridgmanite. *Lines* represent the pressure ranges of spin crossover. *Circles* denote abrupt spin crossover. Data from Badro et al. (2004), Li et al. (2004, 2005, 2006), Jackson et al. (2005), Lin et al. (2005, 2006), Goncharov et al. (2006), Kantor et al. (2006), Speziale et al. (2005), Stackhouse et al. (2006, 2007). After Li (2013), modified

occurs between pressures of 40 and 63 GPa (Badro et al. 2003; Goncharov et al. 2006; Lin et al. 2013), broadening as temperature increases (e.g., Tsuchiya et al. 2006; Lin et al. 2007, 2009; Wentzcovitch et al. 2009).

HL spin transition was experimentally (by a combination of in situ XES and XRD measurements) observed in the B8-structured wüstite (*fe* = 0.96) at ~119 GPa with a volume reduction of ~2.5% at ~120 GPa (Ozawa et al. 2011). While the crystal structure of FeO remains as B8 type across such a volume reduction, the atomic arrangements of Fe and O change from inverse to normal NiAs form with a considerable decrease in the *c*/*a* axial ratio. With recent electrical resistance measurements, this suggests that iron spin crossover, inverse-normal structural change and insulator–metal transition occur concurrently in B8 FeO around 120 GPa.

The summary on HL spin crossover in ferropericlase as a function of depth and composition is presented in Fig. 8.2. The most effective crossover in ferropericlase occurs within 40–63 GPa.

Changes in density and elastic properties. During the course of HL iron spin transition at ~50 GPa, the reduction of the unit cell volume of ferropericlase by 2–4% was experimentally demonstrated, depending on the FeO content in the mineral (Speziale et al. 2005, 2007; Lin and Tsuchiya 2008). According to calculations, after the full HL spin transition in Fe^{2+}, its octahedral sites in ferropericlase reduce in volume by 2–8% (Lin et al. 2005; Fei et al. 2007). Based on static enthalpy differences between LS- and HS-structured ferropericlase, the volume reduction for ferropericlase with *fe* = 0.1875 is 4.2%, and for the compositions with *mg* 0.1875%, the volume reduction is $d\Delta V/dX_{Fe} = -2.22 \times 10^{-2}\, n\, X_{Fe}$ cm^3/mol (Tsuchiya et al. 2006).

With an increase of iron content, the effect of volume reduction increases and a cell of ferropericlase with 39 mol% Fe becomes smaller than a cell of pure MgO under the same *P–T* conditions (Lin et al. 2013 and references therein). In experiments with the HL spin transition in ferropericlase with *fe* = 0.48, using XRD and synchrotron MS at 300 K between 45 and 83 GPa, the cell volume of LS-ferropericlase was ~5% less than for the HS-ferropericlase (Solomatova et al. 2016). The iron content in natural lower-mantle ferropericlase reaches up to 64 mol % (see Sect. 4.3.2). In such a case, the volume reduction should be even greater. Such changes of cell volume are comparable to ~3–7% density changes in olivine → spinel → perovskite structure transitions, responsible for seismic-velocity discontinuities at depths of 410 and 660 km (Speziale et al. 2005).

Wu and Wentzcovitch (2014) confirmed such conclusions with the use of first-principles calculations and suggested that seismic velocity variations in the lower mantle are most likely related not to chemical or phase transitions in the lower mantle, but to the iron HL spin crossover in ferropericlase. However, direct measures of sound velocities under lower-mantle conditions remain controversial due to technical challenges. It was established that the high-pressure, LS state exhibits a much higher bulk modulus and bulk sound velocity than the low-pressure, HS state; the bulk modulus jumps by ~35% and bulk sound velocity increases by ~15% across the transition in ferropericlase $(Mg_{0.83}Fe_{0.17})O$ (Lin et al. 2005). At the same time, the spin transition is associated with a significant reduction of the aggregate V_P/V_S via the aggregate V_P softening because V_S softening does not visibly occur within the transition. Based on thermoelastic modelling along an expected geotherm, the spin crossover in ferropericlase can contribute to a 2% reduction in V_P/V_S in a pyrolite mineralogical model in the mid-lower mantle (Yang et al. 2015).

So far, no clear seismological evidence of this transition exists; and X-ray scattering measurements on ferropericlase across the spin transition area up to 70 GPa showed the HL spin crossover effect was limited to only shear moduli of the elastic tensor (Antonangeli et al. 2011). Single-crystal elastic moduli of ferropericlase, as well as compressional sound velocity (V_p), shear sound velocity (V_s), and bulk sound velocity (V_p) show a monotonic linear increase with depth within the range 0–80 GPa (i.e., up to 1800 km depth), including the area of the HL spin transition (40–63 GPa) (Antonangeli et al. 2011). This explains the absence of deviation in the aggregate seismic velocities and, thus, the lack of a seismic signature of the spin crossover, making the spin crossover in ferropericlase a seismologically transparent transition. The spin state transition does, however, influence shear anisotropy of ferropericlase and is likely to contribute to the seismic shear wave anisotropy of the lower mantle. The different pressure dependence of compressional (and bulk) and shear sound velocities across the HL spin transition implies that the spin transition might be best observed in the V_p/V_s ratio and its pressure (and temperature) derivative (Marquardt et al. 2009).

Magnetism. There are no unpaired electrons and magnetic moments in LS Fe^{2+}, and hence transformation to the LS state also causes a loss of magnetism in ferropericlase. At high pressures and low temperatures, (Mg, Fe)O is antiferromagnetic, with a site-magnetization that increases under pressure. As a result of HL spin transition, new a non-magnetic site appears abruptly at pressures above 30–90 GPa, depending on the composition (Speziale et al. 2005).

Other properties. Upon reaching the HL spin transition of Fe^{2+} in ferropericlase at approximately 60 GPa, enhanced absorption in the mid- and near-infrared spectral range was observed, whereas absorption in the visible ultraviolet was reduced. The observed changes are attributed to *d-d* orbital charge transfer in the Fe^{2+} ion. The results indicate that LS ferropericlase will exhibit lower radiative thermal conductivity than HS ferropericlase, which is something that needs to be considered in geodynamic models (Goncharov et al. 2006).

8.2.3 Iron Spin Crossover in Bridgmanite

The incorporation of iron in orthorhombic bridgmanite is more complex than in cubic ferropericlase and identification of spin transition(s) in bridgmanite has been ambiguous. At ambient conditions, both ions (Fe^{2+} and Fe^{3+}) are in the HS (unpaired electrons) state. Ferrous iron Fe^{2+} with six *3d* electrons substitutes for Mg in the larger, distorted, 8- to 12-coordinated pseudo-dodecahedral A site and can undergo a crossover from the HS to the LS state. Ferric iron (Fe^{3+}) with five *3d* electrons is suggested to enter both the A site and the smaller, relatively undistorted octahedral 6- co-ordinated B site (Fig. 8.1). Studies of bridgmanite demonstrated that bridgmanite can contain a substantial amount of Fe^{3+} in its lattice.

Ferrous iron Fe^{2+} in the A-site was reported to exhibit, in experiments, a spin crossover from the HS to the IS state (Lin et al. 2008; McCammon et al. 2008; Potapkin et al. 2013; Kupenko et al. 2014, 2015). However, all numerous computational studies have found the HS configuration of Fe^{2+} to be the most stable in the entire lower-mantle pressure range (up to 135 GPa), regardless of the iron concentration (Zhang and Oganov 2006; Bengtson et al. 2009; Hsu et al. 2010, 2012; Hsu and Wentzcovitch 2014) and attribute the experimentally observed pressure-induced changes in hyperfine parameters of ferrous iron to small changes in its local environment (Hsu et al. 2012). Experimental results in Al-rich bridgmanite also demonstrate that Fe^{2+} in the A site remains in the HS position in the entire lower mantle, up to pressure of 200 GPa (Fujino et al. 2012).

Ferric iron Fe^{3+} was reported to occupy either exclusively the A-site and to be in the HS state (Vanpeteghem et al. 2006; McCammon et al. 2008; Potapkin et al. 2013; Kupenko et al. 2014) or to occupy both A- and B-sites with Fe_A^{3+} remaining in the HS state at lower-mantle conditions at least to 75 GPa, and Fe_B^{3+} exhibiting HL spin crossover at pressures of 15–50 GPa (Zhang and Oganov 2006; Li et al. 2005; Lin et al. 2007; Catalli et al. 2011; Hsu et al. 2011, 2012). Several studies have additionally proposed that high pressures and high temperatures may promote the site exchange of Fe^{3+} and Al^{3+} by the reaction $\left(\mathrm{HS\,Fe_A^{3+}}\right) + \mathrm{Al_B^{3+}} \rightarrow \left(\mathrm{LS\,Fe_B^{3+}}\right) + \mathrm{Al_B^{3+}}$ (Catalli et al. 2011; Fujino et al. 2012). At the same time, it was demonstrated that the concentration of Al in bridgmanite does not influence the iron spin crossover (Hsu et al. 2012). Although the share of ferric iron in natural bridgmanite $Fe^{3+}/\Sigma Fe$ in the uppermost lower mantle is only 2–8% (Kaminsky 2012), which was confirmed in experimental results (5%; Kupenko et al. 2015), the higher abundance of Fe^{3+} in bridgmanite was proposed to be associated with the Al^{3+}–Fe^{3+} charge-coupled substitution mechanism and self-disproportionation reaction, in which Fe^{2+} transforms to Fe^{3+} and a coexisting metallic iron (Frost et al. 2004; McCammon 2006). Recent experiments demonstrated that high Fe-content promotes the HS–LS transition in bridgmanite at lower pressures; gradual spin transition starts at 50–70 GPa (Dorfman et al. 2015). On the other hand, the low amount of Fe^{3+}_{B} at lower mantle conditions suggests that spin crossover of Fe^{3+} does not affect the properties of bridgmanite in the deep Earth's interior (Kupenko et al. 2015) or does not exist at all even at the lowermost mantle conditions (Mohn and Trønnes 2016).

The summaries on observations on spin crossover in pure and Al-containing bridgmanite are presented in Fig. 8.2 and Table 8.1.

The controversy in the estimate of the iron spin transition in bridgmanite may be explained by the fact that several metastable equilibrium sites for substitutional ferrous iron in bridgmanite exist (Hsu et al. 2010). Their energies are grouped in two different ranges, based on the tilting of Si–O octahedra. In the relevant (lower) energy range, there are two distinct sites for HS, as well as one for LS, and one for IS iron. Because of variable *d*-orbital occupancy across these sites, the two energetically competing HS sites have different iron quadrupole splittings (QS). At low pressure, the HS iron with QS of 2.3–2.5 mm/s is more stable, while the HS iron with QS of 3.3–3.6 mm/s is more favourable at higher pressures. The crossover occurs between 4 and 24 GPa, depending on the iron concentration, the choice of exchange-correlation functional and the inclusion of on-site Coulomb interaction (Hubbard *U*). The calculations performed by Hsu et al. (2010) support the notion that the transition observed in Mössbauer spectra corresponds to an atomic-site change rather than a spin-state crossover.

Additionally, the majority of reports describing the spin state behaviour in bridgmanite at lower mantle conditions are based on analyses of temperature-quenched samples (e.g., Li et al. 2004; Catalli et al. 2011; Fujino et al. 2012; Potapkin et al. 2013). An indication of the temperature effect to stabilize IS Fe^{2+} relative to HS Fe^{2+} was provided by McCammon et al. (2008); however, in their study the temperature was limited to only 1000 K. More recent results based on Nuclear Forward Scattering (NFS) (Kupenko et al. 2014) extended temperatures to lower mantle conditions. However, NFS is not well suited to materials with a large number of components arising from multiple sites, valence states and spin states (which is the case for bridgmanite), leading to spectral complexity and non-uniqueness of fitting models. Energy-domain MS is better suited for the analysis of iron distribution between individual structural sites in bridgmanite, and its oxidation and electronic states (McCammon et al. 2008; Potapkin et al. 2013). However, studies of lower mantle phases at the relevant high pressure and

Table 8.1 Summary of observations of spin transitions in bridgmanite in multi-anvil or DAC experiments using different methods

Character of spin transition	Composition (mol%)	Synthesis	Annealing	Methods*	References
None/no effect	6–40% $FeSiO_3$ (0–80% Fe^{3+})	25–88 GPa (multi-anvil or DAC)	Each step or none	E-SMS, ER, XES, XRD	Potapkin et al. (2013), Dorfman et al. (2015)
Gradual decrease	9–14% $FeSiO_3$; 10% $FeAlO_3$ (15–70% Fe^{3+})	25–35 GPa (multi-anvil or DAC)	None or <1000 K	MS, XES, T-SMS	Li et al. (2004, 2006), McCammon et al. (2008)
Transition below 50 GPa	10% $FeSiO_3$; 10% $FeAlO_3$ (25–70% Fe^{3+})	23–26 GPa (multi-anvil)	None	T-SMS, XES, XRD	Li et al. (2006), Lin et al. (2012)
Gradual transition ends at 50–70 GPa	5–50% $FeSiO_3$ (40% Fe^{3+} or unknown)	26–93 GPa (multi-anvil or DAC)	None	T-SMS, XES, XRD	Jackson et al. (2005), Caracas et al. (2014)
Sharp transition at 50-85 GPa	10–40% $FeSiO_3$; 10% $FeSiO_3$; 10–15% $FeAlO_3$ (10–100% Fe^{3+} or unknown)	25–110 GPa (multi-anvil or DAC)	Each step or none	ER, XES, XRD, T-SMS	Badro et al. (2004), Grocholski et al. (2009), Catalli et al. (2010, 2011) Lin et al. (2010), Ohta et al. (2010), Fujino et al. (2012)
Gradual transition begins at 50–70 GPa	74% $FeSiO_3$ (unknown Fe^{3+})	88 GPa (DAC)	None	XES, XRD	Dorfman et al. (2015)
Sharp transition at 120–135 GPa	10–40% $FeSiO_3$ (10–20% Fe^{3+} or unknown)	30–120 GPa (DAC)	Each step or none	XES, XRD, T-SMS	Badro et al. (2004), Lin et al. (2010), McCammon et al. (2010)

From Dorfman et al. (2015)

Note ER electrical resistivity; *E-SMS* energy-domain synchrotron Mössbauer spectroscopy; *MS* conventional Mössbauer spectroscopy; *T-SMS* time-domain synchrotron Mössbauer spectroscopy; *XES* X-ray emission spectroscopy; *XRD* X-ray diffraction

temperature conditions are not possible using conventional MS with a radioactive source due to the challenge of focusing gamma rays in the laboratory.

The recently developed Synchrotron Mössbauer Source (SMS) allows measurements of high quality and well resolved energy-domain spectra on timescales of only a few hours, combining the advantages of synchrotron radiation and the MS technique (Potapkin et al. 2012). The most recent experimental work, using the in situ SMS technique to elucidate the electronic configuration of iron in bridgmanite in the lower mantle, demonstrated that at lower mantle conditions Fe^{3+} remains predominantly in the HS state, while Fe^{2+} occurs solely in the IS state (Kupenko et al. 2015).

Unit cell volume. The HL iron spin crossover in bridgmanite, like in ferropericlase, results in the reduction of the unit cell volume. However, this reduction, according to theoretical calculations, should be significantly lower, at 0.5–1% (Stackhouse et al. 2007; Tsuchiya and Wang 2013) and is hardly detectable experimentally (Catalli et al. 2010) or in the natural environment. In experiments at low pressure, Fe^{3+} and Al expand the bridgmanite lattice. However, near the pressure range of the abrupt increase in the LS population, the unit cell volume of Fe- and Al-bearing bridgmanite becomes similar to that of pure bridgmanite, while those of Al-free Fe^{3+}-bearing bridgmanite and Al-free Fe^{2+}-bearing bridgmanite remain larger throughout the lower mantle. Consequently, bridgmanite in Al-rich systems should have a lower density in the shallow lower mantle but similar or greater density than bridgmanite in pyrolite in the deep lower mantle (Catalli et al. 2011). Shukla and Wentzcowitch (2016), based on ab initio calculations, concluded that iron spin crossover in Fe^{3+}-rich bridgmanite is accompanied by a clear volume reduction. However, such changes should not significantly influence the density and seismic velocities of bridgmanite. The Fe^{3+} incorporation and its spin transition marginally affect the thermodynamic properties of bridgmanite with a magnesium composition of $mg = 0.0625$ (Tsuchiya and Wang 2013). Effects of the Al_2O_3 incorporation into bridgmanite are also insignificant on both density and bulk modulus at least for the small (4 wt%) concentrations (Tsuchiya and Wang 2013). Only the addition of 25% FeO in bridgmanite noticeably increases its density and bulk sound velocity (Mao et al. 2011), but such a composition does not exist in the natural lower mantle.

Elastic properties of bridgmanite also do not change much during the iron spin crossover. Only marginal differences between isothermal bulk modulus K_T of Fe^{3+}-bearing and pure bridgmanite were found (Tsuchiya and Wang 2013; Shukla and Wentzcowitch 2016). Iron-rich bridgmanite is more compressible than the Mg-end-member below 50 GPa because of the gradual spin transition in the B site together with lattice compression. The completion of the spin transition at 50–60 GPa increases the bulk modulus with no associated change in density (Catalli et al. 2010). The calculations using density functional molecular dynamics (DFT-MD) demonstrated that Fe- and Al-containing bridgmanite, experiencing both Fe^{2+} and Fe^{3+} spin transitions fits the existing 1D seismological model well, down to the D″ layer, in excellent agreement with the average velocities and is also fairly approximate to the density (Zhang et al. 2016).

Optical absorption. Like in ferropericlase, iron spin crossover in bridgmanite optical absorption in the visible and near-infrared spectral range is dominated by O–Fe^{3+} charge transfer and Fe^{3+}–Fe^{2+} intervalence transitions, whereas a contribution from the Fe^{2+} crystal-field transitions is substantially smaller (Goncharov et al. 2008). This in turn, influences the thermal conductivity of the lower mantle and therefore heat flux from the core, controlled by the amount of ferric iron Fe^{3+} (Goncharov et al. 2008).

8.2.4 Iron Spin Crossover in Iron Carbonates

Siderite, magnesite and dolomite form a complete solid solution, where iron is divalent and is located in an octahedral site.

Siderite. The iron spin transition with a spin pairing of *d*-orbital electrons was established in siderite $FeCO_3$ at pressures of ~43–50 GPa by means of XES (Mattila et al. 2007), Raman spectroscopy (Müller et al. 2016), single crystal diffraction (Lavina et al. 2009, 2010a) and visible and near infrared absorption measurements (Lobanov et al. 2015a). It was modeled with first-principles calculations, which showed a volume collapse as a result of the transition, accompanied by increases in bulk modulus and sound velocity (Shi et al. 2008). Unlike in oxides, the iron spin pairing in siderite was established as a sharp event, showing over a narrow pressure range: 50 GPa according to Mattila et al. (2007) and Merlini et al. (2012), 43 GPa according to Lavina et al. (2009), 44–45 GPa, according to Lavina et al. (2010a) and 44 GPa according to Lobanov et al. (2015a).

However, X-ray diffraction (XRD) is a technique sensitive to long-range structures and cannot probe a short-range order such as HS and LS state distributions. Raman, Mössbauer, and X-ray absorption near edge structure (XANES) spectroscopy probe the local structure of neighbouring and groups of atoms, and are therefore sensitive to short-range order features that are not always evident from XRD. Recently, Cerantola et al. (2015), using a combination of Mössbauer, Raman, and XANES spectroscopy in diamond-anvil cells, concluded that at room temperature and under quasi-hydrostatic conditions, spin crossover in siderite takes place over a broad pressure range, between 40 and 47 GPa, in contrast to previous XRD data that described the transition as a sharp volume collapse. Based on these observations, electron spin pairing in siderite may be considered to be a dynamic process, where iron atoms can be either HS or LS in the crossover region. Mode Grüneisen parameters extracted from Raman spectra, collected at pressures below and above spin crossover, showed a drastic change in stiffness of the Fe–O octahedra after the transition, where they become more compact and hence less compressible (Cerantola et al. 2015).

The iron spin transition in siderite causes a volume collapse (and accordingly the density increase) of LS siderite by ~9–10% (Lavina et al. 2009; Lin et al. 2012; Liu et al. 2014). In addition, it was experimentally established that the iron ground state in siderite changes from a low-pressure magnetic state to a high pressure non-magnetic state (Mattila et al. 2007). At the same time, the initially colourless crystals assume an intense green colour after the spin transition, which progressively turns to red above 60 GPa. The colour change is most likely caused by a gain in absorption in the visible range accompanying the spin transition, which has also been described in ferropericlase and bridgmanite, and has implications for the radiative properties of the mantle (Goncharov et al. 2006, 2008).

Magnesiosiderite at pressures above 50 GPa and at a temperature higher than 1400 K obtains an orthorhombic structure, which is stabilized by spin crossover (Liu et al. 2015a).

Ferromagnesite. An iron spin transition was established in ferromagnesite (Fe, Mg)CO_3 under pressure of 42–52 GPa; as a result of the spin transition a volume of ferromagnesite collapses by 6–10% (Lavina et al. 2010b; Lin et al. 2012). Temperature extends the spin crossover region of ferromagnesite (Liu et al. 2014).

The effective ionic radius of Fe^{2+} in the LS configuration is smaller than the radius of Mg^{2+}, and at the spin crossover the volume of LS-siderite is significantly lower than the volume of magnesite. The volume mismatch between siderite and magnesite decreases with pressure, with LS-siderite being less compressible than magnesite, and the volumes of the two carbonates converge at approximately 80 GPa.

The proportion of iron carbonates in the lower-mantle composition is low, and the influence of the spin transition in these minerals on the bulk lower mantle properties is not significant. However, in some local areas that are enriched by these minerals, it may cause minor seismic effects.

8.2.5 Iron Spin Crossover in NAL Phase

Two single-crystal samples (iron-free $Na_{1.14}Mg_{1.83}Al_{4.74}Si_{1.23}O_{12}$ and iron-bearing $Na_{0.71}Mg_{2.05}Al_{4.62}Si_{1.16}Fe^{2+}_{0.09}Fe^{3+}_{0.17}O_{12}$) were recently studied up to 86 GPa at room temperature, using synchrotron NFS and XRD combined with DAC (Wu et al. 2016). A pressure-induced HS–LS transition of the octahedral Fe^{3+} in the iron-bearing NAL was observed at approximately 30 GPa by NFS. Compared with the iron-free NAL, the iron-bearing NAL undergoes a volume reduction of 1.0% at 33–47 GPa, which is associated with the spin transition of the octahedral Fe^{3+}.

Since the NAL phase contains up to 12 mol% iron (see Sect. 5.3.2), the valence and spin states of iron in the NAL phase can potentially affect its elasticity which in turn can affect our understanding of the mineral physics of the middle-ocean ridge basalt (MORB) assemblage in the lower mantle (e.g., Lin et al. 2013). Considering the high volume proportion of the NAL phase in subducted MORB, the distinct elastic properties of the iron-bearing NAL phase across the spin transition may provide an explanation for the observed seismic heterogeneities of subducted slabs in the lower mantle at depths below 1200 km (Wu et al. 2016). As shown above (see Sect. 5.3.2), only K-rich NAL is stable at that level.

8.2.6 Spin Transitions in Elements Other Than Iron

Spin transitions are also known in atoms other than Fe elements, such as Mn (Yoo et al. 2005); they are theoretically predicted in Co (Ohnishi 1978). They have not been well studied to date and are not likely to play a significant role in the lower-mantle structure.

8.3 Changes in Properties of Chemical Elements Under High Pressure and Possible New Mineral Compounds in Deep Earth

Theoretical calculations predict that under high-pressure conditions, chemical elements can obtain dramatic new properties, including the formation of unexpected crystal structures and completely new counter-intuitive compounds. Under high pressure, the orbitals with a higher angular momentum become favourable; hence, atoms typically undergo *s-p* and *s-d* transitions. At the same time, atomic size (volumes) decreases. Electronegativity and chemical hardness of chemical elements can be expected to be highly non-trivial. Concomitantly, new chemical compounds are formed because of the increased reactivity of certain elements and compounds (Zhang et al. 2013). For example, orbital transfer, in particular *s-d* transfer, makes Ni a 'pseudo-noble-gas', Fe and Co strong electron acceptors, while Cu and Zn become active metals (Dong et al. 2015a). In most cases, these transformations are predicted under extremely high-pressure values exceeding 200–500 GPa, but some of them may occur within the lower mantle.

During recent years, the use of evolutionary simulations (mainly with the use of algorithm USPEX; e.g., Lyakhov et al. 2013) experiments under pressures of up to 500 GPa and temperatures corresponding to the lower-mantle and outer-core conditions demonstrated that atoms of some elements change their valences and form new polyatomic anions. As a result, some unexpected compounds may appear. Some of these have been confirmed experimentally.

Alkali halides and fluorides MX, for example, have been viewed as typical ionic compounds, characterized by a 1:1 ratio necessary for charge balance between M1 and X2. It was proposed that group I elements can be oxidized further under high pressure. Sodium chloride, or rocksalt (NaCl) is well characterized at ambient pressure and was identified as the mineral halite as a member of the lower mantle carbonatitic association (Kaminsky et al. 2009, 2013). As a result of the large electronegativity difference between Na and Cl atoms, it has highly ionic chemical bonding (with 1:1 stoichiometry dictated by charge balance) and a B1-type crystal structure. According to the calculated phase diagram, compounds with different stoichiometries were predicted, such as $NaCl_3$, stable above 22 GPa; $NaCl_7$, stable above 142 GPa; and Na_4Cl_3, Na_3Cl_2, Na_2Cl, and Na_3Cl, which are stable above 166, 112, 128, and 68 GPa, respectively (Zhang et al. 2013; Saleh and Oganov

2016). These calculations were confirmed by high-pressure experiments in a laser-heated diamond-anvil cell (LHDAC) on the Na–Cl system in the presence of excess chlorine and sodium, in which an orthorhombic $NaCl_3$ and 2D metallic tetragonal Na_3Cl were synthesized at pressures above 60 GPa (Zhang et al. 2013). These experiments demonstrate that compounds violating chemical intuition can be thermodynamically stable even in simple systems at nonambient conditions.

An analogous Cs-F system that was studied at pressures of up to 100 GPa also demonstrated extremely versatile chemistry. A series of CsF_n ($n \geq 1$) compounds were predicted to be stable already at ambient pressure. Under high pressure, 5*p* electrons of Cs atoms become active, with a growing tendency to form Cs (III) and (V) valence states in fluorine-rich conditions. Although Cs (II) and (IV) are not energetically favoured, the interplay between two mechanisms (polyfluoride anions and polyvalent Cs cations) allows CsF_2 and CsF_4 compounds to be stable under pressure. The estimated defluorination temperatures of CsF_n (n = 2, 3, 5) compounds at atmospheric pressures (218, 150, −15 °C, respectively) are attractive for fluorine storage applications (Zhu et al. 2015).

Alkali hydrides may form analogue compounds. In system Li–H under pressures of 100–165 GPa, compounds LiH_2, LiH_6 and LiH_8 should be more stable than common lithium hydroxide, LiH. The compound LiH_6 (and other LiH_n), may well have extensions beyond the constituent lithium. These hypothetical materials demonstrate that non-traditional stoichiometries can considerably expand the view of chemical combinations under high pressures (Zurek et al. 2009).

Calcium carbides, at pressures from ambient to 100 GPa may form stable compounds Ca_5C_2, Ca_2C, Ca_3C_2, CaC, Ca_2C_3 and CaC_2 on the Ca–C phase diagram (Li et al. 2015a, b). Among these, Ca_2C and Ca_2C_3 were successfully synthesized for the first time in high-pressure experiments with excellent structural correspondence to theoretical predictions. Of particular interest is the base-centred monoclinic phase (space group *C2/m*) of Ca_2C, a quasi-2D metal with layers of negatively charged calcium atoms, and the primitive monoclinic phase (space group $P2_1/c$) of CaC with zigzag C_4 groups (Li et al. 2015a, b). These compounds may be present in the D″ layer.

Lithium and carbon nitrides. A number of new lithium nitrides have been predicted using the ab initio evolutionary structure search at pressures of up to 100 GPa. In addition to the well-known compositions Li_3N, Li_2N_2 and LiN_3, five novel compositions were found, including $Li_{13}N$, Li_5N, Li_3N_2, LiN_2, and LiN_5. Notably, the N–N bonding patterns evolve from isolated N ions to N_2 dumbbells, to linear N_3 groups, infinite nitrogen chains, N_5 rings with increasing N content (Shen et al. 2015). Three new carbon nitrides C_3N_4 are predicted to be stable phases in the range of 68–300 GPa (Dong et al. 2015b). They are of particular interest considering the enrichment of the core–mantle boundary in both carbon and nitrogen (see Sect. 9.3.2).

Noble gases. Unexpected properties occur, under high-pressure conditions, in noble gases. Xenon, which is quite inert under ambient conditions, may become reactive under pressure. Using an ab initio evolutionary algorithm, the existence of thermodynamically stable Xe–O compounds are predicted at high pressures: XeO,

XeO_2 and XeO_3 become stable at pressures above 83, 102 and 114 GPa, respectively (Zhu et al. 2013a). The calculations indicate a large charge transfer in these oxides, suggesting that large electronegativity differences and high pressure are the key factors favouring the formation of xenon compounds. Although xenon oxides are unstable in equilibrium with the metallic iron occurring in the lower mantle, it is possible, however, that xenon atoms may be retained at defects in lower-mantle oxides (Zhu et al. 2013a). The possibility of the formation of stable xenon oxides and silicates in the interior of the Earth could explain the atmospheric missing xenon paradox.

Another noble gas helium, on a par with neon, is the most chemically inert element in the Periodic Table. Due to its extremely stable closed-shell electronic configuration with record-high ionization potential and nearly zero electron affinity, helium is not known to form thermodynamically stable compounds. However, the use of ab initio evolutionary algorithm and subsequent high-pressure synthesis in DAC demonstrate the origin of a thermodynamically stable compound of helium and sodium, Na_2He, which has a fluorite-type structure and is stable at pressures above 113 GPa (Dong et al. 2014). This phase is formed because of increased reactivity of Na under pressure and substantial electronic change of He atoms.

Other compounds. Some other unexpected compounds are available in other systems, such as nitrogen oxides (Li et al. 2015a, b), hafnium oxides (Zhang et al. 2015), and aluminium oxides (Liu et al. 2015b).

More uncommon compounds have been predicted with the use of ab initio evolutionary algorithm (Oganov 2010) to occur at pressures of up to 300–500 GPa (0.3–0.5 TPa) and higher (Niu et al. 2015); however, they cannot occur within the Earth's lower mantle.

Most of these transformations may occur only under specific *P–T* conditions within the lower mantle in specific compositions, which may produce only accessory mineralization. However, they may influence some major lower-mantle phases, such as **ferropericlase**. MgO was experimentally confirmed to be stable up to 250 GPa, i.e., to the core (Duffy et al. 1995; Dorfman et al. 2012). The theoretical calculations with the use of evolutionary algorithm demonstrated that at pressure 116 GPa, magnesium peroxide MgO_2 becomes stable, and contains transitional ions $[O–O]^{2-}$ (Zhu et al. 2013b). Recently these calculations were confirmed experimentally with the use of DAC and synchrotron XRD (Lobanov et al. 2015b). MgO_2 forms as a result of reaction $2MgO + O_2 \rightarrow 2MgO_2$ in the presence of free oxygen at a pressure of 94 GPa and temperature of 2150 K. The resulting magnesium peroxide MgO_2 has a tetragonal structure with the space group *I4/mcm*, which is significantly denser than periclase with the $Fm3m$ structure (Lobanov et al. 2015b). The formation of magnesium peroxide may happen only under high oxygen fugacity conditions.

8.4 Conclusions

HS–LS state transition in Fe atoms is a fundamental physicochemical process, caused by high *P*–*T* conditions in the lower mantle, creating changes in physical, chemical, rheological and transport properties of minerals that can carry significant geophysical and geochemical consequences for our understanding of the deep Earth. This process is most prominent in ferropericlase, where the Fe^{2+} ion gradually obtains LS configuration between pressures of 40 and 63 GPa at ambient lower mantle conditions, corresponding to the depth interval of 1000–1500 km. The high-pressure LS state exhibits a much higher bulk modulus and bulk sound velocity than the low-pressure HS state; the bulk modulus jumps by ~35% and bulk sound velocity increases by ~15% across the transition in ferropericlase. The unit cell volume of ferropericlase decreases by 2–8%, which accordingly affects density and seismic velocities and may potentially cause seismic heterogeneities in the lower mantle. However, the gradual spin transition when HS and LS configurations coexist within ~23 GPa explains the absence of deviation in the aggregate seismic velocities and, thus, the lack of a seismic signature of the spin crossover, making the spin crossover in ferropericlase a 'seismologically transparent transition'.

In bridgmanite, Fe^{2+} in the A site remains in the high-spin state at all mantle conditions. By contrast, Fe^{3+} undergoes a spin transition in the entire range of lower mantle conditions. However, the low amount of Fe^{3+} in bridgmanite does not affect its properties in the deep Earth's interior and the iron spin transition in bridgmanite does not change the existing seismological model down to the D″ layer.

Under very high-pressure conditions, chemical elements can obtain dramatic new properties in the lower mantle, including the formation of unexpected crystal structures and completely new counter-intuitive compounds. Among them are alkali halides and fluorides, alkali hydrides, calcium carbides and lithium and carbon nitrides with uncommon stoichiometry. Noble gases, such as xenon and helium may become reactive under pressure and form compounds with oxygen and sodium. Most of these transformations may occur in specific compositions, which may produce only accessory mineralization. However, they may play a significant role in balancing the Earth's light elements and in the formation of the primordial carbonatitic (or silica-carbonatitic) association, as well as influence some major lower-mantle phases, such as ferropericlase with the formation of magnesium peroxide MgO_2.

References

Antonangeli, D., Siebert, J., Aracne, C. M., Farber, D. L., Bosak, A., Hoesch, M., et al. (2011). Spin crossover in ferropericlase at high pressure: A seismologically transparent transition? *Science, 331*, 64–67.

Badro, J., Fiquet, G., Guyot, F., Rueff, J.-P., Struzhkin, V. V., Vankó, G., et al. (2003). Iron partitioning in Earth's mantle: Toward a deep lower mantle discontinuity. *Science, 300*(5620), 789–791.

Badro, J., Struzhkin, V. V., Shu, J., Hemley, R. J., Mao, H. K., Ruef, J. P., et al. (2004). Electronic transitions in perovskite: Possible nonconvecting layers in the lower mantle. *Science, 305,* 383–386.

Bengtson, A., Li, J., & Morgan, D. (2009). Mössbauer modeling to interpret the spin sate of iron in (Mg, Fe)SiO_3. *Geophysical Research Letters, 36,* L15301.

Bolvin, H., & Kahn, O. (1995). Using model for low-spin high-spin transitions in molecular compounds; within and beyond the mean-field approximation. *Chemical Physics, 192,* 295–305.

Burns, R. G. (1993). *Mineralogical application of crystal field theory* (2nd ed., p. 529). Cambridge, New York, etc: Cambridge University Press.

Cammarano, F., Marquardt, H., Speziale, S., & Tackley, P. J. (2010). Role of iron-spin transition in ferropericlase on seismic interpretation: A broad thermochemical transition in the mid mantle? *Geophysical Research Letters, 37,* L03308. doi:10.1029/2009GL041583

Caracas, R., Ozawa, H., Hirose, K., Ishii, H., Hiraoka, N., Ohishi, Y., et al. (2014). Identifying the spin transition in Fe^{2+}-rich $MgSiO_3$ perovskite from X-ray diffraction and vibrational spectroscopy. *American Mineralogist, 99,* 1270–1276.

Catalli, K., Shim, S.-H., Dera, P., Prakapenka, V. B., Zhao, J., Sturhahn, W., et al. (2011). Effects of the Fe^{3+} spin transition on the properties of aluminous perovskite—new insights for lower-mantle seismic heterogeneities. *Earth and Planetary Science Letters, 310,* 293–302.

Catalli, K., Shim, S. H., Prakapenka, V. B., Zhao, J., Sturhahn, W., Chow, P., et al. (2010). Spin state of ferric iron in $MgSiO_3$ perovskite and its effect on elastic properties. *Earth and Planetary Science Letters, 289,* 68–75.

Cerantola, V., McCammon, C., Kupenko, I., Kantor, I., Marini, C., Wilke, M., et al. (2015). High-pressure spectroscopic study of siderite ($FeCO_3$) with a focus on spin crossover. *American Mineralogist, 100*(11–12), 2670–2681.

Dong, X., Oganov, A. R., Goncharov, A. F., Stavrou, E., Lobanov, S., Saleh, G., et al. (2014). Stable compound of helium and sodium at high pressure. *arXiv: 1309.3827.*

Dong, X., Oganov, A. R., Qian, G., Zhou, X.-F., Zhu, Q., & Wang, H.-T. (2015a) How do chemical properties of the atoms change under pressure? *arXiv: 1503.00230.*

Dong, H., Oganov, A. R., Zhu, Q., & Qian, G.-R. (2015b). The phase diagram and hardness of carbon nitrides. *Scientific Reports, 5,* 9870. doi:10.1038/srep09870

Dorfman, S. M., Badro, J., Rueff, J.-P., Chow, P., Xiao, Y., & Gillet, P. (2015). Composition dependence of spin transition in (Mg, Fe)SiO_3 bridgmanite. *American Mineralogist, 100,* 2246–2253.

Dorfman, S. M., Prakapenka, V. B., Meng, Y., & Duffy, T. S. (2012). Intercomparison of pressure standards (Au, Pt, Mo, MgO, NaCl and Ne) to 2.5 Mbar. *Journal Geophysical Research, 117,* B08210.

Dubrovinsky, L., Narygina, O., & Kantor, I. (2010). Effect of spin transitions in iron on structure and properties of mantle minerals. In E. Boldyreva & P. Dera (Eds.), *High-pressure crystallography: From fundamental phenomena to technological applications* (614 pp). Springer, Dordrecht.doi: 10.1007/978-90-481-9258-8_20

Duffy, T. S., Hemley, R. J., & Mao, H.-K. (1995). Equation of state and shear strength at multimegabar pressures: Magnesium oxide to 227 GPa. *Physical Review Letters, 74,* 1371–1375.

Fei, Y., Zhang, L., Corgne, A., Watson, H. C., Ricolleau, A., Meng, Y., et al. (2007). Spin transition and equations of state of (Mg, Fe)O solid solutions. *Geophysical Research Letters, 34,* L17307. doi:10.1029/2007GL030712

Frost, D. J., Liebske, C., Langenhorst, F., McCammon, C. A., Trønnes, R. G., & Rubie, D. C. (2004). Experimental evidence for the existence of iron-rich metal in the Earth's lower mantle. *Nature, 428,* 409–412.

Fujino, K., Nishio-Hamane, D., Seto, Y., Sata, N., Nagai, T., Shinmei, T., et al. (2012). Spin transition of ferric iron in Al-bearing Mg–perovskite up to 200 GPa and its implication for the lower mantle. *Earth and Planetary Science Letters, 317–318,* 407–412.

Fyfe, W. S. (1960). The possibility of *d*-electron coupling in olivine at high pressures. *Geochimica et Cosmochimica Acta, 19,* 141–143.

Goncharov, A. F., Haugen, B. D., Struzhkin, V. V., Beck, P., & Jacobsen, S. D. (2008). Radiative conductivity in the Earth's lower mantle. *Nature, 456,* 231–234. doi:10.1038/nature07412

Goncharov, A. F., Struzhkin, V. V., & Jacobsen, S. D. (2006). Reduced radiative conductivity of low-spin (Mg, Fe)O in the lower mantle. *Science, 312,* 1205–1208. doi:10.1126/science.1125622

Grocholski, B., Shim, S. H., Sturhahn, W., Zhao, J., Xiao, Y., & Chow, P. C. (2009). Spin and valence states of iron in $(Mg_{0.8}Fe_{0.2})SiO_3$ perovskite. *Geophysical Research Letters, 36,* L24303.

Hsu, H., Blaha, P., Cococcioni, M., & Wentzcovitch, R. M. (2011). Spin-state crossover and hyperfine interactions of ferric iron in $MgSiO_3$ perovskite. *Physical Review Letters, 106,* 118501.

Hsu, H., Umemoto, K., Wentzcovitch, R. M., & Blaha, P. (2010). Spin states and hyperfine interactions of iron in (Mg, Fe)SiO_3 perovskite under pressure. *Earth Planet Sci. Lett., 294,* 19–26.

Hsu, H., & Wentzcovitch, R. M. (2014). First-principles study of intermediate-spin ferrous iron in the Earth's lower mantle. *Physical Review B, 90*(19), 195205.

Hsu, H., Yu, Y. G., & Wentzcovitch, R. M. (2012). Spin crossover of iron in aluminous $MgSiO_3$ perovskite and post-perovskite. *Earth and Planetary Science Letters, 359–360,* 34–39.

Jackson, J. M., Sturhahn, W., Shen, G., Zhao, J., Hu, M. Y., Errandonea, D., et al. (2005). A synchrotron Mössbauer spectroscopy study of (Mg, Fe)SiO_3 perovskite up to 120 GPa. *American Mineralogist, 90,* 199–205.

Kaminsky, F., Wirth, R., Matsyuk, S., Schreiber, A., & Thomas, R. (2009). Nyerereite and nahcolite inclusions in diamond: Evidence for lower-mantle carbonatitic magmas. *Mineralogical Magazine, 73*(5), 797–816.

Kaminsky, F. V., Wirth, R., & Schreiber, A. (2013). Carbonatitic inclusions in deep mantle diamond from Juina, Brazil: New minerals in the carbonate-halide association. *Canadian Mineralogist, 51*(5), 669–688.

Kantor, I. Y., Dubrovinsky, L. S., & McCammon, C. A. (2006). Spin crossover in (Mg, Fe)O: A Mössbauer effect study with an alternative interpretation of x-ray emission spectroscopy data. *Physical Review B, 73,* 100101(R).

Kupenko, I., McCammon, C., Sinmyo, R., Cerantola, V., Potapkin, V., Chumakov, A. I., et al. (2015). Oxidation state of the lower mantle: In situ observations of the iron electronic configuration in bridgmanite at extreme conditions. *Earth and Planetary Science Letters, 423,* 78–86. doi:10.1016/j.epsl.2015.04.027

Kupenko, I., McCammon, C., Sinmyo, R., Prescher, C., Chumakov, A. I., Kantor, A., et al. (2014). Electronic spin state of Fe, Al-containing $MgSiO_3$ perovskite at lower mantle conditions. *Lithos, 189,* 167–172. doi:10.1016/j.lithos.2013.10.022

Lavina, B., Dera, P., Downs, R. T., Prakapenka, V., Rivers, M., Sutton, S., et al. (2009). Siderite at lower mantle conditions and the effects of the pressure-induced spin-pairing transition. *Geophysical Research Letters, 36,* L23306.

Lavina, B., Dera, P., Downs, R. T., Tschauner, O., Yang, W. E., Shebanova, O., et al. (2010a). Effect of dilution on the spin pairing transition in rhom-bohedral carbonates. *High Pressure Research, 30,* 224–229.

Lavina, B., Dera, P., Downs, R. T., Yang, W. G., Sinogeikin, S., Meng, Y., et al. (2010b). Structure of siderite $FeCO_3$ to 56 GPa and hysteresis of its spin-pairing transition. *Physical Review B 82,* 064110.

Li, J. (2013). Electronic transitions and spin states in the lower mantle. In K. Hirose, J. Brodholt, T. Lay, & D. Yuen (Eds.), *Post-Perovskite: The last mantle phase transition. Geophysical monograph 174* (pp. 47–68). Washington, DC: American Geophysical Union.

Li, L., Brodholt, J. P., Stackhouse, S., Weidner, D. J., Alfredsson, M., & Price, G. D. (2005). Electronic spin state of ferric iron in Al-bearing perovskite in the lower mantle. *Geophysical Research Letters, 32,* L17307. doi:10.1029/2005/GL023045

Li, D., Oganov, A. R., Dong, X., Zhou, X.-F., Zhu, Q., Qian, G., et al. (2015a). Nitrogen oxides under pressure: Stability, ionization, polymerization, and superconductivity. *Scientific Reports, 5* 16311. doi:10.1038/srep16311

Li, J., Struzhkin, V. V., Mao, H. K., Shu, J., Hemley, R. J., Fei, Y., et al. (2004). Electronic spin state of iron in lower mantle perovskite. *Proceedings of the National Academy of Sciences of the United States of America, 101,* 14027–14030.

Li, J., Sturhahn, W., Jackson, J. M., Struzhkin, V. V., Lin, J. F., Zhao, J., et al. (2006). Pressure effect on the electronic structure of iron in (Mg, Fe)(Si, Al)O_3 perovskite: a combined synchrotron Mössbauer and X-ray emission spectroscopy study up to 100 GPa. *Physics Chemistry Minerals, 33,* 575–585.

Li, Y.-L., Wang, S.-N., Oganov, A. R., Gou, H., Smith, J. S., & Strobel, T. A. (2015b). Investigation of exotic stable calcium carbides using theory and experiment. *Nature Communications, 6,* 6974. doi:10.1038/ncomms7974

Lin, J. F., Gavriliuk, A. G., Struzhkin, V. V., Jacobsen, S. D., Sturhahn, W., Hu, M., et al. (2006). Pressure-induced electronic spin transition of iron in magnesiowüstite-(Mg, Fe)O. *Physical Review B, 73,* 113107.

Lin, J.-F., Gavriliuk, A. G., Sturhahn, W., Jacobsen, S. D., Zhao, J., Lerche, M., et al. (2009). Synchrotron Mössbauer spectroscopic study of ferropericlase at high pressures and temperatures. *American Mineralogist, 94,* 594–599.

Lin, J.-F., Jacobsen, S. D., & Wentzcovitch, R. M. (2007). Electronic spin transition of iron in the Earth's deep mantle. *Eos Transactions American Geophysical Union 88*(2), 13, 17.

Lin, J.-F., Liu, J., Jacobs, C., & Prakapenka, V. B. (2012). Vibrational and elastic properties of ferromagnesite across the electronic spin-pairing transition of iron. *American Mineralogist, 97,* 583–591.

Lin, J.-F., Speciale, S., Mao, Z., & Marquardt, H. (2013). Effects of the electronic spin transitions of iron in lower mantle minerals: Implications for deep mantle geophysics and geochemistry. *Reviews of Geophysics, 51*(2), 244–275.

Lin, J.-F., Speziale, S., Prakapenka, V., Dera, P., Lavina, B., & Watson, H. (2010). High-pressure X-ray diffraction and X-ray emission studies on iron-bearing silicate perovskite under high pressures. *High Pressure Research, 30,* 230–237.

Lin, J.- F., Struzhkin, V. V., Jacobsen, S. D., Hu, M. Y., Chow, P., King, J., et al. (2005) Spin transition of iron in magnesiowüstite in Earth's lower mantle. *Nature 436*, 377–380

Lin, J.-F., & Tsuchiya, T. (2008). Spin transition of iron in the Earth's lower mantle. *Physics of the Earth and Planetary Interiors, 170,* 248–259.

Lin, J.-F., Watson, H., Vankó, G., Alp, E. E., Prakapenka, V. B., Dera, P., et al. (2008). Intermediate-spin ferrous iron in lowermost mantle post-perovskite and perovskite. *Nature Geoscience, 1,* 688–691.

Lin, J.-F., & Wheat, A. (2012). Electronic spin transition of iron in the Earth's lower mantle. *Hyperfine Interactions, 207,* 81–88.

Liu, J., Lin, J. F., Mao, Z., & Prakapenka, V. B. (2014). Thermal equation of state and spin transition of magnesiosiderite at high pressure and temperature. *American Mineralogist, 99,* 84–93.

Liu, J., Lin, J.-F., & Prakapenka, V. B. (2015a). High-pressure orthorhombic ferromagnesite as a potential deep-mantle carbon carrier. *Scientific Reports, 5,* 7640.

Liu, Y., Oganov, A. R., Wang, S., Zhu, Q., Dong, X., & Kresse, G. (2015b). Prediction of new thermodynamically stable aluminum oxides. *Scientific Reports, 5,* 9518. doi:10.1038/srep09518

Lobanov, S. S., Goncharov, A. F., & Litasov, K. D. (2015a). Optical properties of siderite ($FeCO_3$) across the spin transition: Crossover to iron-rich carbonates in the lower mantle. *American Mineralogist, 100,* 1059–1064.

Lobanov, S. S., Zhu, Q., Holtgrewe, N., Prescher, C., Prakapenka, V. B., Oganov, A. R., et al. (2015b). Stable magnesium peroxide at high pressure. *Scientific Reports, 5,* 13582. doi:10.1038/srep13582

Lyakhov, A. O., Oganov, A. R., Stokes, H. T., & Zhu, Q. (2013). New developments in evolutionary structure prediction algorithm USPEX. *Computer Physics Communications, 184,* 1172–1182.

Mao, Z., Lin, J. F., Scott, H. P., Watson, H. C., Prakapenka, V. B., Xiao, Y., et al. (2011). Iron-rich perovskite in the Earth's lower mantle. *Earth and Planetary Science Letters, 309,* 179–184. doi:10.1016/j.epsl.2011.06.030

Marquardt, H., Speziale, S., Reichmann, H. J., Frost, D. J., Schilling, F. R., & Garnero, E. J. (2009). Elastic shear anisotropy of ferropericlase in Earth's lower mantle. *Science, 324* (5924), 224–226.

Mattila, A., Pylkkanen, T., Rueff, J. P., Huotari, S., Vanko, G., Hanfland, M., et al. (2007). Pressure induced magnetic transition in siderite $FeCO_3$ studied by X-ray emission spectroscopy. *Journal of Physics-Condensed Matter, 19,* 386206.

McCammon, C. (2006). Microscopic properties to macroscopic behavior: The influence of iron electronic states. *Journal of Mineralogical and Petrological Sciences, 101,* 130–144.

McCammon, C. A., Dubrovinsky, L., Narygina, O., Kantor, I., Wu, X., Glazyrin, K., et al. (2010). Low-spin Fe^{2+} in silicate perovskite and a possible layer at the base of the lower mantle. *Physics of the Earth and Planetary Interiors, 180,* 215–221.

McCammon, C., Kantor, I., Narygina, O., Roquette, J., Ponkratz, U., Sergueev, I., et al. (2008). Stable intermediate spin ferrous iron in lower mantle perovskite. *Nature Geoscience, 1*(10), 684–687. doi:10.1038/ngeo309

Merlini, M., Crichton, W., Hanfland, M., Gemmi, M., Müller, H., & Kupenko, I. (2012). Structures of dolomite at ultrahigh pressure and their influence on the deep carbon cycle. *Proceedings of the National Academy of Sciences of the U.S.A. 109,* 13509–13514.

Mohn, C. E., & Trønnes, R. G. (2016). Iron spin state and site distribution in $FeAlO_3$-bearing bridgmanite. *Earth and Planetary Science Letters, 440,* 178–186. doi:10.1016/j.epsl.2016.02.010

Müller, J., Speziale, S., Efthimiopoulos, I., Jahn, S., & Koch-Müller, M. (2016). Raman spectroscopy of siderite at high pressure: Evidence for a sharp spin transition. *American Mineralogist, 101,* 2638–2644. doi:10.2138/am-2016-5708

Narigina, O., Mattesini, M., Kantor, I., Pascarelli, S., Wu, X., Aquilanti, G., et al. (2009). High-pressure experimental and computational XANES studies of (Mg, Fe)(Si, Al)O_3 perovskite and (Mg, Fe)O ferropericlase as in the Earth's lower mantle. *Physical Review B, 79,* 174115.

Niu, H., Oganov, A. R., Chen, X.-Q., & Li, D. (2015). Prediction of novel stable compounds in the Mg-Si-O system under exoplanet pressures. *Scientific Reports, 5,* 18347. doi:10.1038/srep18347

Oganov, A. R. (Ed.). (2010). *Modern methods of crystal structure prediction*. Berlin: Wiley-VCN. ISBN 978-3-527-40939-6.

Ohnishi, S. (1978). A theory of the pressure-induced high-spin–low-spin transition of transition metal oxides. *Physics of the Earth and Planetary Interiors, 17,* 130–139.

Ohta, K., Hirose, K., Shimizu, K., Sata, N., & Ohishi, Y. (2010). The electrical resistance measurements of (Mg, Fe)SiO_3 perovskite at high pressures and implications for electronic spin transition of iron. *Physics of the Earth and Planetary Interiors, 180,* 154–158.

Ozawa, H., Hirose, K., Ohta, K., Ishii, H., Hiraoka, N., Ohishi, Y., et al. (2011). Spin crossover, structural change, and metallization in NiAs-type FeO at high pressure. *Physical Review B, 84,* 134417.

Potapkin, V., Chumakov, A. I., Smirnov, G. V., Celse, J. P., Rüffer, R., McCammon, C., et al. (2012). The ^{57}Fe synchrotron Mössbauer source at the ESRF. *Journal of Synchrotron Radiation, 19,* 559–569. doi:10.1107/S0909049512015579

Potapkin, V., McCammon, C., Glazyrin, K., Kantor, A., Kupenko, I., Prescher, C., et al. (2013). Effect of iron oxidation state on the electrical conductivity of the Earth's lower mantle. *Nature Communications, 4,* 1427. doi:10.1038/ncomms2436

Saleh, G., & Oganov, A. R. (2016). Alkali subhalides: High-pressure stability and interplay between metallic and ionic bonds. *Physical Chemistry Chemical Physics: PCCP, 18,* 2840–2849. doi:10.1039/c5cp06026e

Shen, Y., Oganov, A. R., Qian, G., Zhang, J., Dong, H., Zhu, Q., et al. (2015). Novel lithium-nitrogen compounds at ambient and high pressures. *Scientific Reports, 5,* 14204. doi:10.1038/srep14204

Shi, H., Luo, W., Johansson, B., & Ahuja, R. (2008). First-principles calculations of the electronic structure and pressure-induced magnetic transition in siderite $FeCO_3$. *Physical Review B, 78,* 155119.

Shukla, G., & Wentzcovitch, R. M. (2016). Spin crossover in (Mg, Fe^{3+})(Si, Fe^{3+})O^3 bridgmanite: Effects of disorder, iron concentration, and temperature. *Physics of the Earth and Planetary Interiors, 260,* 53–61. doi:10.1016/j.pepi.2016.09.003

Solomatova, N. V., Jackson, J. M., Sturhahn, W., Wicks, J. K., Zhao, J., Toellner, T. S., et al. (2016). Equation of state and spin crossover of (Mg, Fe)O at high pressure, with implications for explaining topographic relief at the core-mantle boundary. *American Mineralogist, 101,* 1084–1093. doi:10.2138/am-2016-5510

Speziale, S., Lee, V. E., Clark, S. M., Lin, J. F., Pasternak, M. P., & Jeanloz, R. (2007). Effects of Fe spin transition on the elasticity of (Mg, Fe)O magnesiowuustites and implications for the seismological properties of the Earth's lower mantle. *Journal of Geophysical Research, 112,* B10212. doi:10.1029/2006JB004730

Speziale, S., Milner, A., Lee, V. E., Clark, S. M., Pasternak, M. P., & Jeanloz, R. (2005). Iron spin transition in Earth's mantle. *Proceedings of the National Academy of Sciences, 102,* 17918–17922.

Stackhouse S., Brodholt, J., Dobson, D. P., & Price, G. D. (2006). Electronic spin transitions and the seismic properties of ferrous iron-bearing $MgSiO_3$ post-perovskite. *Geophysical Research Letters, 33,* L12S03. doi 10.1029/2005GL025589

Stackhouse, S., Brodholt, J. P., & Price, G. D. (2007). Electronic spin transitions in iron-bearing $MgSiO_3$ perovskite. *Earth and Planetary Science Letters, 253,* 282–290.

Tsuchiya, T., & Wang, X. (2013). Ab initio investigation on the high-temperature thermodynamic properties of Fe^{3+}-bearing $MgSiO_3$ perovskite. *Journal of Geophysical Research, 118,* 83–91. doi:10.1029/2012JB009696

Tsuchiya, T., Wentzcovitch, R. M., da Silva, C. R. S., & de Gironcoli, S. (2006). Spin transition in magnesiowüstite in Earth's lower mantle. *Physical Review Letters, 96,* 198501. doi:10.1029/2012JB009696

Umemoto, K., Wentzcovitch, R. M., Yu, Y. G., & Requist, R. (2008). Spin transition in (Mg, Fe) SiO_3 perovskite under pressure. *Earth and Planetary Science Letters, 276,* 198–206.

Vanpeteghem, C., Angel, R., Ross, N., Jacobsen, S., Dobson, D., Litasov, K., et al. (2006). Al, Fe substitution in the $MgSiO_3$ perovskite structure: A single-crystal X-ray diffraction study. *Physics of the Earth and Planetary Interiors, 155*(1–2), 96–103.

Wentzcovitch, R. M., Justo, J. F., Wu, Z., da Silva, C. R. S., Yuen, D. A., & Kohlstedt, D. (2009). Anomalous compressibility of ferropericlase throughout the iron spin cross-over. *Proceedings of the National Academy of Sciences of the United States of America, 106,* 8447–8452.

Wu, Z., Wentzcovitch, R. M. (2014). Spin crossover in ferropericlase and velocity heterogeneities in the lower mantle. *Proceedings of the National Academy of Sciences of the U.S.A, 111,* 10468–10472.

Wu, Y., Wu, X., Lin, J.-F., McCammon, C. A., Xiao, Y., Chow, P., et al. (2016). Spin transition of ferric iron in the NAL phase: Implications for the seismic heterogeneities of subducted slabs in the lower mantle. *Earth and Planetary Science Letters, 434,* 91–100.

Yang, J., Tong, X., Lin, J.-F., Okuchi, T., & Tomioka, N. (2015). Elasticity of ferropericlase across the spin crossover in the Earth's lower mantle. *Scientific Reports, 5,* 17188. doi:10.1038/srep17188

Yoo, C. S., Maddox, B., Klepeis, J.-H. P., Iota, V., Evans, W., McMahan, A., et al. (2005). First-order isostructural Mott transition in highly compressed MnO. *Physical Review Letters, 94,* 115503.

Zhang, S., Cottaar, S., Liu, T., Stackhouse, S., & Militzera, B. (2016). High-pressure, temperature elasticity of Fe-and Al-bearing $MgSiO_3$: Implications for the Earth's lower mantle. *Earth and Planetary Science Letters, 434,* 264–273. doi:10.1016/j.epsl.2015.11.030

Zhang, F., & Oganov, A. R. (2006). Valence state and spin transitions of iron in Earth's mantle silicates. *Earth and Planetary Science Letters, 249,* 436–443.

Zhang, W., Oganov, A. R., Goncharov, A. F., Zhu, Q., Boulfelfel, S. E., Lyakhov, A. O., et al. (2013). Unexpected stable stoichiometries of sodium chlorides. *Science, 342,* 1502–1506.

Zhang, J., Oganov, A. R., Li, X., Xue, K.-H., Wang, Z., & Dong, H. (2015). Pressure-induced novel compounds in the Hf–O system from first-principles calculations. *Physical Review B, 92,* 184104. doi:10.1103/PhysRevB.92.184104

Zhu, Q., Jung, D. Y., Oganov, A. R., Glass, C. W., Gatti, C., & Lyakhov, A. O. (2013a). Stability of xenon oxides at high pressures. *Nature Chemistry, 5,* 61–65.

Zhu, Q., Oganov, A. R., & Lyakhov, A. O. (2013b). Novel stable compounds in the Mg–O system under high pressure. *Physical Chemistry Chemical Physics, 15,* 7696–7700.

Zhu, Q., Oganov, A. R., & Zeng, Q. (2015). Formation of stoichiometric CsF_n compounds. *Scientific Reports, 5,* 7875. doi:10.1038/srep07875

Zurek, E., Hoffmann, R., Ashcroft, N. W., Oganov, A. R., & Lyakhov, A. O. (2009). A little bit of lithium does a lot for hydrogen. *PNAS, 106*(42), 17640–17643. doi:10.1073/pnas.0908262106

Chapter 9
D″ Layer: Transition from the Lower Mantle to the Earth's Core

Abstract The D″ layer is a ~200 km layer at the bottom of the lower mantle (at ~2700–2900 km depth). It has low *S*-wave velocity gradients and increased scatter in travel times and amplitudes. Compositionally, there are two sources for the D″ layer: the oxide lower mantle and the outer core. Oxides from the lower mantle experience phase and physical changes within the D″ layer; the outer core delivers the metallic part to the D″ layer composition. The transformation of bridgmanite in post-perovskite, creating a 1–1.5% shear velocity increase, is the major effect for distinguishing of the D″ layer. Post-perovskite may incorporate 1–2 wt% H_2O and thus may store significant amounts of hydrogen within the D″ layer. The transformation of bridgmanite in post-perovskite is accompanied with the transition of the orthorhombic $CaCl_2$-structured SiO_2 in seifertite, α-PbO_2-structured SiO_2, resulting in a slight decrease in bulk sound speed by ~0.4% and shear wave decrease. Iron-rich liquid metal from the outer core (containing 5–10% light elements), namely C, N, O and Si infiltrates into the lowermost mantle and forms a series of native Fe^0, iron carbides and nitrides, and silicon carbide. Of particular importance is the presence of Fe_7C_3 and Fe_7N_3. When in association with diamond, these are the first solidus phases crystallizing from the metallic liquid in the D″ layer. The presence of iron nitrides in the D″ layer is closely related to their suggested presence in the inner core and helps to solve the problem of 'missing nitrogen' in the Earth's nitrogen balance.

9.1 General

As previously noted, the D″ layer is a ~200 km layer at the bottom of the lower mantle (at depths of ~2700–2900 km), characterized by anti-correlated bulk and shear wave velocities. There is a laterally variable thickness of between 0 and >300 km, with an added complicated seismic anisotropy structure, and large regions of generally low shear wave velocities. The anomalous seismic properties of the D″ region in the form of low S-wave velocity gradients and increased scatter in travel times and amplitudes, were recognized by Bullen (1949). Its lower border

F.V. Kaminsky, *The Earth's Lower Mantle*, Springer Geology,
DOI 10.1007/978-3-319-55684-0_9

with the outer core (the 'Gutenberg discontinuity') is strongly marked by zero shear-wave velocity and drastic changes in compressional wave velocity and density in the PREM and other models (Dziewonski and Anderson 1981; Kennett and Engdahl 1991). Bullen (1942) proposed the 'D layer' name for the entire lower mantle. However, by 1949 he had recognized that the lowermost region of the lower mantle (a layer of thickness 200–300 km) had properties that made it distinct from the overlying and much thicker layer of this region. He was, therefore, compelled by the evidence to split the D layer into two parts, which he labelled D′ and D″ (Bullen 1949, 1950). The former of these regions we simply refer to as the lower mantle, but the latter has retained the name originally given by Bullen. Chao (2000) suggested, on this basis, that this layer might be more reasonably referred to as the 'Bullen layer', a suggestion that ra a great deal of interesting historical comments, in which it was recalled that the discovery of a seismically distinct layer immediately above the core–mantle boundary (CMB) was made much earlier by Cornelius Damn in his 1934 doctoral dissertation at St. Lois University (Peltier 2007; Kisslinger 2000).

Initially the D″ layer was considered to be a thermal layer resulting from an expected large temperature drop from the core across D″, up to 1500 °C (Stacey and Loper 1983; Williams 1998). Based on geophysical and geochemical data, the following years produced a chemically different composition model of the D″ layer (compared with the rest of the lower mantle), which gained popularity (Garnero 2000; Trønnes 2010; Cobden et al. 2015). Because of the extremely large density discontinuity at the CMB, there is very good reason to conclude that the interface between the outer core and the mantle is a boundary at which chemically distinct phases with density intermediates between the core and the lower mantle might concentrate (Peltier 2007).

A significant polarization anisotropy can be observed within the D″ layer, especially under the circum-Pacific regions, where the horizontally polarized S-wave velocity (V_{SH}) is faster by 1–3% compared with the vertically polarized S-wave velocity (V_{SV}) (Lay et al. 1998; Panning and Romanowicz 2004). Since such polarization anisotropy is present only below the D″ discontinuity, it is naturally expected that the origin of S-wave anisotropy is related to the preferred orientation of post-perovskite (Tsuchiya et al. 2004a; Hirose 2007; Wookey and Kendall 2007).

In addition to anisotropy, the D″ layer is heterogeneous (observations from seismologic data), exhibits radial and lateral heterogeneity in seismic velocities, anomalous seismic velocity gradients, discontinuities in both shear- and compressional-wave velocities, and strong scattering (Knittle and Jeanloz 1991; Tsuchiya et al. 2004a; Lay and Garnero 2004 and references therein). It is the final resting place for the subducting slabs (van der Hilst et al. 1997; Ohta et al. 2008) and, at the same time, may be the source region for volcanic hot spots (Williams 1998).

The upper border of the D″ layer fluctuates between ~100 and 450 km above the CMB (Cobden et al. 2015). In the upper part of the D″ layer and partly above it (up to 1000 km above the CMB) in two areas (beneath Africa and beneath the central Pacific), large low shear velocity provinces (LLSVP) with V_s reduced by

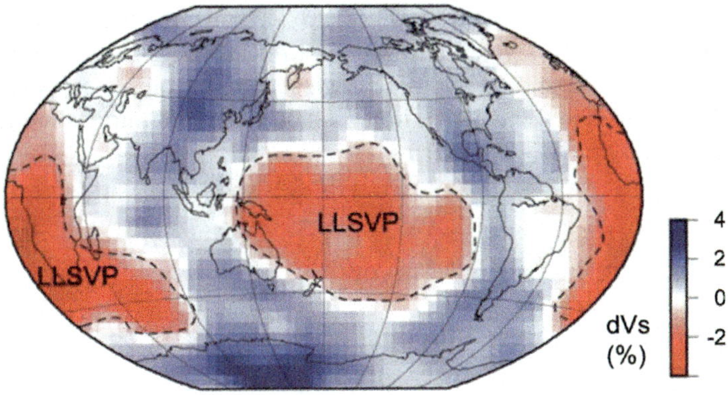

Fig. 9.1 Variations of *S*-wave velocity in the lowermost 200-km of the mantle, in the seismic tomography model of Grand (2002). From Ohta et al. (2008) © Elesevier

3–5% were discovered in the 1990s (Su et al. 1994; Li and Romanowicz 1996) and studied with the use of seismic tomography (Masters et al. 2000; Grand 2002; Trampert et al. 2004; He and Wen 2012) (Fig. 9.1). Seismic tomography also revealed in the LLSVPs, the elevated bulk sound speed, hotter temperature, elevated density and sharp boundaries (Lay and Garnero 2004; Trampert et al. 2004; Davies et al. 2015; Li et al. 2015a).

The composition of LLSVPs is currently extensively debated and it is most likely to be different to the surrounding lower mantle. The primordial remnants of mantle differentiation processes (Lee et al. 2010), the accumulation of subducted ancient oceanic crust (Davies 2008), the mixture of MORB and pyrolitic material (Ohta et al. 2008), and the presence of metallic Fe–Ni–S liquid, trapped during the crystallization of a dense basal magma ocean (Zhang et al. 2016) are among the hypotheses discussed regarding such a composition.

There is a thin (4–13 km thick) core–mantle transitional zone (CMTZ) at the base of the D″ layer, in which small-scale patches with strongly reduced velocities (up to 10% in V_p and 20% in V_s lower than average) and increased densities, known as ultra-low velocity zones (ULVZ) occur (Garnero 2004; Garnero and Helmberger 1995, 1996; Williams and Garnero 1996; McNamara et al. 2010). Its nature is now debatable (Gassner et al. 2015). Two explanations have been proposed: 10–30% of partial melt (Williams and Garnero 1996; Garnero 2004; Lay et al. 2004) or some type of iron-enriched solid. In the latter case, it may be the result of the infiltration of an iron-nickel alloy from the outer core (Kanda and Stevenson 2006; Otsuka and Karato 2012) or residual Fe-enriched mantle material as a final result of an evolved basal magma ocean (Labrosse et al. 2007; Nomura et al. 2011).

The entire D″ layer should be considered as a transitional zone between the mantle and the core, which is presumably very geodynamically active (as, for example, the source of mantle plumes), and is not yet fully understood. Compositionally, there are two sources for the D″ layer: the outer core and the lower mantle. The lower mantle's input is the major oxides, which experience phase and physical changes within the D″ layer (the oxide part); the outer core delivers the metallic part to the D″ layer composition. Mineral physics currently provides central knowledge for the oxide part; geological samples presented in recent years provide information about the native iron, iron carbides and nitrides in the D″ layer (the 'metallic part').

9.2 Oxide Part of the D″ Layer

9.2.1 General

Major lower-mantle oxides experience structural and physical changes at the upper border of the D″ layer. Bridgmanite obtains a post-perovskite structure, and SiO_2 becomes seifertite. Ferropericlase-magnesiowüstite and CaSi-perovskite remain structurally unchanged.

9.2.2 Post-perovskite

The transformation of bridgmanite in post-perovskite is the major effect for distinguishing the D″ layer (Duffy 2005). There is a confluence of evidence suggesting that the explanation of the existence of the D″ layer is primarily due to the bridgmanite to post-perovskite phase transformation (Peltier 2007).

This phase transition was predicted by Sidorin et al. (1999), who noted that an exothermic phase change above the CMB would explain the seismic D″ discontinuity better than a thermally or chemically distinct layer. The transformation of the bridgmanite *Pbnm* structure in a post-perovskite *Cmcm* ($CaIrO_3$-type) structure (with the same formula ABO_3 and composition (Mg, Fe)(Si, Al)O_3) was experimentally discovered with the use of LHDAC and in situ XRD by Murakami et al. (2004) at ~125 GPa and 2500 K, followed almost simultaneously by other independent experiments at 124 GPa and 2500 K (Mao et al. 2004; Shim et al. 2004; Oganov and Ono 2004; Ono and Oganov 2005) and theoretical calculations (Iitaka et al. 2004; Oganov and Ono 2004; Tsuchiya et al. 2004b; Dorogokupets et al. 2015). These pressure–temperature (*P–T*) conditions match the depth of the D″ discontinuity. In recent years, the post-perovskite formation and its properties, along with its geophysical and geodynamic implications, have been extensively studied both theoretically and experimentally (e.g., Hirose et al. 2013 and references therein).

The post-perovskite phase transformation also occurred in perovskites formed in natural peridotites and basaltic compositions (Murakami et al. 2005; Ono et al. 2005; Ohta et al. 2008). In peridotite, the transformation from bridgmanite to post-perovskite was observed at 113 GPa and 2500 K (Murakami et al. 2005) and at 116–121 GPa and 2500 K (Ohta et al. 2008), which is consistent with the experimental transition pressure observed in pure $MgSiO_3$ (119 GPa at 2500 K;

Hirose et al. 2006) and in $(Mg_{0.9}Fe_{0.1})SiO_3$ composition (Shieh et al. 2006). The synthesized peridotite post-perovskite has a composition similar to bridgmanite, but contains a lower amount of iron (1.9 wt%; Murakami et al. 2005). The experiments on iron partitioning between bridgmanite and post-perovskite demonstrate that post-perovskite is indeed poorer in iron than coexisting bridgmanite (Hirose et al. 2008).

Synchrotron diffraction experiments have shown that the presence of iron may lower the transition boundary in $(Mg, Fe)SiO_3$ compositions (Mao et al. 2004, 2007); however, this effect may be balanced by the addition of Al (Mao et al. 2007).

The post-perovskite transition occurs within a 5 GPa pressure interval in pyrolitic material, corresponding to a depth range of 90 km in the lowermost mantle (Ohta et al. 2008). It is larger than the estimates for the sharpness of the D″ seismic discontinuity, which have inferred a depth extent of velocity increase of less than 30 km (Lay and Young 1989), or not more than 50–75 km (Revenaugh and Jordan 1991; Weber et al. 1996). This discrepancy may be due to the errors in pressure determinations in the experiments (Ohta et al. 2008).

The phase boundary between bridgmanite *Pbnm* and post-perovskite *Cmcm* phases has a large positive Clapeyron slope, varying, according various experiments and calculations, from 7.5–8 MPa/K to 13–16 MPa/K (Tsuchiya et al. 2004b; Oganov and Ono 2004; Ono and Oganov 2005; Ohta et al. 2008; Li et al. 2015a) (Fig. 9.2).

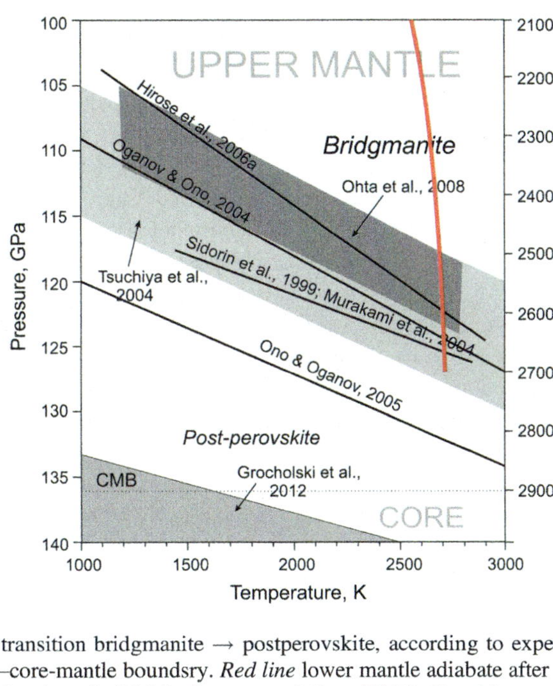

Fig. 9.2 Phase transition bridgmanite → postperovskite, according to experimental and theoretical data. *CMB*—core-mantle boundsry. *Red line* lower mantle adiabate after Katsura et al. (2010). Data from Murakami et al. (2004), Oganov and Ono (2004), Tsuchiya et al. (2004b), Ono and Oganov (2005), Hirose et al. (2006), Ohta et al. (2008), Grocholski et al. (2012)

The chemical composition of post-perovskite is as variable as that of bridgmanite, with the magnesium index *mg* varying in the same range (0.84–0.96) and with substantial admixtures of Al, Cr, Ca and Na (see Sect. 4.2.2). In addition, the same charge-coupled substitution $Al^{3+} + Fe^{3+} \leftrightarrow Si^{4+} + Mg^{2+}$ may be expected in post-perovskite. However, some compositional changes, particularly in the Fe concentration, may be suggested.

The partition coefficient between bridgmanite and post-perovskite is high (2–4; Murakami et al. 2005; Ono and Oganov 2005; Hirose et al. 2008; Andrault et al. 2010), indicating the depletion of post-perovskite in iron. Experimental Fe–Mg partition coefficient between coexisting magnesiowüstite and the post-perovskite phase is also large (3–8), confirming that iron is strongly depleted in the post-perovskite phase compared with the perovskite phase at lower pressures (Kobayashi et al. 2005; Murakami et al. 2005; Hirose et al. 2008). On the other hand, Mao et al. (2004), from analysis of XRD data, suggested that the post-perovskite phase is enriched in iron. It was suggested later that the decrease in bulk iron content causes controversies in the Fe–Mg partitioning between post-perovskite and ferropericlase, and that iron prefers ferropericlase over post-perovskite (Sakai et al. 2010). However, the decrease in bulk iron content is not a case for the lowermost lower mantle. The reverse was demonstrated for natural ferropericlase and bridgmanite pairs: both minerals are enriched in Fe as a consequence of the bulk compositional stratification (Kaminsky and Lin 2017; see Sect. 4.4). Within the D″ layer, further enrichment in Fe may be expected as a result of interference with the outer core. Experimental works demonstrate that post-perovskite can accommodate a significant part of iron, up to *fe* = 0.40 (Mao et al. 2005, 2006).

In the case of the HS–LS transition in post-perovskite, iron partitioning should significantly change, and the ferric iron part $Fe^{3+}/\Sigma Fe_{tot}$ is expected to increase up to 0.65 (Sinmyo et al. 2006). This conclusion, however, is debated (e.g., Murakami et al. 2005; Kobayashi et al. 2005; Nishio-Hamane and Yagi 2009; Lin et al. 2013). First-principles calculations predicted equal occupancies of HS Fe^{3+} at the A-site and LS Fe^{3+} at the B-site for Fe_2O_3-bearing post-perovskite (Yu et al. 2012). In LHDAC experiments with Fe_2O_3-bearing post-perovskite, nearly equal occupancies of HS Fe^{3+} at the A-site and LS Fe^{3+} at the B-sites were found for 128–138 GPa, i.e., for the entire range of D″ layer (Catalli et al. 2010).

First-principles calculations demonstrate a possibility for post-perovskite to incorporate 1–2 wt% H_2O at D″ pressures (Townsend et al. 2015). In this case, the formula of stable hydrous post-perovskite should be $Mg_{1-x}SiH_{2x}O_3$. During the course of water-containing bridgmanite → post-perovskite structural transformation, water favours bridgmanite over post-perovskite by a factor of approximately 5:1 for the Al-free system. In the Al-bearing system, in reverse, water favours post-perovskite over bridgmanite by a factor of approximately 3:1 at ambient mantle conditions (Townsend et al. 2016). Since bridgmanite in deep lower mantle is always Al-rich (see Sect. 4.4), post-perovskite and the entire D″ layer may store significant amounts of hydrogen.

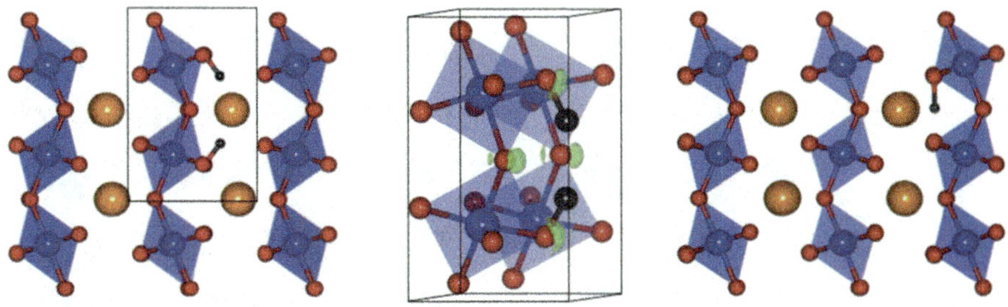

Fig. 9.3 Predicted stable hydrogen defect structure of hydrous post-perovskite at 120 GPa. *Left* view of hydrous bridgmanite down the *a*-axis. *Right* electron localisation function 93-percentile iso-surface (*green*) of a Mg vacancy site in hydrous bridgmanite with super-imposed hydrogen groups. Atom positions are shown as O^{2-} (*red*), Si^{4+} (*blue*), Mg^{2+} (*brown*), and H^+ (*black*). From Townsend et al. (2015). © Elesevier

The structure of post-perovskite was established as an orthorhombic $CaIrO_3$-type with the space group *Cmcm* (Murakami et al. 2004; Tsuchiya et al. 2004b; Oganov and Ono 2004). It contains alternating layers of edge-sharing SiO_6 octahedra and Mg polyhedra in eight coordination to oxygen. The structure is anisotropic being, more compressible in the *b* direction. Both oxygen sites in the structure of post-perovskite are slightly under-bonded, and thus the structure is potentially more accommodating of hydrogen than bridgmanite (Hernández et al. 2013). The stable structure of hydrous post-perovskite at 120 GPa is illustrated in Fig. 9.3. The hydrogen bond in hydrous post-perovskite is considerably more non-linear than in hydrous bridgmanite (Townsend et al. 2015). In hydrous post-perovskite, the O–H bond distance increases slightly with pressure as the hydrogen bond distance $d(O\cdots O)$ shortens, which is consistent with general trends among hydrogen bonds in minerals (Libowitzky 1999).

Single-crystal structure determinations in post-perovskite demonstrated that pure and low-Fe (*fe* = 0.10) post-perovskites have similar *Cmcm* structures (Zhang et al. 2013). High-Fe post-perovskite (*fe* = 0.40) crystallizes in a structure with the *Pmcm* (*Pmma*) space group, which is energetically more favourable than the *Cmcm*, as was shown in experimental works and theoretical calculations (Yamanaka et al. 2012) (Fig. 9.4). The *Pmcm* (*Pmma*) structure has a significantly higher density ($\rho = 6.119\ g/cm^3$) than the *Cmcm* ($\rho = 5.694\ g/cm^3$) due to both the larger amount of iron and the smaller ionic radius of Fe^{2+} as a result of an electronic spin transition observed by X-ray emission spectroscopy (XES). However, as demonstrated above, high-Fe post-perovskite in the D″ layer is unlikely, and only *Cmcm* structure may be suggested for this mineral phase.

Ab initio calculations on post-perovskite with Fe = 0.0625–0.1250 demonstrated that, under pressures of up to 180 GPa at 0 K, iron remains in a HS state (Stackhouse et al. 2005, 2006), and no spin transition in post-perovskite should occur (Zhang and Oganov 2006; Yu et al. 2012). However, in experiments with iron-containing Al-post-perovskite, both iron ions Fe^{2+} (in A-site) and Fe^{3+} (in B-site) experience the HS–LS transition at 123–135 GPa within the D″ layer (Fujino et al. 2014).

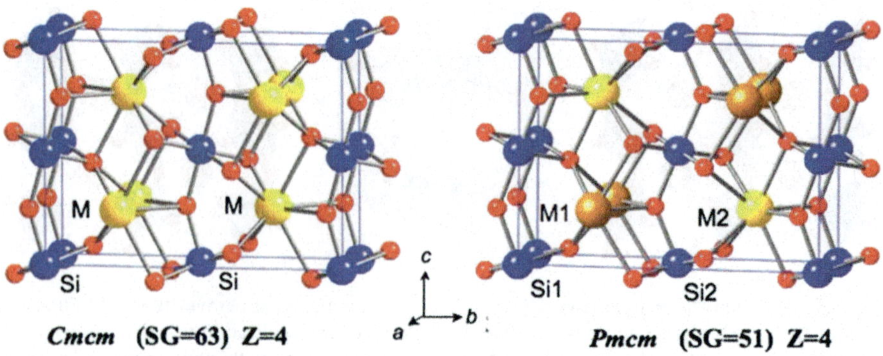

Fig. 9.4 Crystal structures of post-perovskite with Fe = 0.10 composition and *Cmcm* space group (*left*) and with Fe = 0.40 m and *Pmcm* space group (*right*). Atom positions are shown as O^{2-} (*red*), Si^{4+} (*blue*), Mg^{2+} (*yellow*), Fe^{2+} (*brown*). Used with permission of the PNAS from Yamanaka et al. (2012)

Elastic properties. The bridgmanite → post-perovskite structure transformation is accompanied by a density increase of 0.9–1.6% (Murakami et al. 2004; Shieh et al. 2006; Shim et al. 2008), which theoretically causes a 1–1.5% shear velocity increase (Stackhouse et al. 2005; Tsuchiya and Tsuchiya 2006; Wentzcovitch et al. 2006), although the incorporation of Fe and Al reduces the velocity contrast (Caracas and Cohen 2005; Tsuchiya and Tsuchiya 2006). Further experiments show that the enrichment of Fe and Al in post-perovskite can produce an increase in its density, but substantially reduce the bulk sound velocity (V_ϕ) across the phase transition (by 2.4%; Shim et al. 2008). As a result, the combined effect of Fe and Al leads to an anti-correlation between the enhanced density and the reduced bulk sound velocity at the pressure condition of the D″ layer, and (Fe,Al)-rich post-perovskite existing in the D″ region would be shown as a relatively high-density and low-velocity region in deep-mantle seismic observations (Mao et al. 2014).

The hydration of post-perovskite at D″ conditions should decrease wave velocities in the mineral, but the bulk sound velocity (V_ϕ) is insensitive to the presence of hydrogen. At 120 GPa, V_ϕ is reduced by less than 0.1% per wt% H_2O. In contrast to the bulk sound velocity, both P-wave (V_P) and S-wave (V_S) velocities decrease by 1.1% per wt% H_2O at 120 GPa. Of particular note is that at 120 GPa, the S-wave velocity of hydrous post-perovskite is slower than that of post-perovskite, implying that a very hydrous post-perovskite, if present in the D″ layer, may be difficult to distinguish from post-perovskite using bulk sound velocity and/or shear-wave profiles alone (Townsend et al. 2015).

Post-perovskite is a strongly anisotropic mineral, and its presence may thus explain the anisotropy observed in the D″ layer (Wookey et al. 2005, 2007). The seismic anisotropy of pure post-perovskite $MgSiO_3$ is 15% for V_P and 26% for V_S. The addition of Fe^{2+} to post-perovskite results in anisotropic expansion of the structure. This expansion is more pronounced along the direction perpendicular to the octahedral layers, the *b* lattice parameter (Caracas and Cohen 2007).

9.2.3 SiO_2

The transition of the orthorhombic $CaCl_2$-structured SiO_2, characteristic of the deep part of the lower mantle in seifertite α-PbO_2-structured SiO_2 is described in Sect. 4.6.2. It occurs (as per experimental data) at 113–130 GPa for pure SiO_2 (Murakami et al. 2003; Shieh et al. 2005; Grocholski et al. 2013); an admixture of Al lowers the transition by 5–10 GPa (Hirose et al. 2005; Andrault et al. 2014). On average, this transition takes place near the D″ layer upper border (Fig. 4.26).

Experiments with pure silica carried out in the LHDAC, using insulating compressed noble gas media, demonstrated that seifertite has the smallest molar volume among all known silica polymorphs. Therefore, having the highest density at 1.5% denser than the $CaCl_2$-structured SiO_2, resulting in a slight decrease in bulk sound speed of ~0.4% at 130 GPa and shear wave decrease (Grocholski et al. 2013). In such a case, this silica transition may provide an alternative explanation for the discontinuities observed with a shear wave velocity decrease near the D″ discontinuity.

9.2.4 Ferropericlase-Magnesiowüstite

There is no structural transformation in ferropericlase-magnesiowüstite within the entire lower mantle (see Sect. 4.3.3) and no seismic effect may be caused by this mineral within or near the D″ layer. However, (Mg,Fe)O, like post-perovskite is expected to have both azimuthal and polarization anisotropies. For horizontal flow, (Mg,Fe)O shows $V_{SH} > V_{SV}$ anisotropy, while perovskite will produce $V_{SV} > V_{SH}$ anisotropy. V_{SH}/V_{SV} anisotropy caused by post-perovskite depends on the elastic anisotropy and the dominant glide plane, both of which are not well constrained. For this glide plane, the magnitude of the anisotropy of (Mg,Fe)O is much larger than that for post-perovskite; therefore, (Mg,Fe)O plays a more important role than post-perovskite for the interpretation of seismic anisotropy within the D″ layer (Yamazaki and Karato 2007).

The total thermal conductivity of ferropericlase, in contrast, plays much smaller role than that of post-perovskite at the core/mantle boundary because of the effects of strong iron impurity scattering and spin crossover (Ohta et al. 2017). This data indicates that post-perovskite is the best heat conductor at the CMB.

9.2.5 Other Minerals

It was demonstrated in Sect. 8.3 that, at pressures corresponding to D″ layer conditions (126–137 GPa), chemical elements can obtain dramatic new properties, including the formation of unexpected crystal structures and completely new counter-intuitive compounds. In the D″ layer, one may expect the appearance of the following compounds: Na_3Cl_2 and Na_2Cl, CsF_2 and CsF_4, LiH_6, XeO_3, Na_2He and others, some of which may play a significant role in the Earth's balance of light elements and in the formation of the primordial carbonatitic (or silica-carbonatitic) association.

9.3 Metal-Carbide-Nitride Part of the D″ Layer

9.3.1 General

One of the major models for the formation of the D″ layer is that its substance was dissolved in a liquid core, fractionated, separated and uplifted during its crystallization (Buffett 2003). A possibility of periodic infiltrations of Fe-rich liquid metal from the outer core into the lowermost mantle was suggested as the result of the deformation-induced mechanical instabilities at the CMB, caused by the accumulation of strain deformations in high-stress zones associated with downwelling material (Petford et al. 2007). Such processes are supported in experiments with chondrites, in which metallic layers were infiltrating silicate chondritic material (Rushmer et al. 2005). Another mechanism of the infiltration of liquid metal into the lower mantle (based on experiments) was proposed as a result of a 'morphological instability', caused by a contact of ferropericlase with iron-rich liquids and resulting in the penetration of blobs of liquids for 50–100 km away from the CMB (Otsuka and Karato 2012).

Disequilibrium between core and mantle compositions produces chemical interactions at the CMB. However, the process of the interaction of liquid metal with the oxide part of the D″ layer remains unclear. Mao et al. (2006), after studying a Fe-enriched system at pressures of up to 170 GPa, suggested the presence of Fe-rich post-perovskite at the CMB as the result of the reaction of oxide with a liquid iron alloy from the core. It was experimentally demonstrated that the addition of 25% Fe in bridgmanite increases its density and bulk sound velocity (V_{Φ}) by 4–6% and 6–7%, respectively, at lower mantle pressures of up to 135 GPa. This effect can be explained by the occurrence of the low-spin Fe^{3+} and the extremely high-quadrupole component of Fe^{2+}. It explains the seismic observations of reduced shear-wave velocity in regions interpreted as dense, stiff piles in the lower mantle (Mao et al. 2011).

On the other hand, a series of experiments demonstrated that bridgmanite and ferropericlase are strongly depleted in Fe, whereas the coexisting metal is enriched in Si and O. In early experiments involving the interaction of bridgmanite and metallic iron, formation of SiO_2, FeO and FeSi was reported (Knittle and Jeanloz 1991; Goarant et al. 1992). Subsequently, Takafuji et al. (2005), using a combination of LHDAC and TEM, studied the solubility of O and Si in liquid iron in equilibrium with bridgmanite. The results demonstrated that the dissolution of both O and Si into molten iron was significantly enhanced with increasing pressure. The quenched liquid iron contained 5.3 wt% O and 2.8 wt% Si at 97 GPa and 3150 K, and the coexisting bridgmanite was depleted in FeO. Reaction products such as

SiO_2, FeO, and FeSi were not observed as stable phases, but FeO and SiO_2 were found as quench crystals formed from (O, Si)-rich molten iron. Based on the partitioning data of oxygen and silicon between perovskite and molten iron, Takafuji et al. (2005) suggested that the bottom of the mantle in contact with the outer core should be extremely depleted in iron (less than 1% FeO) if the local equilibrium is attained at the CMB. Later, Sakai et al. (2006) performed experiments on the interaction between iron and post-perovskite at higher-pressure conditions, corresponding to the bottom of the D″ layer (130–139 GPa; 2000–3000 K). They confirmed the results of previous experiments and demonstrated that, at such conditions, even higher amounts of O (6.3 wt%) and Si (up to 4.0 wt%) can be dissolved into molten iron; however, no reaction occurred between the phases. Both phases of ferropericlase and bridgmanite are low-Fe (Ozawa et al. 2008, 2009).

These experiments were performed with the use of pure iron, undersaturated with O and Si. Other models, in reverse, propose the enrichment of the outer core with light elements that, in turn, cause the extraction of these elements from the outer core to the bottom of the lower mantle (Buffett et al. 2000). Light elements, including C, N, O and Si, comprise 5–10 wt% of the Earth's core (Stixrude et al. 1997; McDonough 2003). Geological samples demonstrate the presence of compounds containing these elements in the D″ layer.

Light elements penetrate into the bottom of the lower mantle along with iron. It was experimentally demonstrated that iron and other siderophile elements are mobile in the grain boundaries of polycrystalline periclase, obtained by quantifying the extent of alloy formation between initially pure metals separated by 1 mm of polycrystalline MgO (Hayden and Watson 2007, 2008). Grain-boundary diffusion resulted in a significant alloying of sink and source particles, enabling grain-boundary fluxes to be calculated. The computed diffusivities were high enough to allow the transport of a number of siderophile elements over geologically significant lengths (tens of kilometres) during the age of Earth. Such grain-boundary diffusion is a potential fast pathway for chemical communication between the core and lower mantle.

9.3.2 Native Iron, Iron Carbides and Nitrides, and Silicon Carbide

Native iron, in association with iron carbides, nitrides and silicon carbide, presupposed to originate in the D″ layer, was discovered in natural samples as microinclusions in several diamonds from the Rio Soriso area in Brazil (Kaminsky and Wirth 2011; Kaminsky et al. 2015).

Native iron, in this association, was found as an intergrowth with iron carbide and magnetite (Fig. 9.5) and is almost pure in its composition. A minor nitrogen impurity was observed in the EDX spectrum, as well as a minor admixture of Ni.

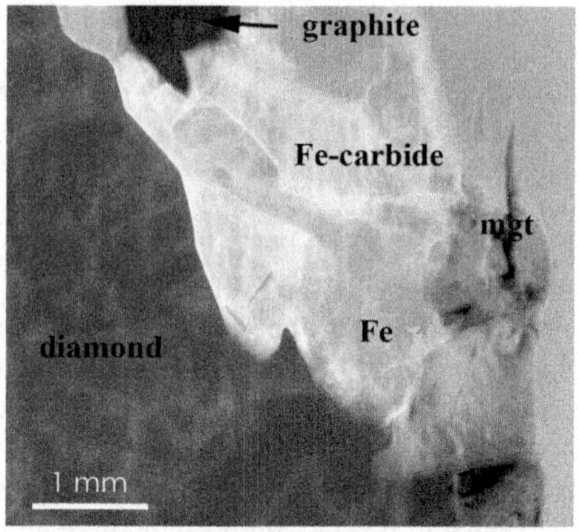

Fig. 9.5 TEM image of an inclusion in diamond composed of native iron (Fe, *dark grey*), iron carbide (*light grey*), magnetite (*mgt*), and graphite. The iron carbide phase is an aggregate of grains, 0.5–1 μm each, having different orientations, which are expressed in various shades of *grey* in the image. *Dark areas* in magnetite are pores. Note the high density of dislocations in host diamond. From Kaminsky and Wirth (2011). Used with a permission of the Canadian Mineralogist

Table 9.1 Chemical compositions of carbides and nitrocarbides (at.%). From Kaminsky and Wirth (2011)

Composition	$Fe_5(C,N)_3$	Fe_2C	Yarlongite Fe_9C_4	$Fe_7(C,N)_3$	Haxonite $Fe_{23}C_6$
Sample No.	2053-13	2053-7	2053-1	2053-14	2053-3
Fe	62.2	66.1	68.2	69.9	78.2
C	30.5	33.9	31.8	22.1	21.8
N	7.3	–	–	8.0	–
C/(C + N)	0.81	–	–	0.73	–
Fe/(C + N)	1.65	1.95	2.14	2.32	3.59

Carbides, nitrocarbides, carbonitrides and nitrides form a series of stoichiometric compositions: $Fe_3(C, N)$ (cohenite and nitrocohenite), $Fe_9(C, N)_4$ (yarlongite and nitroyarlongite), $Fe_7(C, N)_3$, $Fe_{23}(C, N)_6$ (haxonite and nitrohaxonite). Their chemical compositions are presented in Table 9.1. The most peculiar feature of their compositions is the variable concentration of nitrogen, with N/(N + C) varying from 0 to 100, up to pure nitrides (Kaminsky and Wirth 2017).

Iron carbide Fe_7C_3. Of principal importance is the presence of Fe_7C_3. The possibility has not been excluded that this mineral is 'chalypite'; it was identified in 1854 as Fe_2C from meteorite (Shepard 1867), but was later deleted from the mineral list (Buchwald 1977). Lately, it has been repeatedly identified in lower-mantle associations with the use of both composition and electron diffraction analyses (Kaminsky and Wirth 2011). In the binary Fe–C system, Fe_7C_3 is the first subsolidus mineral, stable within the range of 50–130 GPa; at pressure above 135 GPa

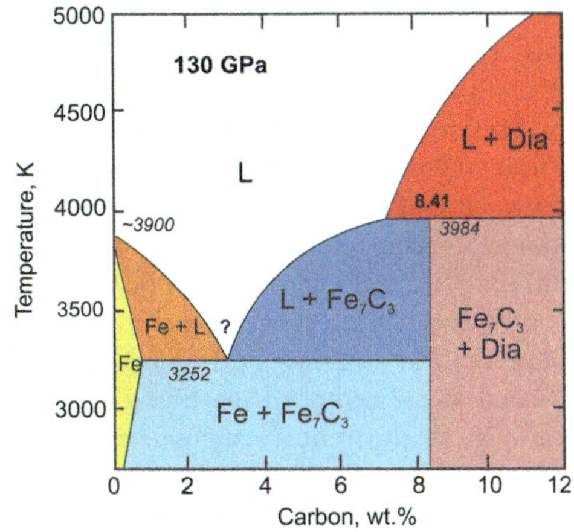

Fig. 9.6 The iron-rich portion of the Fe–C phase diagram at 130 GPa. *L* liquid; *Dia* diamond. Values in *italics* represent melting temperatures and those in **bold** represent compositions. From Lord et al. (2009). © EPSL

it forms a eutectic relation with Fe (Lord et al. 2009). At lower temperatures, cohenite joins Fe_7C_3, but only at higher pressures, corresponding to the D″ layer conditions native iron and diamond associate with Fe_7C_3 in subsolidus (Fig. 9.6). This is the reason that it is considered to be most likely that Fe_7C_3 is formed within the D″ layer, with the source of iron from the outer core.

In the Fe–C system under extreme *P*–*T* conditions (starting from the core), diamond crystallizes first and can associate in the subsolidus with Fe_7C_3 if the carbon concentration is sufficient (Fig. 9.6). It was suggested earlier (based on thermodynamic calculations with the use of EoS data) that iron carbide Fe_3C might be a constituent of the Earth's inner core that solidified from the carbon-rich liquid outer core (Wood 1993). However, later high-pressure experiments on the Fe–C system have shown that Fe_7C_3 rather than Fe_3C, might be stabilized under core pressures and temperatures (Lord et al. 2009; Nakajima et al. 2009); and Wood et al. (2013) agreed with such conclusions in their later work.

Until recently, the enigmatic low shear wave velocity in the inner core (Dziewonski and Anderson 1981) remained unexplained in terms of the pure iron composition of this layer. Partial melting of the outer layer of the inner core, resulting from heat flux variations in the CMB, was suggested as a cause for this effect (Gubbins et al. 2011). Another explanation can be offered with respect to the presence of Fe_7C_3 in the inner core. The calculated density for Fe_7C_3 provides a good explanation for the density of the Earth's inner core obtained from seismological observations (Nakajima et al. 2011; Chen et al. 2012). (The density of Fe_7N_3 provides the same results; see below.) A further density change may also be explained by the HL spin transition, which produces remarkable shear softening in the LS-Fe_7C_3 and may reproduce the obtained V_s value for the inner core (Chen et al. 2014).

Partial melting of the outer layer of the inner core was unable to explain another inner core feature, its high Poisson's ratio, which is practically uniform throughout the entire inner core. The presence of Fe_7C_3, which has a Poisson's ratio similar to that of the Earth's inner core, however, can explain it well. Thus, the presence of Fe_7C_3 can explain all anomalous elastic properties of the Earth's core (Prescher et al. 2015).

The structure of Fe_7C_3 was established as hexagonal with the space group *P6₃mc* (Herbstein and Snyman 1964) (Fig. 9.7a). The iron atoms form a ditrigonal prism around the central carbon atom (black). Three such prisms are combined in a triangular arrangement. There are three distinct Fe sites: FeI (Wykcoff site 2b, green in Fig. 9.7), FeII (Wykcoff site 6c, golden), and FeIII (Wykcoff site 6c, red). There is only one site for the carbon atom, C (Wykcoff site 6c). Each ditrigonal prism has three FeIII atoms, two FeII atoms, and one FeI atom (Fig. 9.7b). The FeI atom is located at the centre of the triangle (Fig. 9.7a). This structure remains stable (according to theoretical calculations) up to inner core conditions of 300 GPa (Bazhanova et al. 2012), where it is the most likely mineral phase (Mookherjee et al. 2011; Chen et al. 2012).

Other carbides, such as cohenite, yarlongite, haxonite, and, possibly other stoichiometries correspond to a series of Fe, Co, Ni, Cr, and Mn carbides, which are well studied in technical applications. They belong to the so-called intermediate transition-metal carbides, which have features of both the interstitial and the salt-like carbides. Like all interstitial carbides, iron carbides are metal-like compounds that have no ion connection in their crystal lattice; in their structures, carbon atoms occupy interstices between closely packed metal atoms. Different positions of metal atoms in a crystal lattice lead to multiple stoichiometries, such as $M_{23}C_6$, M_3C, M_7C_3, M_3C_2, M_4C_3, MC, MC_2 with variable (but normally fixed for each stoichiometry) M/C ratios (e.g., Kosolapova 1971; Cottrell 1995). Within each

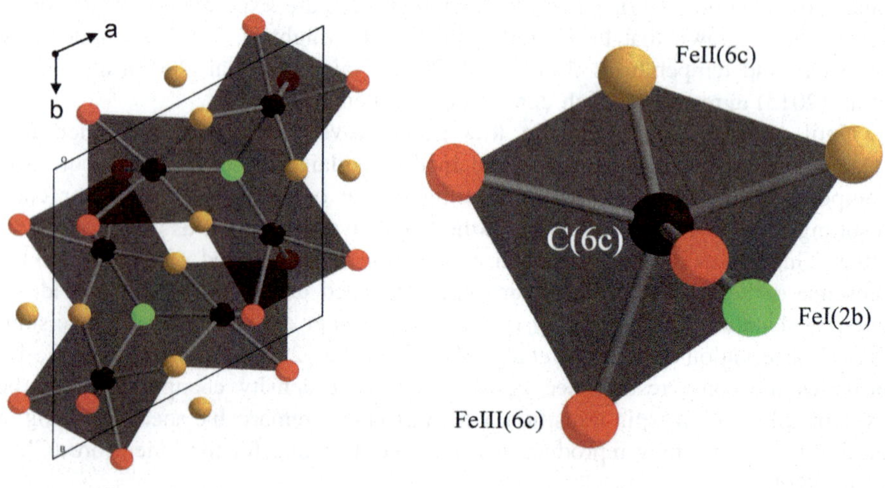

Fig. 9.7 Crystal structure of Fe_7C_3 with hexagonal symmetry (space group $P6_3mc$). *Left* General view along *c* axis. *Right* Detailed structure of ditrigonal carbon prism. Three such prisms form a triangular arrangement. See details in the text. From Mookherjee et al. (2011). © Journal of Geophysical Research

stoichiometry, significant variations in the Fe/C ratio may occur (Walker et al. 2013; Fang et al. 2016). For example, in experiments at 7 GPa Fe/C ratio in Fe_7C_3 varies from 1.78 to 2.44, which is explained by compositional zonation between Fe and C during the process of mineral growth (Walker et al. 2013). In natural Fe_7C_3, Fe/C ratio varies from 1.65 to 2.32.

In addition to intermediate transition-metal carbides, one may expect to meet calcium carbides, most likely of Ca_2C and/or CaC, both monoclinic (Li et al. 2015b) in the D″ layer.

Nitrogen in carbides. Nitrides. In the first discoveries of lower mantle Fe_7C_3 and other carbides, (with the use of EDX and EEL spectra) the presence of nitrogen in the amount of 7.3–9.1 at.% was found, which corresponds to the anion ratio N/(C + N) = 0.19–0.27 (Kaminsky and Wirth 2011), i.e., these minerals may be referred to as nitrocarbides. Subsequently, Fe_7C_3 with 12.8–18.42 at.% of nitrogen (i.e., N/(C + N) = 0.40–0.63) was found, which may be referred to as nitrocarbide or carbonitride and indexed as $Fe_7(C, N)_3$ or $Fe_7(N, C)_3$ (Kaminsky et al. 2015). Recently, nitroyarlongite $Fe_9(N_{0.8}C_{0.2})_4$ and pure nitrides Fe_2N and Fe_3N, in association with Fe_7C_3 were found (Kaminsky and Wirth 2017). The EDX spectrum of native iron, associated with $Fe_7(C, N)_3$, also has a N peak (Kaminsky and Wirth 2011). This means that the entire natural Fe–C system from which the minerals were formed is enriched in nitrogen and is, in fact, Fe–C–N. Iron carbides and nitrides have impurities of Ni (0.08–0.93 at.%), Cr (0.03–1.8 at.%) and Mn (up to 1.22 at.%); some show the noticeable presence of Si (5.1–5.7 at.%) (Kaminsky and Wirth 2017). No traces of sulphur were found, confirming the immiscibility of the Fe–C system with Fe–S under high pressures (Dasgupta et al. 2009).

The presence of iron nitrides in the D″ layer is closely related to their suggested presence in the inner core (Adler and Williams 2005). Two iron nitrides, cubic γ-Fe_4N (known from iron meteorites as roaldite) and the hexagonal ε-Fe_7N_3 are stable at high pressures; the latter nitride should be more stable at the inner core conditions and, in this case, the calculated inner core density corresponds to the mixture of Fe_7N_3 + 8Fe. The elastic parameters of ε-Fe_7N_3 are indistinguishable from those of ε-Fe, and Fe_7N_3 fits the PREM no worse than Fe_7C_3. The presence of Fe_7N_3 in the inner core solves another enigma—anisotropy in this layer. It is well established that the inner core exhibits significant anisotropy, with *P*-wave velocities along the polar axis $\sim$3% faster than in the equatorial plane (Oreshin and Vinnik 2004 and references therein). Several hypotheses were offered to explain this, but none of them seem satisfactory (Vocadlo 2009). The anisotropy of ε-Fe_7N_3 (which is greater than of pure ε-Fe) corresponds to the seismologically observed anisotropy in the inner core (Adler and Williams 2005). This is in contrast to Fe_7C_3, in which elastic anisotropy is negligible (Leineweber et al. 2015). These features make ε-Fe_7N_3 the likely component of the core and explain the presence of nitrides in the D″ layer as a result of the core–mantle interaction. One may expect the HS–LS transition in Fe_7N_3, like in iron carbides (Chen et al. 2014). Although this effect has not been studied to date, one may expect the spin-associated shear softening behaviour of compressed Fe_7N_3.

The identification of transitional nitrocarbides and carbonitrides in geological samples from the D″ layer solves the alternative of Fe_7N_3 or Fe_7C_3 in the inner core: neither pure nitride nor pure carbide but intermediate compounds should be expected in the inner core.

9.4 Conclusions

The D″ layer, occupying the lowermost 200 km of the lower mantle (at a depth of ~2700–2900 km) is compositionally different to the other parts of the lower mantle. It has two compositional sources: the lower mantle and the outer core. The lower mantle's input is the major oxides, which experience phase and physical changes within the D″ layer (the oxide part); the outer core delivers the metallic part to the D″ layer composition.

The major oxide component is post-perovskite, structurally transformed from bridgmanite. The phase transformation from the bridgmanite *Pbnm* structure to the post-perovskite *Cmcm* structure occurs at 124–125 GPa and 2500 K is the major effect for distinguishing the D″ layer. The transformation creates the mineral density increase by 0.9–1.6%, which theoretically causes a 1–1.5% shear velocity increase. It is accompanied by the transition to the orthorhombic $CaCl_2$-structured SiO_2, characteristic for the lower part of the lower mantle, in seifertite, α-PbO_2 -structured SiO_2, with 1.5% density increase, resulting in a slight decrease in bulk sound speed by ~0.4% and shear wave decrease.

Iron-rich liquid metal from the outer core, containing 5–10% light elements (C, N, O and Si) infiltrates into the lowermost mantle and forms a series comprising native Fe^0, iron carbides and nitrides, and silicon carbide. Of particular importance is the presence of Fe_7C_3 and Fe_7N_3, which, in association with diamond, are the first solidus phases, crystallized from the metallic liquid in the D″ layer. The presence of iron nitrides in the D″ layer is closely related to their suggested presence in the inner core.

References

Adler, J. F., & Williams, Q. (2005). A high-pressure X-ray diffraction studies of iron nitrides: Implications for Earth's core. *Journal of Geophysical Research, 110,* B011203. doi:10.1029/2004JB003103

Andrault, D., Munoz, M., Bolfan-Casanova, N., Guignot, N., Perrillat, J., Aquilanti, G., et al. (2010). Experimental evidence for perovskite and post-perovskite coexistence throughout the whole D″ region. *Earth Planet Sci Lett, 293,* 90–96.

Andrault, D., Pesce, G., Bouhifd, M. A., Bolfan-Casanova, N., Hénot, J.-M., & Mezouar, M. (2014). Melting of subducted basalt at the core-mantle boundary. *Science, 344*(6186), 892–895.

Bazhanova, Z. G., Oganov, A. R., & Gianola, O. (2012). Fe–C and Fe–H systems at pressures of the Earth's inner core. *Uspekhi Fizicheskikh Nauk, 182*(5), 521–530. (in Russian).

Stackhouse S., Brodholt, J., Dobson, D. P., & Price, G. D. (2006). Electronic spin transitions and the seismic properties of ferrous iron-bearing $MgSiO_3$ post-perovskite. *Geophysical Research Letters, 33*, L12S03. doi:10.1029/2005GL025589

Buchwald, V. F. (1977). Mineralogy of iron meteorites. *Philosophical Transactions of the Royal Society of London: Mathematical and Physical Sciences, 286,* 453–491.

Buffett, B. A. (2003). The thermal state of Earth's core. *Science, 299,* 1675–1677.

Buffett, B. A., Garnero, E. J., & Jeanloz, R. (2000). Sediments at the top of Earth's core. *Science, 290,* 1338–1343.

Bullen, K. E. (1942). The density variation of the Earth's central core. *Bulletin of the Seismological Society of America, 30,* 235–250.

Bullen, K. E. (1949). Compressibility-pressure hypothesis and the Earth's interior. *Monthly Notices of the Royal Astronomical Society, Geophysical Supplements, 5,* 355–368.

Bullen, K. E. (1950). An Earth model based on compressibility-pressure hypothesis. *Monthly Notices of the Royal Astronomical Society, Geophysical Supplements, 6,* 50–59.

Caracas, R., & Cohen, R. E. (2005). Effect of chemistry on the stability and elasticity of the perovskite and post-perovskite phase in the $MgSiO_3$–$FeSiO_3$–Al_2O_3 system and implications for the lowermost mantle. *Geophysical Research Letters, 32,* L16310. doi:10.1029/2005GL023164

Caracas, R., & Cohen, R.E. (2007). Effect of chemistry on the physical properties of Perovskite and Post-Perovskite. In K. Hirose, J. Brodholt, T. Lay, & D. Yuen (Eds.), *Post-Perovskite: The Last Mantle Phase Transition*. Geophysical Monograph 174 (pp. 115–128). Washington, D.C.: American Geophysical Union. doi:10.1029/174GM10

Catalli, K., Shim, S. H., Prakapenka, V. B., Zhao, J., Sturhahn, W., Chow, P., et al. (2010). Spin state of ferric iron in $MgSiO_3$ perovskite and its effect on elastic properties. *Earth and Planetary Science Letters, 289,* 68–75.

Chao, B. F. (2000). *Renaming D double prim.* EOS 81, p. 46, February 1 2000.

Chen, B., Gao, L., Lavina, B., Dera, P., Alp, E. E., & Li, J. (2012). Magneto-elastic coupling in compressed Fe_7C_3 supports carbon in Earth's inner core. *Geophysical Research Letters, 39,* L18301. doi:10.1029/2012GL052875

Chen, B., Li, Z., Zhang, D., Liu, J., Hu, M. Y., Zhao, J., et al. (2014). Hidden carbon in Earth's inner core revealed by shear softening in dense Fe_7C_3. *PNAS, 111*(50), 17755–17758. doi:10.1073/pnas.1411154111

Cobden, L., Thomas, C., & Trampert, J. (2015). Seismic detection of post-perovskite inside the Earth. In A. Khan, F. Deschamps (Eds.), *The Earth's heterogeneous mantle* (pp. 391–440). Berlin: Springer. doi:10.1007/978-3-319-15627-9_13

Cottrell, A. H. (1995). *Chemical bonding in transition metal carbides* (p. 97). London: Institute of Metals.

Dasgupta, R., Buono, A., Whelan, G., & Walker, D. (2009). High-pressure melting relations in Fe–C–S systems: Implications for formation, evolution, and structure of metallic cores in planetary bodies. *Geochimica et Cosmochimica Acta, 73*(21), 6678–6691. doi:10.1016/j.gca.2009.08.001

Davies, G. F. (2008). Episodic layering of the early mantle by the 'basalt barrier' mechanism. *Earth and Planetary Science Letters, 275*(3–4), 382–392.

Davies, D. R., Goes, S., & Lau, H. C. P. (2015). Thermally dominated deep mantle LLSVPs: A review. In: A. Khan, A, F. Deschamps (Eds.), *The Earth's heterogeneous mantle. A geophysical, geodynamical, and geochemical perspective* (pp. 441–478). Berlin: Springer.

Dorogokupets, P. I., Dymshits, A. M., SokolovaT, S., Danilov, B. S., & Litasov, K. D. (2015). The equations of state of forsterite, wadsleyite, ringwoodite, akimotoite, $MgSiO_3$-perovskite, and postperovskite and phase diagram for the Mg_2SiO_4 system at pressures of up to 130 GPa. *Russian Geology and Geophysics, 56,* 172–189. doi:10.1016/j.rgg.2015.01.011

Duffy, T. S. (2005). Synchrotron facilities and the study of the Earth's deep interior. *Reports on Progress in Physics, 68,* 1811–1859.

Dziewonski, A., & Anderson, D. (1981). Preliminary reference Earth model. *Physics of Earth and Planetary Interiors, 25,* 297–356. doi:10.1016/0031-9201(81)90046-7

Fang, C. M., van Huis, M. A., & Sluiter, M. H. F. (2016). Formation, structure and magnetism of the g-(Fe, M)23C6 (M = Cr, Ni) phases: A first-principles study. *Acta Materialia, 103,* 273–279. doi:10.1016/j.actamat.2015.08.078

Fujino, K., Nishio-Hamane, D., Nagai, T., Seto, Y., Kuwayama, Y., Whitaker, M., et al. (2014). Spin transition, substitution, and partitioning of iron in lower mantle minerals. *Physics of the Earth and Planetary Interiors, 228,* 186–191.

Garnero, E. J. (2000). Heterogeneity of the lowermost mantle. *Annual Review of the Earth and Planetary Sciences, 28,* 509–537.

Garnero, E. J. (2004). A new paradigm for Earth's core-mantle boundary. *Science, 304,* 834–836.

Garnero, E. J., & Helmberger, D. V. (1995). A very slow basal layer underlying large-scale low-velocity anomalies in the lower mantle beneath the Pacific: evidence from core phases. *Physics of the Earth and Planetary Interiors, 91*(1–3), 161–176.

Garnero, E. J., & Helmberger, D. V. (1996). Seismic detection of a thin laterally varying boundary layer at the base of the mantle beneath the central-Pacific. *Geophysical Research Letters, 23,* 2777–2780.

Gassner, A., Thomas, C., Krüger, F., & Weber, M. (2015). Probing the core–mantle boundary beneath Europe and Western Eurasia: A detailed study using PcP. *Physics of the Earth and Planetary Interiors, 246,* 9–24.

Goarant, F., Guyot, F., Peyronneau, J., & Poirier, J. P. (1992). High-pressure and high-temperature reactions between silicates and liquid iron alloys, in the diamond anvil cell, studied by analytical electron microscopy. *Journal of Geophysical Research, 97,* 4477–4487.

Grand, S. P. (2002). Mantle shear-wave tomography and the fate of subducted slabs. *Philosophical Transactions of the Royal Society of London A, 360,* 2475–2491.

Grocholski, B., Catalli, K., Shim, S.-H., & Prakapenka, V. (2012). Mineralogical effects on the detectability of the postperovskite boundary. *PNAS, 109*(7), 2275–2279. doi:10.1073/pnas.1109204109

Grocholski, B., Shim, S. H., & Prakapenka, V. B. (2013). Stability, metastability, and elastic properties of a dense silica polymorph, seifertite. *Journal of Geophysical Research. Solid Earth, 118*(9), 4745–4757. doi:10.1002/jgrb.50360

Gubbins, D., Sreenivasan, B., Mound, J., & Rost, S. (2011). Melting of the Earth's inner core. *Nature, 473,* 361–363. doi:10.1038/nature10068

Hayden, L. A., & Watson, E. B. (2007). A diffusion mechanism for core-mantle interaction. *Nature, 450*(7170), 709–711. doi:10.1038/nature06380

Hayden, L. A., & Watson, E. B. (2008). Grain boundary mobility of carbon in Earth's mantle: A possible carbon flux from the core. *Proceedings of the National Academy of Sciences of the USA, 105*(25), 8537–8541. doi:10.1073/pnas.0710806105

He, Y., & Wen, L. (2012). Geographic boundary of the pacific anomaly and its geometry and transitional structure in the north. *Journal of Geophysical Research: Solid Earth, 117.* doi:10.1029/2012JB009436

Herbstein, H., & Snyman, J. A. (1964). Identification of Eckstrom-Adcock iron carbide as Fe_7C_3. *Inorganic Chemistry, 3,* 894–896.

Hernández, E. R., Alfe, D., & Brodholt, J. (2013). The incorporation of water into lowermantle perovskites: A first-principles study. *Earth and Planetary Science Letters, 364,* 37–43. doi:10.1016/j.epsl.2013.01.005

Hirose, K. (2007). Discovery of post-perovskite phase transition and the nature of D″ layer. In K. Hirose, J. Brodholt, T. Lay, D. Yuen (Eds.), *Post-Perovskite: The Last Mantle Phase Transition.* Geophysical Monograph 174 (pp. 19–35). Washington, D.C.: American Geophysical Union.

Hirose, K., Brodholt, J., Lay, T., & Yuen, D. (Eds.). (2013). *Post-Perovskite: The last mantle phase transition.* Geophysical Monograph 174 (287 p). Washington, D.C.: American Geophysical Union.

Hirose, K., Sinmyo, R., Sata, Y., & Ohishi, Y. (2006). Determination of postperovskite phase transition boundary in $MgSiO_3$ using Au and MgO pressure standards. *Geophysical Research Letters, 33,* L01310. doi:10.1029/2005GL024468

Hirose, K., Takafuji, N., Fujino, K., Shieh, S. R., & Duffy, T. S. (2008). Iron partitioning between perovskite and post-perovskite: A transmission electron microscope study. *American Mineralogist, 93,* 1678–1681. doi:10.2138/am.2008.3001

Hirose, K., Takafuji, N., Sata, N., & Ohishi, Y. (2005). Phase transition and density of subducted MORB crust in the lower mantle. *Earth and Planetary Science Letters, 237,* 239–251.

Iitaka, T., Hirose, K., Kawamura, K., & Murakami, M. (2004). The elasticity of the $MgSiO_3$ post-perovskite phase in the Earth's lowermost mantle. *Nature, 430,* 442–445.

Kaminsky, F. V., & Lin, J.-F. (2017). Iron partitioning in natural lower-mantle minerals: Toward a chemically heterogeneous lower mantle. *American Mineralogist, 102*(4), 824–832. doi:10.2138/am-2017-5949

Kaminsky, F. V., & Wirth, R. (2011). Iron carbide inclusions in lower-mantle diamond from Juina, Brazil. *Canadian Mineralogist, 49*(2), 555–572.

Kaminsky, F., & Wirth, R. (2017). *Nitride, carbonitride and nitrocarbide inclusions in lower-mantle diamonds: A key to the balance of nitrogen in the Earth.* European Geosciences Union General Assembly Abstract No. EGU2017-1751, Vienna, Austria.

Kaminsky, F. V., Wirth, R., & Schreiber, A. (2015). A microinclusion of lower-mantle rock and some other lower-mantle inclusions in diamond. *Canadian Mineralogist, 53*(1), 83–104. doi:10.3749/canmin.1400070

Kanda, R. V. S., & Stevenson, D. J. (2006). Silicon mechanism for iron entrainment into the lower mantle. *Geophysical Research Letters, 33,* L02310. doi:10.1029/2005GL025009

Katsura, T., Yoneda, A., Yamazaki, D., Yoshino, T., & Ito, E. (2010). Adiabatic temperature profile in the mantle. *Physics of the Earth and Planetary Interiors, 183,* 212–218. doi:10.1016/j.pepi.2010.07.001

Kennett, B. L. N., & Engdahl, E. R. (1991). Traveltimes for global earthquake location and phase identification. *Geophysical Journal International, 105,* 429–465.

Kisslinger, C. (2000). "Dahm" layer: Comments on renaming of D″. *EOS, 81,* 210 (May 9, 2000).

Knittle, E., & Jeanloz, R. (1991). Earth's core-mantle boundary: Results of experiments at high pressures and temperatures. *Science, 251*(5000), 1438–1443.

Kobayashi, Y., Kondo, T., Ohtani, E., Hirao, N., Miyajima, N., Yagi, T., et al. (2005). Fe–Mg partitioning between (Mg, Fe)SiO_3 post-perovskite, perovskite, and magnesiowüstite in the Earth's lower mantle. *Geophysical Research Letters, 32,* L19301. doi:10.1029/2005GL023257

Kosolapova, T. Y. (1971). *Carbides: Properties, production, and applications* (p. 298). New York-London: Plenum Press.

Labrosse, S., Hemlund, J. V., & Coltice, N. (2007). A crystallising dense magma ocean at the base of the Earth's mantle. *Nature, 450,* 866–869.

Lay, T., & Garnero, E. J. (2004). Core–mantle boundary structures and processes. In: R. S. J. Sparks, C. J. Hawkesworth (Eds.), *The state of the planet: Frontiers and challenges in geophysics*. Geophys. Monogr. (Vol. 150, pp. 25–41). Washington, D.C.: American Geophysical Union.

Lay, T., Garnero, E. J., & Williams, Q. (2004). Partial melting in a termo-chemical boundary layer at the base of the mantle. *Physics of the Earth and Planetary Interiors, 146,* 441–467.

Lay, T., Williams, Q., Garnero, E. J., Kellogg, L., & Wysession, M. F. (1998). Seismic wave anisotropy in the D″ region and its implications. In M. Gurnis et al. (Eds.), *The core-mantle boundary region* (pp. 299–318). Washington, D.C.: AGU.

Lay, T., & Young, C. J. (1989). Waveform complexity in teleseismic broadband SH displacements: Slab diffractions or deep mantle reflections? *Geophysical Research Letters, 16,* 605–608.

Lee, C.-T. A., Luffi, P., Höink, T., Li, J., Dasgupta, R., & Hernlund, J. (2010). Upside-down differentiation and generation of a 'primordial' lower mantle. *Nature, 463*(7283), 930–933.

Leineweber, A., Shang, S. L., & Liu, Z. K. (2015). Elastic anisotropy of iron carbides with trigonal-prismatic coordination of C by Fe. *Journal of Alloys and Compounds, 633,* 390–394. doi:10.1016/j.jallcom.2015.02.040

Li, Y., Deschamps, F., & Tackley, P. J. (2015a). Effects of the post-perovskite phase transition properties on the stability and structure of primordial reservoirs in the lower mantle of the Earth. *Earth and Planetary Science Letters, 432,* 1–12.

Li, X.-D., & Romanowicz, B. (1996). Global mantle shear velocity model developed using nonlinear asymptotic coupling theory. *Journal of Geophysical Research, 101,* 22245–22272. doi:10.1029/96JB01306

Li, Y.-L., Wang, S.-N., Oganov, A. R., Gou, H., Smith, J. S., & Strobel, T. A. (2015b). Exotic stable calcium carbides. *Nature Communications, 6,* 6974. doi:10.1038/ncomms7974

Libowitzky, E. (1999). Correlation of O–H stretching frequencies and O–HO hydrogen bond lengths in minerals. *Monatschefte Chem, 130,* 1047–1059.

Lin, J.-F., Speciale, S., Mao, Z., & Marquardt, H. (2013). Effects of the electronic spin transitions of iron in lower mantle minerals: Implications for deep mantle geophysics and geochemistry. *Reviews of Geophysics, 51*(2), 244–275.

Lord, O. T., Walter, M. J., Dasgupta, R., Walker, D., & Clark, S. M. (2009). Melting in the Fe–C system to 70 GPa. *Earth and Planetary Science Letters, 284,* 157–167.

Mao, W. L., Campbell, A. J., Prakapenka, V. B., Hemley, R. J., & Mao, H.-K. (2007). Effect of iron on the properties of Post-Perovskite Silicate. In: K. Hirose, J. Brodholt, T. Lay, D. Yuen (Eds.), *Post-Perovskite: The last mantle phase transition*. Geophysical Monograph 174 (pp. 37–46). Washington, D.C.: American Geophysical Union.

Mao, Z., Lin, J. F., Scott, H. P., Watson, H. C., Prakapenka, V. B., Xiao, Y., et al. (2011). Iron-rich perovskite in the Earth's lower mantle. *Earth and Planetary Science Letters, 309,* 179–184. doi:10.1016/j.epsl.2011.06.030

Mao, Z., Lin, J. F., Yang, J., Bian, H., Liu, J., Watson, H. C., et al. (2014). Fe, Al bearing postperovskite in the Earth's lower mantle. *Earth and Planetary Science Letters, 403,* 157–163.

Mao, W. L., Mao, H.-K., Sturhahn, W., Zhao, J., Prakapenka, V. B., Meng, Y., et al. (2006). Iron-rich post-perovskite and the origin of ultralow-velocity zones. *Science, 312,* 564–565. doi:10.1126/science.1123442

Mao, W. L., Meng, Y., Shen, G., Prakapenka, V. B., Campbell, A. J., Heinz, D. L., et al. (2005). Iron-rich silicates in the Earth's D″ layer. *Proceedings of the National Academy of Sciences, 102,* 9751–9753. doi:10.1073/pnas.0503737102

Mao, W. L., Shen, G., Prakapenka, V. B., Meng, Y., Cambell, A. J., Heinz, D., et al. (2004a). Ferromagnesian postperovskite silicates in the D″ layer of the earth. *Proceedings of the National Academy of Sciences, 101,* 15867–15869.

Mao, W. L., Shen, G. Y., Prakapenka, V. B., Meng, Y., Campbell, A. J., Heinz, D. L., et al. (2004b). Ferromagnesian post-perovskite silicates in the D″ layer of the Earth. *Proceedings of the National Academy of Sciences, 101,* 15867–15869.

Masters, G., Laske, G., Bolton, H., & Dziewonski, A. (2000). The relative behavior of shear velocity, bulk sound speed, and compressional velocity in the mantle: Implications for chemical and thermal structure. In *Earth's deep interior: Mineral physics and tomography from the atomic to the global scale* (pp. 63–87).

McDonough, W. F. (2003). Compositional model for the Earth's core. In H. D. Holland, & K. K. Turekian (Eds.), *Treatise on Geochemistry* (pp. 547–568). Pergamon: Oxford (To see in the 2nd edition 2013).

McNamara, A. K., Garnero, E. J., & Rost, S. (2010). Tracking deep mantle reservoirs with ultra-low velocity zones. *Earth and Planetary Science Letters, 299,* 1–9.

Mookherjee, M., Nakajima, Y., Steinle-Neumann, G., Glazyrin, K., Wu, X., Dubrovinsky, L., et al. (2011). High-pressure behavior of iron carbide (Fe_7C_3) at inner core conditions. *Journal of Geophysical Research, 116,* B04201. doi:10.1029/2010JB007819

Murakami, M., Hirose, K., Kawamura, K., Sata, N., & Ohishi, Y. (2004). Post-perovskite phase transition in $MgSiO_3$. *Science, 304,* 855–858.

Murakami, M., Hirose, K., Ono, S., & Ohishi, Y. (2003). Stability of $CaCl_2$-type and α-PbO_2-type SiO_2 at high pressure and temperature determined by in situ X–ray measurements. *Geophysical Research Letters, 30,* 1207. doi:10.1029/2002GL016722

Murakami, M., Hirose, K., Sata, N., & Ohishi, Y. (2005). Post-perovskite phase transition and mineral chemistry in the pyrolitic lowermost mantle. *Geophysical Research Letters, 32,* L03304. doi:10.1029/2004GL021956

Nakajima, Y., Takahashi, E., Sata, N., Nishihara, Y., Hirose, K., Funakoshi, K.-I., et al. (2011). Thermoelastic property and high-pressure stability of Fe_7C_3: Implication for iron-carbide in the Earth's core. *American Mineralogist, 96*(7), 1158–1165.

Nakajima, Y., Takahashi, E., Suzuki, T., & Funakoshi, K. (2009). "Carbon in the core" revisited. *Physics of the Earth and Planetary Interiors, 174,* 202–211.

Nishio-Hamane, D., & Yagi, T. (2009). Equations of state for postperovskite phases in the $MgSiO_3$–$FeSiO_3$–$FeAlO_3$ system. *Physics of the Earth and Planetary Interiors, 175,* 145–150.

Nomura, R., Ozawa, H., Tateno, S., Hirose, K., Hernlund, J., Muto, S., et al. (2011). Spin crossover and iron-rich silicate melt in the Earth's deep mantle. *Nature, 473,* 199–203.

Oganov, A. R., & Ono, S. (2004). Theoretical and experimental evidence for a post-perovskite phase of $MgSiO_3$ in Earth's D″ layer. *Nature, 430,* 445–448.

Ohta, K., Hirose, K., Lay, T., Sata, N., & Ohishi, Y. (2008). Phase transitions in pyrolite and MORB at lowermost mantle conditions: Implications for a MORB-rich pile above the core–mantle boundary. *Earth and Planetary Science Letters, 267,* 107–117.

Ohta. K., Yagi, T., Hirose, K., & OhishI, Y. (2017) Thermal conductivity of ferropericlase in the Earth's lower mantle. *Earth and Planetary Science Letters*, 465, 29–37. doi:10.1016/j.epsi. 2017.02.030.

Ono, S., & Oganov, A. R. (2005). In situ observations of phase transition between perovskite and $CaIrO_3$-type phase in $MgSiO_3$ and pyrolitic mantle composition. *Earth and Planetary Science Letters, 236,* 914–932.

Ono, S., Ohishi, Y., Isshiki, M., & Watanuki, T. (2005). In situ X-ray observations of phase assemblages in peridotite and basalt compositions at lower mantle conditions: Implications for density of subducted oceanic plate. *Journal of Geophysical Research, 110,* B02208.

Oreshin, S. I., & Vinnik, L. P. (2004). Heterogeneity and anisotropy of seismic attenuation in the inner core. *Geophysical Research Letters, 31,* L02613.

Otsuka, K., & Karato, S.-I. (2012). Deep penetration of molten iron into the mantle caused by a morphological instability. *Nature, 492,* 243–246. doi:10.1038/nature11663

Ozawa, H., Hirose, K., Mitome, M., Bando, Y., Sata, H., & Ohishi, Y. (2008). Chemical equilibrium between ferropericlase and molten iron to 134 GPa and implications for the iron content at the bottom of the mantle. *Geophysical Research Letters, 35,* L05308. doi:10.1029/ 2007GL032648

Ozawa, H., Hirose, K., Mitome, M., Bando, Y., Sata, H., & Ohishi, Y. (2009). Experimental study of reaction between perovskite and molten iron to 146 GPa and implications for chemically distinct buoyant layer at the top of the core. *Physics and Chemistry of Minerals, 36,* 355–363. doi:10.1007/s00269-008-0283-x

Panning, M., & Romanowicz, B. (2004). Inferences on flow at the base of Earth's mantle based on seismic anisotropy. *Science, 303,* 351–353.

Peltier, W. R. (2007). Mantle dynamics and the D″ layer: Impact of the post perovskite phase. In: K. Hirose, J. Brodholt, T. Lay, & D. Yuen (Eds.), *Post-perovskite: The last mantle phase transition* (pp. 217–227). Geophysical Monograph Series 174. doi:10.1029/174GM15

Petford, N., Rushmer, T., & Yuen, D. A. (2007). Deformation-induced mechanical instabilities at the core-mantle boundary. In K. Hirose, J. Brodholt, T. Lay, & D. Yuen (Eds.), *Post-Perovskite: The last mantle phase transition*. Geophysical Monograph 174 (pp. 271–287). Washington, D.C.: American Geophysical Union. doi:10.1029/174GM18

Prescher, C., Dubrovinsky, L., Bykova, E., Kupenko, I., Glazyrin, K., Kantor, A., et al. (2015). High Poisson's ratio of Earth's inner core explained by carbon alloying. *Nature Geoscience, 8,* 220–223.

Revenaugh, J., & Jordan, T. H. (1991). Mantle layering from *ScS* reverberations. 3. The upper mantle. *Journal of Geophysical Research, 96*(B12), 19781–19810.

Rushmer, T., Petford, N., Humayun, M., & Campbell, A. J. (2005). Fe-liquid segregation in deforming planetesimals: Coupling core forming compositions with transport phenomena. *Earth and Planetary Science Letters, 239,* 185–202.

Sakai, T., Kondo, T., Ohtani, E., Terasaki, H., Endo, N., Kuba, T., et al. (2006). Interaction between iron an post-perovskite at core-mantle boundary and core signature in plume source region. *Geophysical Research Letters, 33,* L15317. doi:10.1029/2006GL026868

Sakai, T., Ohtani, E., Terasaki, H., Miyahara, M., Nishijima, M., Hirao, N., et al. (2010). Fe–Mg partitioning between post-perovskite and ferropericlase in the lowermost mantle. *Physics and Chemistry of Minerals, 37,* 487–496.

Shepard, C. U. (1867). New classification of meteorites with an enumeration of meteoric species. *American Journal of Science and Arts, 43,* 22–28. (Second Series).

Shieh, S. R., Duffy, T. S., Kubo, A., Shen, G., Prakapenka, V. B., Sata, N., et al. (2006). Equation of state of the post-perovskite phase synthesized from a natural (Mg, Fe)SiO_3 orthopyroxene. *Proceedings of the National Academy of Sciences, 103,* 3039–3043.

Shieh, S. R., Duffy, T. S., & Shen, G. (2005). XRD study of phase stability in SiO_2 at deep mantle conditions. *Earth and Planetary Science Letters, 235,* 273–282.

Shim, S.-H., Catalli, K., Hustoft, J., Kubo, A., Prakapenka, V. B., Caldwell, W. A., et al. (2008). Crystal structure and thermoelastic properties of $(Mg_{0.91}Fe_{0.09})SiO_3$ postperovskite up to 135 GPa and 2700 K. *Proceedings of the National Academy of Sciences of the United States of America, 105,* 7382–7386.

Shim, S. H., Duffy, T. S., Jeanloz, R., & Shen, G. (2004). Stability and crystal structure of $MgSiO_3$ perovskite to the core-mantle boundary. *Geophysical Research Letters, 31,* 110603. doi:10.1029/2004GLOI9639

Sidorin, I., Gurnis, M., & Helmberger, D. V. (1999). Dynamics of a phase change at the base of the mantle consistent with seismological observations. *Journal of Geophysical Research: Solid Earth, 104,* 15005–15023.

Sinmyo, R., Hirose, K., O'Neill, H. S. C., & Okunishi, E. (2006). Ferric iron in Al-bearing post-perovskite. *Geophysical Research Letters, 33,* L12S13.

Stacey, F. D., & Loper, D. E. (1983). The thermal boundary-layer interpretation of D″ and its role as a plume source. *Physics of the Earth and Planetary Interiors, 33,* 45–55.

Stackhouse, S., Brodholt, J. P., Wookey, J., Kendall, J. M., & Price, G. D. (2005). The effect of temperature on the seismic anisotropy of the perovskite and postperovskite polymorphs of $MgSiO_3$. *Earth and Planetary Science Letters, 230,* 1–10.

Stixrude, L., Wasserman, E., & Cohen, R. E. (1997). Composition and temperature of Earth's inner core. *Journal of Geophysical Research, 102*(B11), 24729–24739.

Su, W.-J., Woodward, R. L., & Dziewonski, A. M. (1994). Degree 12 model of shear velocity heterogeneity in the mantle. *Journal of Geophysical Research: Solid Earth, 99,* 6945–6980. doi:10.1029/93JB03408

Takafuji, N., Hirose, K., Mitome, M., & Bando, Y. (2005). Solubilities of O and Si in liquid iron in equilibrium with (Mg, Fe)SiO_3 perovskite and the light elements in the core. *Geophysical Research Letters, 32,* L06313. doi:10.1029/2005GL022773

Townsend, J. P., Tsuchiya, J., Bina, C. R., & Jacobsen, S. D. (2015). First-principles investigation of hydrous post-perovskite. *Physics of the Earth and Planetary Interiors, 244,* 42–48.

Townsend, J. P., Tsuchiya, J., Bina, C. R., & Jacobsen, S. D. (2016). Water partitioning between bridgmanite and postperovskite in the lowermost mantle. *Earth and Planetary Science Letters, 454,* 20–27. doi:10.1016/j.epsl.2016.08.009

Trampert, J., Deschamps, F., Resovsky, J., & Yuen, D. (2004). Probabilistic tomography maps chemical heterogeneities throughout the lower mantle. *Science, 306,* 853–856. doi:10.1126/science.1101996

Trønnes, R. G. (2010). Structure, mineralogy and dynamics of the lowermost mantle. *Mineralogy and Petrology, 99,* 243–261. doi:10.1007/s00710-009-0068-z

Tsuchiya, T., & Tsuchiya, J. (2006). Effect of impurity on the elasticity of perovskite and post-perovskite: Velocity contrast across the post-perovskite transition in (Mg, Fe, Al)(Si, Al) O_3. *Geophysical Research Letters, 33*, L12S04. doi:10.1029/2006GL025706

Tsuchiya, T., Tsuchiya, J., Umemoto, K., & Wentzcovitch, R. M. (2004a). Elasticity of post-perovskite $MgSiO_3$. *Geophysical Research Letters, 31,* L14603.

Tsuchiya, T., Tsuchiya, J., Umemoto, K., & Wentzcovitch, R. M. (2004b). Phase transition in $MgSiO_3$ perovskite in the earth's lower mantle. *Earth and Planetary Science Letters, 224,* 241–248.

van der Hilst, R. D., Widlyantoro, S., & Engdahl, E. R. (1997). Evidence for deep mantle circulation from global tomography. *Nature, 386,* 578–584.

Vocadlo, L. (2009). Mineralogy of the Earth—The Earth's core: Iron and iron alloys. In G. D. Price (Ed.), *Treatise on Geophysics. Mineral Physics* (pp. 91–120). Elsevier.

Walker, D., Dasgupta, R., Li, J., & Buono, A. (2013). Nonstoichiometry and growth of some Fe carbides. *Contributions to Mineralogy and Petrology, 166,* 935–957.

Weber, M., Davis, J. P., Thomas, C., Kruger, F., Sherbaum, F., Schlittenhardt, J., et al. (1996). The structure of the lowermost mantle as determined from using seismic arrays. In E. Boschi, G. Ekstrom, & A. Morelli (Eds.), *Seismic modeling of the Earth's structure* (pp. 399–442). Rome: Istituto Nazionale di Geophysica.

Wentzcovitch, R. M., Tsuchiya, T., & Tsuchiya, J. (2006). $MgSiO_3$ postperovskite at D″ conditions. *Proceedings of the National Academy of Sciences of the United States of America, 103,* 543–546. doi:10.1073/pnas.0506879103

Williams, Q. (1998). The temperature contrast across D9. In M. Gurnis, M. E. Wysession, E. Knittle, & B. A. Buffett (Eds.), *The core-mantle boundary* (pp. 73–81). Washington, D.C.: AGU.

Williams, Q., & Garnero, E. J. (1996). Seismic evidence for partial melt at the base of the mantle. *Science, 273,* 1528–1530.

Wood, B. J. (1993). Carbon in the core. *Earth and Planetary Science Letters, 117,* 593–607.

Wood, B. J., Jie Li, J., & Shahar, A. (2013). Carbon in the core: Its influence on the properties of core and mantle. *Reviews in Mineralogy and Geochemistry, 75,* 231–250.

Wookey, J., & Kendall, J.-M. (2007). Seismic anisotropy of post-perovskite and the lowermost mantle. In K. Hirose, J. Brodholt, T. Lay, D. Yuen (Eds.), *Post-Perovskite: The last mantle phase transition.* Geophysical Monograph 174 (pp. 171–188). Washington, D.C.: American Geophysical Union. doi:10.1029/174GM13

Wookey, J., Kendall, J. M., & Rümpker, G. (2005). Lowermost mantle anisotropy beneath the north Pacific from differential S-ScS splitting. *Geophysical Journal International, 161,* 829–838.

Yamanaka, T., Hirose, K., Mao, W. L., Menge, Y., Ganesh, P., Shulenburger, L., et al. (2012). Crystal structures of (Mg1-X, FeX)SiO_3 postperovskite at high pressures. *Proceedings of the National Academy of Sciences of the United States of America, 109,* 1035–1040.

Yamazaki, D., & Karato, S.-I. (2007). Lattice-preferred orientation of lower mantle materials and seismic anisotropy in the D″ layer. In K. Hirose, J. Brodholt, T. Lay, & D. Yuen (Eds.), *Post-Perovskite: The last mantle phase transition.* Geophysical Monograph 174 (pp. 69–78). Washington, D.C.: American Geophysical Union.

Yu, Y. G., Hsu, H., Cococcioni, M., & Wentzcovitch, R. M. (2012). Spin states and hyperfine interactions of iron incorporated in $MgSiO_3$ post-perovskite. *Earth and Planetary Science Letters, 331–332*(2012), 1–7. doi:10.1016/j.epsl.2012.03.002

Zhang, Z., Dorfman, S. M., Labidi, J., Zhang, S., Li, M., Manga, M., et al. (2016). Primordial metallic melt in the deep mantle. *Geophysical Research Letters, 43*(8), 3693–3699. doi:10.1002/2016GL068560

Zhang, L., Meng, Y., Dera, P., Yang, W., Mao, W. L., & Mao, H.-K. (2013). Single-crystal structure determination of (Mg, Fe)SiO_3 postperovskite. *Proceedings of the National Academy of Sciences, 110,* 6292–6295.

Zhang, F., & Oganov, A. R. (2006). Valence state and spin transitions of iron in Earth's mantle silicates. *Earth and Planetary Science Letters, 249,* 436–443.

Chapter 10
Seismic Heterogeneities and Their Nature in the Lower Mantle

Abstract Over the last 20 years, global seismology has made significant progress in mapping the deep interior of the Earth. Tomographic studies identified variations in lower-mantle chemistry and phase transitions with depths of observed seismic heterogeneities occupying the entire range of the lower mantle. Three major zones of seismic heterogeneities can be outlined. The upper (shallow) zone from 660–1300 km includes ~70% of all heterogeneities, observed almost equally near subduction zones and beneath the tectonic plates. The middle zone, from 1300 to 1900 km, includes ~20% of all heterogeneities, which are observed entirely near subduction zones. The lower zone, from 1900 km to the border of D″ layer at 2700 km includes only a small number of heterogeneities. The deepest seismic heterogeneities, identified within the central parts of the Eurasian and North American plates, are located at depths of ~2630 km and ~2400 km. No correlation between the observed seismic heterogeneities and major mineral phase transitions and spin crossover were identified. The seismic heterogeneities, most likely, reflect local and regional chemical variations within the lower mantle.

10.1 General

To date, the most reliable (and almost the only) information about the structure of deep Earth comes from the analysis of seismic waves generated by large earthquakes around the world. Based on averaged profiles of the compressional velocity (V_p), shear velocity (V_s) and density (ρ) with depth, the general seismic model of the Earth's interior is presented in Fig. 10.1.

From 660–2700 km, the lower mantle exhibits smooth variations in seismic velocities and densities with depth. Single-dimensional interpretations of deep-seated earthquakes, carried out in the 1970s–1980s, failed to show any heterogeneities within the lower mantle up to the D″ layer at 2700 km (Dziewonski and Anderson 1981; Kennett et al. 1995). Areas of very low seismic velocities (ultra-low velocity zones; ULVZ), which may be areas of partial melting at the base of the mantle and/or chemical reactions between the mantle and the core, mantle

F.V. Kaminsky, *The Earth's Lower Mantle*, Springer Geology,
DOI 10.1007/978-3-319-55684-0_10

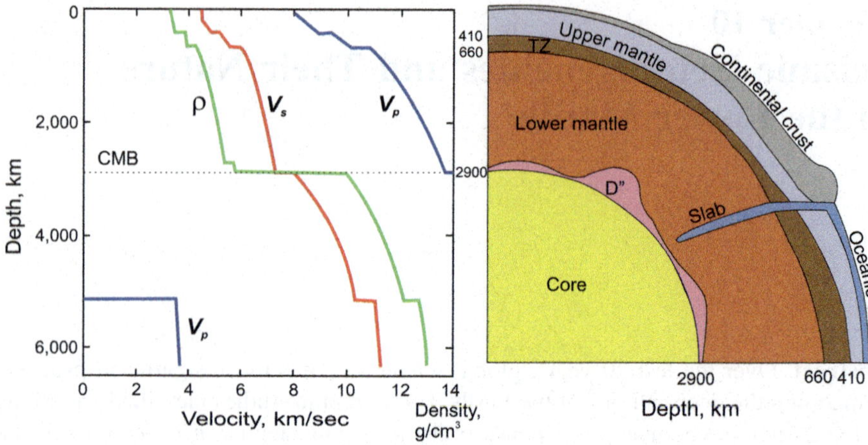

Fig. 10.1 Global average variations of compressional velocity (V_p), shear velocity (V_s) and density (ρ) with depth from the Earth model AK135. Also shown (*right*) is a cutaway view of the Earth, showing a region of continental crust, upper mantle, transition zone (TZ), lower mantle and a subducted lithospheric slab extending into the lower mantle. The D″ layer is irregular in thickness. Modified after Helffrich and Wood (2001). Used with permission of the Nature

plumes, and relics of subducting slabs significantly complicate the ideal model of the Earth's interior.

In the late 1960s, Johnson (1969) completed a study of 212 deep-seated earthquakes and identified global anomalous gradients in V_p at depths of 830, 1000, 1230, 1540, 1910 and 2370 km. He suggested that these anomalies might be caused by changes in the composition and density of the lower mantle. The existence of regional seismic heterogeneities as strong reflectors within the lower mantle was established later, beneath the northwest margin of the Pacific and other areas (Muirhead and Hales 1980; Souriau 1986; Revenaugh and Jordan 1991; Petersen et al. 1993; Wicks and Richards 1993; Revenaugh and Sipkin 1994).

Over the last 20 years, global seismology has made significant progress in mapping the deep interior of the Earth (e.g., Deuss et al. 2013; Trampert and Fichtner 2013). Two complementary approaches have been used to this end. Combining long-period seismic normal modes and surface waves with shorter-period body wave observations has resulted in tomographic models showing the existence of smoothly varying, large-scale heterogeneity (e.g., Trampert and Fichtner 2013). Tomographic studies, using the effect of temperature on velocity variations, helped to identify variations in lower-mantle chemistry and phase transitions, as well as some geodynamic features, such as sinking lithospheric slabs and rising plumes. In addition to the smooth structure, the Earth's mantle is delineated by sharp changes in seismic properties. These sharp heterogeneities (reflectors or scatterers) can be studied using seismic body waves, which are reflected, converted or refracted at these boundaries.

Seismic heterogeneities in the lower mantle are more difficult to distinguish compared with the upper mantle. However, modern methods of interpretation with the use of high-resolution seismic array techniques, coupled with the analysis of different types of scattered waves, mostly of short ($\sim$1s) period (mainly with *S*-to-*P* scattering; as well as with *P*-to-*P* scattering prior to *PP* waves; *P*-to-*P* scattering prior to *P′P′*; *P*-to-*P* scattering in *P* coda waves; and *P*-to-*P* scattering prior to *PKP*) (Kaneshima 2016; Jenkins et al. 2017), and in combination with 3D seismic tomography, demonstrated the presence of seismic heterogeneities at depths from 720 km to the D″ layer beneath the circum-Pacific area (Izu-Bonin-Mariana Arc, Kermadec-Tonga-Fiji Trench, Indonesian Arc, Japan-Kuril-Kamchatka Zone) and other areas (Petersen et al. 1993; Kawakatsu and Niu 1994; Le Stunff et al. 1995; Wicks et al. 1996; Niu and Kawakatsu 1997; Vinnik et al. 1998, 2001, 2010; Kaneshima and Helffrich 1998, 1999, 2003, 2010; Castle and Creager 1999; Krüger et al. 2001; Castle and van der Hilst 2003; Kaneshima et al. 2003; Liu et al. 2003; Niu et al. 2003; Shen et al. 2003; Vanacore et al. 2006; Cao and Romanowicz 2007; Courtier and Revenaugh 2008; Rost et al. 2008; Deuss et al. 2013; Kaneshima 2013; Li and Yuen 2014; Niu 2014; Ma et al. 2016). Technical aspects of seismic array studies can be found in Rost et al. (2008) and in Kaneshima (2016).

Seismic transitions for such heterogeneities are sharp, not exceeding 5 km (i.e., half the size of the 660 km discontinuity) (Niu 2014). This excludes a possibility of a purely thermic nature of such transitions (Shen et al. 2003). Usually, lower layers in heterogeneities demonstrate a higher density of the reflecting layer by 2–9% and the 2–12% lower V_s under constant V_p (Kaneshima and Helffrich 1999; Niu et al. 2003; Kaneshima 2013; Li and Yuen 2014; Niu 2014). These anomalies are frequently greater than the calculated 2–4% decrease of the V_s for the pyrolite/basalt transmission in the lower mantle (Tsuchiya 2013).

10.2 Seismic Heterogeneities Within the Lower Mantle

Most seismic heterogeneities within the lower mantle are related to modern or ancient subduction zones, particularly in the circum-Pacific area. They extend from the Aleutian Arc and Kamchatka (Ma et al. 2016) to the Kuril Islands (Petersen et al. 1993; Kaneshima 2009; Kaneshima and Helffrich 2010), Japan (Niu and Kawasaki 1997; Li and Yuen 2014), Izu-Bonin and Mariana trenches (Wicks et al. 1996; Niu and Kawasaki 1997; Kaneshima and Helffrich 1998, 1999, 2003, 2010; Krüger et al. 2001; Kaneshima 2003; Niu et al. 2003; Vinnik et al. 2001; Rost et al. 2008; Deuss et al. 2013) and occur in the southern Pacific in the Fiji-Tonga-Kermadec area (Kawakatsu and New 1994; Vinnik et al. 2001; Rost et al. 2008; Kaneshima and Helffrich 2010; Deuss et al. 2013). Eastern Pacific heterogeneities are observed in areas of subduction of the Pacific Plate beneath the North American Plate (Deuss et al. 2013) and the Nazca plate beneath the South American Plate (Liu et al. 2003; Kaneshima and Helffrich 2010; Deuss et al. 2013). Numerous seismic heterogeneities also occur beneath the Indonesian Arc (Niu and Kawasaki 1997; Vinnik et al. 1998,

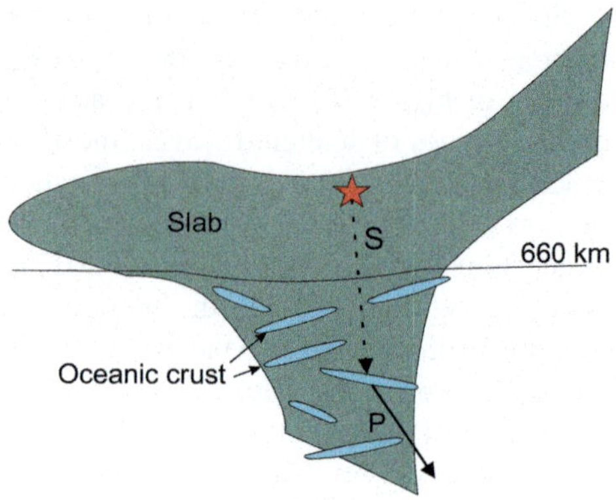

Fig. 10.2 Model of an ancient stagnant slab with a local penetration into the lower mantle beneath NE China. The later *P* arrival observed here is generated at one of the many fragments of former ocean crust with a favourable orientation. From Niu (2014). © Elsevier

2001). Recently, these data on heterogeneities caused by basaltic bodies in the mid-lower mantle were reviewed by Kaneshima (2016).

In all areas, the dipping of the seismic heterogeneities coincides with the dipping of the subducted lithospheric plates. The thicknesses of the observed seismic heterogeneities are 5–12 km (Castle and Greager 1999; Kaneshima and Helffricn 1999; Niu et al. 2003; Kaneshima 2013; Niu 2014) and confirm their lithospheric nature. However, estimations of thickness is not always possible. After passing through the 660 km discontinuity, some lithospheric slabs continue dipping at angles of 12–45°, while others experience breakage, folding and sink (owing to their negative buoyancy; Tsuchiya 2011) as separate reflecting/scattering fragments to the core–mantle boundary (CMB) (Fig. 10.2).

Recently, seismic heterogeneities at depths of 930–1120 km were identified beneath north eastern China (Li and Yuen 2014; Niu 2014). Like more southern anomalies (beneath Japan, Izu-Bonin and Mariana arches), they were also interpreted as remnants of subducted oceanic plates, although they are located several hundred kilometres away from modern subduction zones. The most plausible tectonic explanation for the positions of these and some other reflectors/scatterers is they are remnants of former subducted and folded oceanic crust, which preceded the current subduction of the plates. Li and Yuen (2014) suggested that these large surfaces (reaching almost 1000 km in size) are relics of the ancient (now non-existent) Izanagi Plate, which began subducting beneath the Okhotsk section of the North American Plate during 130–100 Ma. This implies that kilometre scale heterogeneity persisted for billions of years despite stirring due to convection and, therefore, is ubiquitous in the lower mantle.

Seismic heterogeneities were found to not only be related to modern/ancient subduction zones, but also to under lithospheric plates at depths of 785–1200 km beneath South Africa (Le Stunff et al. 1995; Vanacore et al. 2006; Vinnik et al. 2010), 940–1530 km beneath the North American Plate (Courtier and Revenaugh 2008; Deuss et al. 2013), 850–1100 km beneath the Indo-Australian Plate (Courtier and Revenaugh 2008; Deuss et al. 2013), 1240–1270 km beneath Western Europe (Vinnik et al. 2010), and 780–1200 km beneath the Eurasian Plate (Vinnik et al.

2001; Deuss et al. 2013). Recently, a detailed analysis of mid-mantle heterogeneity beneath Western Europe was performed by Jenkins et al. (2017). It was shown that seismic velocity jumps beneath Europe and the North-Atlantic within the depth range of 800–1400 km, predominantly between depths of 975 and 1050 km. The heterogeneity has a linear trend across Western Europe from Iceland in the northwest to Eifel in the southeast, which coincides with areas of active upwelling (under Iceland) and an elongate lateral low-velocity anomaly. The heterogeneity is believed to be caused either by a phase change within chemically distinct plume material itself, or by small-scale chemical heterogeneities entrained within the upwelling plume (Jenkins et al. 2017).

The depths of seismic heterogeneities occupy the entire range of the lower mantle. Figure 10.3 shows a histogram detailing the depths of the identified seismic heterogeneities, determined by all seismic methods, from the data published between 1993 and 2016. (In some cases, depth determinations of refractory surfaces made from the same earthquakes in different laboratories vary by several hundreds kilometres; e.g., data on the Izu-Bonin arch made by Kaneshima and Helffrich 1998 and Krüger et al. 2001.)

Three major zones of seismic heterogeneities can be outlined: (1) The zone from 660–1300 km includes ~70% of all heterogeneities, observed almost equally near subduction zones and beneath the tectonic plates; (2) The zone from 1300–1900 km includes ~20% of all heterogeneities, which are observed entirely near subduction zones; and (3) The zone from 1900 to the border of the D″ layer at 2700 km includes only a small number of heterogeneities. Conventionally, these zones may be called the shallow lower mantle, the middle lower mantle, and the deep lower mantle.

Within the shallow part of the lower mantle, heterogeneities form two maxima at depths of 800–900 and 1100–1200 km. Two more saddle maxima exist within the middle lower mantle in the ranges 1500–1600 and 1800–1900 km. The latter coincides with the results on *S*-to-*P* scatterers obtained by Kaneshima (2016) for the western Pacific. Of particular interest are the deeper seismic heterogeneities, identified within central parts of the Eurasian and North American plates. The

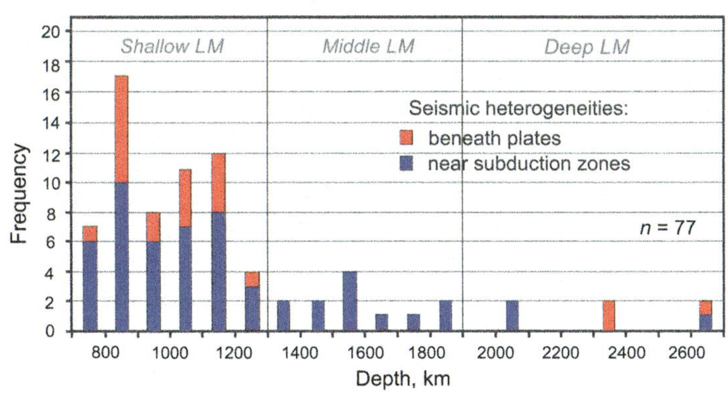

Fig. 10.3 Depths of seismic heterogeneities within the lower mantle from 1993–2017 publications

seismic migration of shear waves reflecting from the seismic discontinuity method (which is widely used in oil exploration seismology for steep topography layers) enabled two extensively undulating reflectors to be identified just above the CMB beneath north-western Siberia (Thomas et al. 2004). They are located at depths of 55–85km and 206–316 km above the CMB, i.e., at depths of ~2630 and ~2400 km, respectively. The upper reflector is characterized by the increase of seismic velocities while the lower one is characterized by a decrease, demonstrating the vicinity of the low velocity zone in the D″ layer. They were interpreted as a result of the subduction of unknown ancient slabs to the CMB. Another seismic heterogeneity was identified beneath the central part of the North American plate, extending from southern Ontario to northern Saskatchewan in Canada within depth ranges of 2270–2470 km (Cao and Romanowicz 2007). It was suggested to be a 'fossil slab', but there is no tectonic indication of such a slab or subduction zone in this area or in the vicinity for several thousand kilometres.

A summary of the seismic heterogeneities within the lower mantle is presented in Table 10.1 and Fig. 10.4.

10.3 Causes of Seismic Heterogeneities in the Lower Mantle

10.3.1 General

The nature of lower-mantle seismic heterogeneities is still disputable. Based on the 3D tomography data, most of them that are spatially related to subduction zones are undoubtedly caused by subducted slabs. Among them, there are both subhorizontal and flat and steep dipping surfaces (Niu and Kawakatsu 1997; Castle and Greager 1999; Niu et al. 2003); some are layered (Kaneshima and Helffrich 2003). This indicates the existence of compositional changes within the lower mantle both laterally and vertically (e.g., Kellogg et al. 1999). The most prominent scatterers for subducted slabs were established at ~1500–1600-km depth (Kaneshima and Helffrich 1998, 1999, 2003), pointing to the presence of a single seismic inhomogeneity at this depth beneath the entire Pacific area.

Seismic heterogeneities beneath lithospheric plates are distributed within the entire depth range including the lowermost parts of the lower mantle, and their genetic nature needs clarification. These heterogeneities are mostly local, only several hundred kilometres long. Some of them, such as those within mantle plumes in Hawaii and Iceland (Shen et al. 2003), as well as beneath the South-African and Zagros rift zones (Le Stunff et al. 1995, 2015; Deuss et al. 2013), cannot be attributed to the process of the lithospheric plate subduction.

Some sub-horizontal seismic heterogeneities were suggested as having a global character (Le Stunff et al. 1995; Vinnik et al. 2001). The absence of such seismic heterogeneities globally is probably because they are below the resolution limit of

Table 10.1 Seismic heterogeneities established within the lower mantle in 1993–2016

No.	Location	Depth, km	Lateral size, km	Dipping	Authors' interpretation	Reference
1	Japan-Kuril-Kamchatka subduction zones	900			Regional feature related to processes in the Japan-Kuril-Kamchatka subduction zone	Petersen et al. (1993)
2	Tonga subduction zone	~920		Horizontal	May be global	Kawakatsu and Niu (1994)
3	Southern African rift	785			Global	Le Stunff et al. (1995)
4	Mariana subduction zone	1200		Steep		Wicks et al. (1996)
5	Indonesian Arc	940–1080		West		Niu and Kawasaki (1997)
6	Japan Sea	945		Sub-horizontal		
7	Izu-Bonin Trench	970		Sub-horizontal		
8	Izu-Bonin Trench	1590	30–150		Relic of MZ Java subducted slabs or upwelling mantle heterogeneity	Kaneshima and Helffrich (1998)
9		1850				
8a		720			Penetration of a slab through 660-km discontinuity	Krüger et al. (2001)
9a		790				
10	Sunda Arch, Indonesia	860		N	Dipping plate	Vinnik et al. (1998)
11		1070				
12		1170				
13	Izu-Bonin Trench	1000	200–300	West steeply	Subducted ancient slab	Castle and Creager (1999)
14	East of Mariana and Izu-Bonin subduction zones	1400–1600	500 × 300	S 30–40°	Subducted ancient slab	Kaneshima and Helffrich (1999)

(continued)

Table 10.1 (continued)

No.	Location	Depth, km	Lateral size, km	Dipping	Authors' interpretation	Reference
15	Mariana subduction zone	720; 790			Penetration of the slab through the 660 km discontinuity	Krüger et al. (2001)
16	Izu-Bonin region	860–880			Global	Vinnik et al. (2001)
17	Kermadec-Fiji-Tonga zone	1250				
18	Eurasian plate	1200				
19	Mariana Trench	1670				
20	Sunda Arch	1070				
21	Sunda Arch	900				
22	Mariana Trench	~1600	Local			Castle and van der Hilst (2003)
23	Mariana Trench	710; 860				Kaneshima (2003)
23a	NE of the Mariana subduction zone	1000–1400	>500 × 300	30–40°	Remnant of subducted lithospheric plate	Kaneshima and Helffrich (2003)
24		1300–1650				
25		1600–1900				
26	Nazca subduction zone	820–850			Regional (not global) feature	Liu et al. (2003)
27	Nazca subduction zone	900–938			Regional (not global) feature	
28	Mariana subduction zone	1115	100 × 100	SW ~ 20°	A piece of sank subducted ancient slab	Niu et al. (2003)
29	Hawaii	1050			Unknown phase transitions	Shen et al. (2003)
30	Iceland	1050				
31	NW Siberia	2400			Whole-mantle convection?	Thomas et al. (2004)
32		2630				
33	Kaapvaal Craton	1080				Vanacore et al. (2006)
34		1200				

(continued)

Table 10.1 (continued)

No.	Location	Depth, km	Lateral size, km	Dipping	Authors' interpretation	Reference
35	North America	2270–2470	>1000	NW	Fossil slabs	Cao and Romanowicz (2007)
36	Indo-Australian Plate	852; 1108				Courtier and Revenaugh (2008)
37	North-American Plate	940–1130; 1380–1530				
38	Mariana Trench	600–1000		Dipping	Basaltic crust penetrating into lower mantle	Rost et al. (2008)
39	Tonga subduction zone	600–1000		Dipping		
40	Mariana Zone	2100		Dipping		
41	Tonga Zone	2100		Dipping		
42	Izu-Bonin (BX)	1190–1240		Dipping	Remnants of subducted and folded former oceanic crust	Kaneshima and Helffrich (2010)
43	Izu-Bonin (BY)	1350				
44	Izu-Bonin (BZ)	1600				
45	North Kuril islands	1285–1350				
46	Northern Fiji (FX)	1120–1160				
47	Northern Fiji (FY)	1245–1230				
48	Northern Fiji (FZ)	1710–1900				
49	Peru (SX)	1300–1500				
50	Peru (SY)	1600				
51	South Africa	1170			Global	Vinnik et al. (2010)
52	Western Europe	1240–1270				
53	Western USA	830				Deuss et al. (2013)
54	Chile	915				
55	Persian Gulf	820				

(continued)

Table 10.1 (continued)

No.	Location	Depth, km	Lateral size, km	Dipping	Authors' interpretation	Reference
56	China	810				
57	Northern Australia	980				
58	Japan	900				
59	Japan Sea	1200				
60	Izu-Bonin	870				
61	Fiji	920				
62	Fiji	1050				
63	Fiji	1080				
64	Fiji	1200				
65	Hudson Bay	810				
66	Mid-Atlantic	900				
67	East Siberia	880				
68	Mongolia?	780				
69	NE China and Japan Sea	930–1120	~800		Subduction of ancient Izanagi plate	Li and Yuen (2014)
70	NE China	987	200 × 50	N ~ 17°	Subducted parts of the ocean crust	Niu (2014)
71	NE China	1043	200 × 50	N ~ 17°		
72	Aleutian Islands and Kamchatka	2500–2900	>1000		Remnants of ancient subducted slabs	Ma et al. (2016)
73	Western Europe	975–1050	Linear > 1000		Plume material	Jenkins et al. (2017)

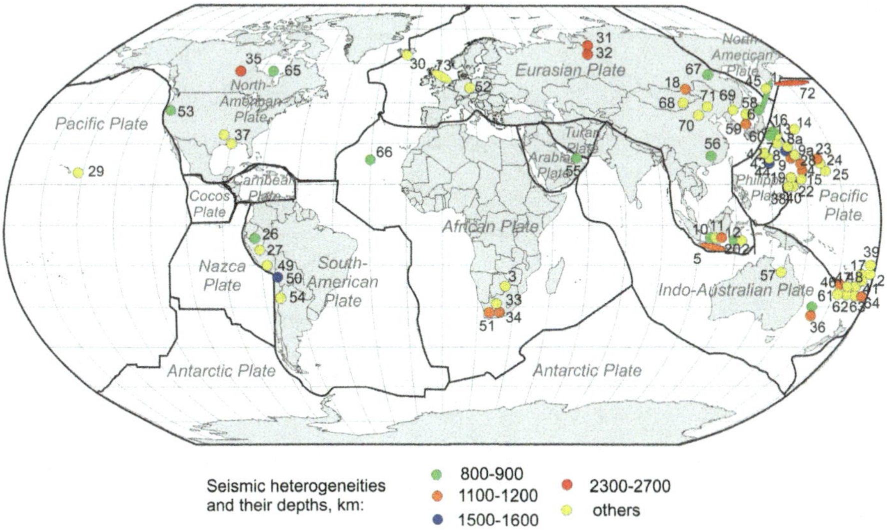

Fig. 10.4 Global distribution of seismic heterogeneities at different depths. Numbers in the map correspond to numbers in Table 10.1

present methods. It should be also noted that the most common technique of detecting seismic heterogeneities with the use of near-source *S*-to-*P* converted waves in short-period seismograms from deep-seated earthquakes is limited to the areas located directly near the hypocentre. Such earthquakes usually occur within zones of deep subduction, while plates and particularly their central parts are seismically passive, which determines the identification of seismic heterogeneities within the lower mantle to mainly beneath subduction zones, leaving subcratonic areas unstudied.

Summarizing data on seismic heterogeneities within the lower mantle, Deuss et al. (2013) established the wide distribution of seismic heterogeneities—beneath both oceans and continents—within two major ranges, 800–900 and 1000–1200 km; however, the results remain undefined. These ranges are close to those established in this work at 800–900 and 1100–1200 km (Fig. 10.3). The heterogeneities (as well as the 1500–1600 km one) have not yet been explained. Considering large velocity contrasts and the sharpness of the observed lower-mantle seismic heterogeneities, they should have a chemical rather than thermal origin.

10.3.2 Structure of the Lower Mantle

Two different structural models were proposed for the interpretation of seismic heterogeneities (Fig. 10.5). The first model suggests the presence of a seismic border at depth with considerable topography, dividing the lower mantle into two

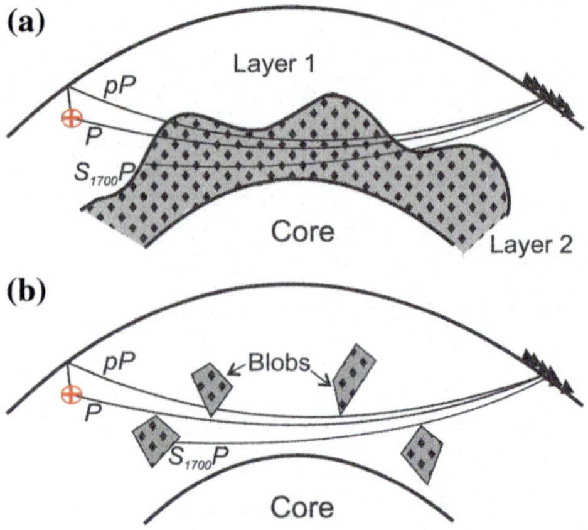

Fig. 10.5 Two different models of seismic heterogeneity within the mantle. **a** Two-layer model with a strongly undulated border at a depth of 1600–1700 km (Kellogg et al. 1999). **b** Homogeneous whole mantle containing chemically heterogeneous blobs of varying size (Becker et al. 1999). Superimposed on both models are the schematic ray paths of the *P*, *pP*, and $S_{1700}P$ for a 500 km deep earthquake (*red cross*) recorded at 80° distance. After Castle and van der Hilst (2003)

layers (Kellogg et al. 1999) (Fig. 10.5a). The lower layer is denser than the upper mantle. It includes juvenile ^{40}Ar and ^{3}He, which did not migrate to the atmosphere and generates the excess heat, explaining the current observed heat flow. This layer may have originated during the early differentiation of the Earth and may be related to deep mantle ocean development (Kellogg et al. 1999; Dai et al. 2013).

It has been suggested that the crystallization of bridgmanite from basal magma ocean evolves such that the composition of residual magma is poor in SiO_2, and rich in MgO, FeO and other incompatible elements such as Na_2O and K_2O (Labrosse et al. 2007; Nomura et al. 2011). Such residual melts are denser than any of the typical lower mantle minerals below depths of 1800 km and therefore may have accumulated at the base of the mantle. As a consequence, the late-stage crystallization from the basal magma ocean could have formed chemically anomalous solid patches enriched in ferropericlase and the CF phase in the lowermost mantle. Both ferropericlase and CF exhibit substantially lower sound velocities than those of bridgmanite (Dai et al. 2013). The enrichment in ferropericlase and the CF phase therefore causes a particularly large seismic anomaly and a remarkably low shear velocity in the deep lower mantle.

Another plausible cause for the layered structure of the lower mantle is the increase of iron content in the lower part of the lower mantle, which is different to that of a pyrolitic composition. It has been recently established, based on the Fe partitioning in the bridgmanite + ferropericlase association, that the deep lower mantle is iron-rich and differs markedly from a pyrolitic composition (Kaminsky and Lin 2017; see Section 4.4). This should cause changes in sound velocity in the lowermost mantle.

According to the second model, local seismic borders at different depths within the lower mantle are caused by the occurrence of large blobs composed of juvenile undepleted lower-mantle compositions (Becker et al. 1999). The occurrence of dense, undepleted zones in the lower third of the mantle is suggested from seismic data (van der Hilst and Kárason 1999), free-air gravity constraints (Tromp and Ishii 1998) and is supported by thermochemical convection models in 3D spherical geometry (Li et al. 2014). They can persist for relatively long periods of time (billions of years), even in a mantle undergoing 'whole mantle' convection, and could thus represent regions of the mantle that are largely unsampled geochemically. Such primitive blobs within the lower mantle can exist from the early stages of Earth's history (van der Hilst and Kárason 1999), or form as a result of partial melting during the initial stages of whole-mantle convection (Lee et al. 2010).

The second model is shown in more detail in Fig. 10.6. A significant part of the lower mantle is comprised of blobs of the primary, undifferentiated, undepleted, dense material, which is composed of the predominantly ferropericlase and bridgmanite (Davies 1984; Becker et al. 1999). Such compositional model coincides with observed geological data, according to which ferropericlase comprises 70–75% of all inclusions in lower-mantle diamonds (instead of 16–20%) in the model composition of the uppermost lower mantle (Kaminsky 2012). This model was additionally confirmed by experiments on the rheological properties of a mixture of bridgmanite and magnesiowüstite under high pressures and temperatures (Girard et al. 2016). The controlled generation of measurements from stress and strain under high pressure and temperature conditions (24–27.5 GPa and 2000–2150 K) demonstrate that bridgmanite is substantially stronger than magnesiowüstite, and that magnesiowüstite largely accommodates the strain. These

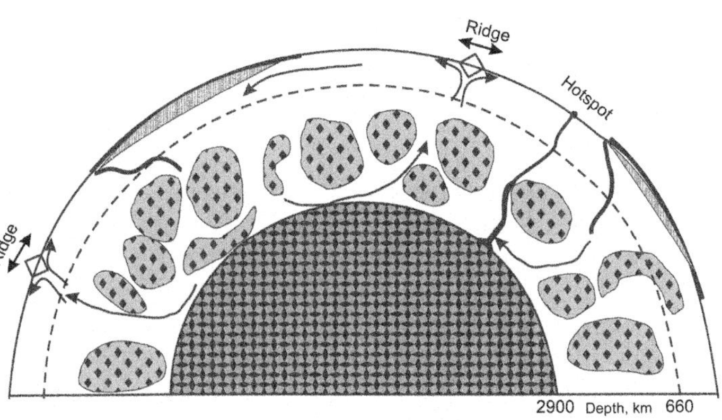

Fig. 10.6 The blob model of the lower mantle. Convection occurs in the whole mantle mode with varying morphology of slab penetration through the 660-km-transition zone. The blobs reside mainly in the cores of the convective cells and represent the primitive reservoir. Surrounding material and especially the upper mantle region are depleted and degassed by melting at the ridges and earlier continent-formation. Blobs are sampled by rising plumes that entrain material and lead to a heterogeneous OIB isotope source. From Becker et al. (1999). © Elsevier

results suggest that strain weakening and resultant shear localisation probably occurs in the lower mantle. Furthermore, this would explain the preservation of long-lived geochemical reservoirs and the lack of seismic anisotropy in the majority of the lower mantle (Girard et al. 2016).

In the blob model, the convection encompasses the entire mantle, both its lower and upper parts. It does not restrict the penetration of lithospheric plates through the 660 km discontinuity into the lower mantle and down to the CMB. The blobs reside mainly in the cores of the convective cells and represent the primitive reservoir. Surrounding material, and especially the upper mantle region, are depleted and degassed by melting at the ridges and earlier continent formation. Manga (1996) studied the effects of viscosity variations associated with compositional heterogeneities on mixing in the mantle and found out that such highly viscous blobs can persist relatively undisturbed for long periods of time without being substantially deformed or mixed with the surrounding flow; they have a tendency to aggregate, leading to the formation of large scale heterogeneities. Numerical studies suggest that bodies of up to a kilometre in size can persist for 1 Ga (Kellogg and Turcotte 1990), or up to 2 Ga (Gurnis and Davies 1986), or even for 4 Ga (Becker et al. 1999), remaining as isolated blobs, exempt from recycling in the case of whole mantle convection. To satisfy geochemical constraints, blobs would have to fill 30–65% of the mantle (Becker et al. 1999).

Geochemical model calculations confirm the chemical heterogeneity of the lower mantle. Ryabchikov and Kaminsky (2013, 2014) demonstrated that, in addition to fertile areas (similar in composition to pyrolite), more refractory rocks (close to a harzburgitic composition) may occur in the lower mantle. The only microxenolith (known to date) of a lower-mantle rock that originated at depth of more than 1900 km is composed of ferropericlase and two new orthorhombic oxides, $MgCr_2O_4$ and $CaCr_2O_4$ (Kaminsky et al. 2015).

10.3.3 Phase Transitions and Spin Crossover

Among other possible reasons for lower-mantle heterogeneities, mineral phase transitions and Fe–Mg redistributions between iron containing minerals as a result of spin effect have also been suggested (Vinnik et al. 2010).

The most plausible phase transition causing seismic effects is the transition of stishovite into the $CaCl_2$-structured SiO_2, occurring in experiments at 70–80 GPa pressure at mantle adiabatic conditions, which may cause the reduction of seismic wave velocities, particularly V_s (by 12–25%) (Nomura et al. 2010). According to experimental data, it should occur within the 200 km range above and below the phase transition at depths of ~1600–1800 km (see Sect. 4.6.2). However, no significant seismic heterogeneities were recorded within this depth range (Figs. 9.3 and 9.7).

Another phase transition, which was suggested to be responsible for seismic heterogeneities, is the transition of wüstite into rhombohedral structure at depths of

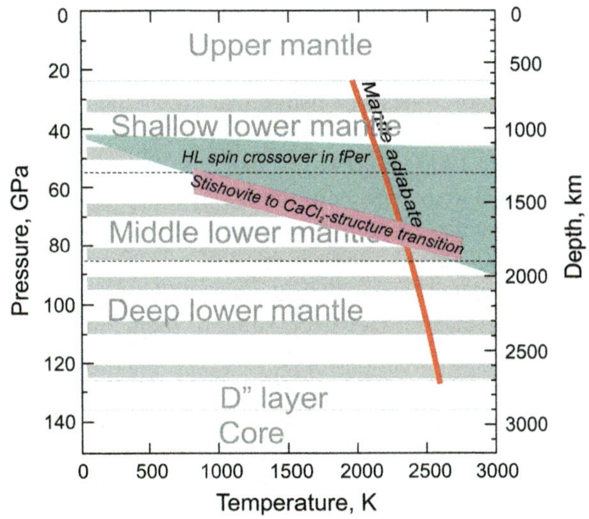

Fig. 10.7 Seismic heterogeneities in the lower mantle (*grey*) and major mineral phase transitions: stishovite to $CaCl_2$-structured SiO_2 (*pink*) and HL spin crossover in ferropericlase (*blue*) and lower mantle adiabate after Katsura et al. (2010) (*red*)

1700 km (Bina 1998). First, like the stishovite → $CaCl_2$-structured SiO_2 transition, this process is not reflected in seismic scatterers (Figs. 10.3 and 10.7); second, it may be realized at temperatures much lower than the mantle adiabatic (Fig. 4.13); and third, natural lower-mantle associations include not wüstite but ferropericlase and magnesiowüstite (see Sect. 4.3.2).

HL iron spin crossover in ferropericlase occurs within a wide range of pressures between 45 and 60–80 GPa at ambient lower-mantle conditions, which corresponds to the depth interval of 1100–1800 km. The LS state exhibits a much higher bulk modulus and bulk sound velocity than the HS state; the bulk modulus increases by ~35% and bulk sound velocity increases by ~15% across the transition in ferropericlase. As a result, the unit cell volume of ferropericlase decreases by 2–8%, which accordingly affects density and seismic velocities and may potentially cause seismic heterogeneities in the lower mantle (see Sect. 8.2.2). However, the spin transition occurs within shallow and middle parts of the lower mantle, where most seismic heterogeneities are related to zones of subduction (Figs. 9.3 and 9.7). HL spin transition in bridgmanite (if it occurs) has no significant physical effect (see Sect. 8.2.3).

The lack of observed seismic effects, corresponding to both stishovite → $CaCl_2$-structure in SiO_2 and HL iron spin crossover in ferropericlase, may be explained by the gradual nature of these transitions. This situation demonstrates the absence of strong stratigraphic discontinuities within the lower mantle, which is comparable to the 660 km or D″ discontinuities. The seismic heterogeneities, most likely, reflect local and regional chemical differences within the lower mantle.

10.4 Conclusions

Three major zones of seismic heterogeneities can be outlined. The shallow lower mantle zone from 660–1300 km includes ~70% of all heterogeneities. The middle zone from 1300–1900 km includes ~20% of all heterogeneities, and the deep zone from 1900 to the border of the D″ layer at 2700 km includes only a small number of heterogeneities. The major scatters are at depths of 800–900, 1100–1200, 1500–1600, 1800–1900, 2000–2100, 2300–2400, and 2600–2700 km. Most of the scatters are related to subduction zones and are caused by lithospheric slabs sinking in the lower mantle. However, some of them, such as the ones within mantle plumes and beneath rift zones cannot be attributed to the process of lithospheric plate subduction. One of the plausible causes for the layered structure of the lower mantle is the increase of iron content in the lower part of the lower mantle, which is different to that of a pyrolitic composition. Some sub-horizontal seismic heterogeneities are suggested as having a global character. However, currently neither mineral phase transition nor HL iron spin crossover can be considered as the cause for lower-mantle heterogeneity.

References

Becker, T. W., Kellog, J. B., & O'Connell, R. J. (1999). Thermal constraints on the survival of primitive blobs in the lower mantle. *Earth and Planetary Science Letters, 171*, 351–365.

Bina, C. R. (1998). Lower mantle mineralogy and the geophysical perspective. *Reviews in Mineralogy, 37*, 205–239.

Cao, A., & Romanowicz, B. (2007). Locating scatterers in the mantle using array analysis of PKP precursors from an earthquake doublet. *Earth and Planetary Science Letters, 255*, 22–31. doi:10.1016/j.epsl.2006.12.002

Castle, J. C., & Creager, K. C. (1999). A steeply dipping discontinuity in the lower mantle beneath Izu-Bonin. *Journal of Geophysical Research, 104*(B4), 7279–7292.

Castle, J. C., & van der Hilst, R. D. (2003). Searching for seismic scattering off mantle interfaces between 800 km and 2000 km depth. *Journal of Geophysical Research, 108*(B2), 2095. doi:10.1029/2001JB000286

Courtier, A. M., & Revenaugh, J. (2008). Slabs and shear wave reflectors in the midmantle. *Journal of Geophysical Research, 113*, B08312. doi:10.1029/2007JB005261

Dai, L., Kudo, Y., Hirose, K., Murakami, M., Asahara, Y., Ozawa, H., et al. (2013). Sound velocities of Na0.4Mg0.6Al1.6Si0.4O4 NAL and CF phases to 73 GPa determined by Brillouin scattering method. *Physics and Chemistry of Minerals, 40*, 195–201.

Davies, G. F. (1984). Geophysical and isotopic constraints on mantle convection: An interim synthesis. *Journal of Geophysical Research. Solid Earth 89*, 6017–6040. doi:10.1029/JB089iB07p06017

Deuss, A., Andrews, & J., Day, E. (2013). Seismic observations of mantle discontinuities and their mineralogical and dynamical interpretation. In: S.-I. Karato, (Ed.), *Physics and Chemistry of Deep Earth*. (pp.287–323). USA: Wiley.

Dziewonski, A. M., & Anderson, D. L. (1981). Preliminary reference Earth model. *Physics of the Earth and Planetary Interiors, 25*(4), 297–356.

Girard, J., Amulele, G., Farla, R., Mohiuddin, A., & Karato, S.-I. (2016). Shear deformation of bridgmanite and magnesiowüstite aggregates at lower mantle conditions. *Science, 351*(6269), 144–147. doi:10.1126/science.aad3113

Gurnis, M., & Davies, G. F. (1986). Mixing in numerical-models of mantle convection incorporating plate kinematics, mixing in the mantle and the possible survival of primitive mantle. *Journal of Geophysical Research, 91*, 6375–6395.

Helffrich, G. R., & Wood, B. J. (2001). The Earth's mantle. *Nature, 412*(6846), 501–507.

Jenkins, J., Deuss, A., & Cottaar, S. (2017). Converted phases from sharp 1000 km depth mid-mantle heterogeneity beneath Western Europe. *Earth and Planetary Science Letters, 459*, 196–207. doi:10.1016/j.epsl.2016.11.031

Johnson, L. R. (1969). Array measurements of *P* velocities in the lower mantle. *Bulletin of the Seismological Society of America, 59*(2), 973–1008.

Kaminsky, F. V. (2012). Mineralogy of the lower mantle: A review of 'super-deep' mineral inclusions in diamond. *Earth-Science Reviews, 110*(1–4), 127–147.

Kaminsky, F. V., & Lin, J.-F. (2017). Iron partitioning in natural lower-mantle minerals: Toward a chemically heterogeneous lower mantle. *American Mineralogist, 102*(4), 824–832. doi:10.2138/am-2017-5949.

Kaminsky, F. V., Wirth, R., & Schreiber, A. (2015). A microinclusion of lower-mantle rock and some other lower-mantle inclusions in diamond. *Canadian Mineralogist, 53*(1), 83–104. doi:10.3749/canmin.1400070

Kaneshima, S. (2003). Small-scale heterogeneity at the top of the lower mantle around the Mariana slab. *Earth and Planetary Science Letters, 209*, 85–101. doi:10.1016/S0012-821X(03)00048-7

Kaneshima, S. (2009). Seismic scatterers at the shallowest lower mantle beneath subducted slabs. *Earth and Planetary Science Letters, 286*, 304–315. doi:10.1016/j.epsl.2009.06.044

Kaneshima, S. (2013). Lower mantle seismic scatterers below the subducting Tonga slab: Evidence for entrainment of transition zone material. *Physics of the Earth and Planetary Interiors, 222*, 35–46. doi:10.1016/j.pepi.2013.07.001

Kaneshima, S. (2016). Seismic scatterers in the mid-lower mantle. *Physics of the Earth and Planetary Interiors, 257*, 105–114. doi:10.1016/j.pepi.2016.05.004

Kaneshima, S., & Helffrich, G. (1998). Detection of lower mantle scatterers northeast of the Marianna subduction zone using short-period array data. *Journal of Geophysical Research, 103*, 4825–4838.

Kaneshima, S., & Helffrich, G. (1999). Dipping low-velocity layer in the mid-lower mantle: Evidence for geochemical heterogeneity. *Science, 283*, 1888–1891.

Kaneshima, S., & Helffrich, G. (2003). Subparallel dipping heterogeneities in the mid-lower mantle. *Journal of Geophysical Research 108*(B5), 2272. doi:10.1029/2001JB001596

Kaneshima, S., & Helffrich, G. (2010). Small scale heterogeneity in the mid-lower mantle beneath the circum-Pacific area. *Physics of the Earth and Planetary Interiors, 183*, 91–103. doi:10.1016/j.pepi.2010.03.011

Katsura, T., Yoneda, A., Yamazaki, D., Yoshino, T., & Ito E. (2010). Adiabatic temperature profile in the mantle. *Physics of the Earth and Planetary Interiors, 183*, 212–218. doi: 10.1016/j.pepi.2010.07.001

Kawakatsu, H., & Niu, F. L. (1994) Seismic evidence for a 920-km discontinuity in the mantle, *Nature, 371*, 301–305.

Kellogg, L. H., & Turcotte, D. L. (1990) Mixing and the distribution of heterogeneities in a chaotically convecting mantle. *Journal of Geophysical Research: Solid Earth, 95*(B1), 421–432. doi:10.1029/JB095iB01p00421

Kellogg, L. H., Hager, B. H., & van der Hilst, R. D. (1999). Compositional stratification in the deep mantle, *Science, 283*, 1881–1884.

Kennett, B., Engdahl, E., & Buland, R. (1995). Constraints on seismic velocities in the Earth from traveltimes. *Geophysical Journal International, 122*, 108–124.

Krüger, F., Banumann, M., Scherbaum, F., & Weber, M. (2001). Mid mantle scatterers near the Mariana slab detected with a double array method. *Geophysical Research Letters, 28*, 667–670.

Labrosse, S., Hernlund, J. W., & Coltice, N. (2007). A crystallizing dense magma ocean at the base of Earth's mantle. *Nature, 450*, 866–869.

Lee, C.-T. A., Luffi, P., Höink, T., Li, J., Dasgupta, R., & Hernlund, J. (2010). Upside-down differentiation and generation of a 'primordial' lower mantle. *Nature, 463*, 930–933.

Le Stunff, Y., Wicks, C. W., & Jr., Romanowicz, B. (1995). Evidence for mid-mantle reflectors. *Science, 270*(5233), 74–77.

Le Stunff, Y., Wicks, C. W., & Jr., Romanowicz , B. (2015) P′P′ Precursors under Africa: Evidence for mid-mantle reflectors. *Science, 270*(5233), 74–77.

Li, J., & Yuen, D. A. (2014). Mid-mantle heterogeneities associated with Izanagi plate: Implications for regional mantle viscosity. *Earth and Planetary Science Letters, 385*, 137–144.

Li, Y., Deschamps, F., & Tackley, P.J. (2014). The stability and structure of primordial reservoirs in the lower mantle: insights from models of thermochemical convection in three-dimensional spherical geometry. *Geophysical Journal International, 199*, 914–930. doi:10.1093/gji/ggu295

Liu, K., Gao, S., Silver, P., & Zhang, Y. (2003). Mantle layering across central South America. *Journal of Geophysical Research, 108*(B11), 2510. doi:10.1029/2002JB002208

Ma, X., Sun, X., Wiens, D.A., Wend, L., Nyblade, A., Anandakrishnan, S., et al. (2016). Strong seismic scatterers near the core–mantle boundary north of the Pacific Anomaly. *Physics of the Earth and Planetary Interiors, 253*, 21–30. doi:10.1016/j.pepi.2016.01.007

Manga, M. (1996). Mixing of heterogeneities in the mantle: Effect of viscosity differences. *Geophysical Research Letters, 23*(4), 403–406.

Muirhead, K. J., & Hales, A. L. (1980). Evidence for P wave velocity discontinuities at depths greater than 650 km in the mantle. *Physics of the Earth and Planetary Interiors, 23*, 304–313.

Niu, F. (2014). Distinct compositional thin layers at mid-mantle depths beneath northeast China revealed by the US Array. *Earth and Planetary Science Letters, 402*, 305–312. doi:10.1016/j.epsl.2013.02.015

Niu, F., & Kawakatsu, H. (1997). Depth variation of the mid-mantle seismic discontinuity. *Geophysical Research Letters, 24*, 429–432.

Niu, F., Kawakatsu, H., & Fukao, Y. (2003). Seismic evidence for a chemical heterogeneity in the midmantle: A strong and slightly dipping seismic reflector beneath the Mariana subduction zone. *Journal of Geophysical Research, 108*(B9), 2419. doi:10.1029/2002JB002384

Nomura, R., K. Hirose, N. Sata, and Ohishi Y. (2010). Precise determination of post-stishovite phase transition boundary and implications for seismic heterogeneities in the mid-lower mantle. *Physics of the Earth and Planetary Interiors, 183*, 104–109, doi:10.1016/j.pepi.2010.08.004

Nomura R., Ozawa H., Tateno S., Hirose K., Hernlund H., Muto S., et al. (2011). Spin crossover and iron-rich silicate melt in the Earth's deep mantle. *Nature, 473*, 199–203.

Petersen, N., Gossler, J., Kind, R, Stammler, K., & Vinnik, L. (1993). Precursorsto SS and the structure of transition zone of the north-western Pacific. *Geophysical Research Letters, 20*(4), 281–284.

Revenaugh, J., & Jordan, T. H. (1991). Mantle Layering From *ScS* Reverberations. 3. The Upper Mantle. *Journal of Geophysical Research, 96*(B12) 19781–19810.

Revenaugh, J., & Sipkin, S. A. (1994). Mantle discontinuity structure beneath China. *Journal of Geophysical Research, 99*(B11), 21911–21927.

Rost, S., Garnero, E. J., & Williams, Q. (2008). Seismic array detection of subducted oceanic crust in the lower mantle. *Journal of Geophysical Research, 113*, B06303. doi:10.1029/2007JB005263

Ryabchikov, I. D., & Kaminsky, F. V. (2013). The composition of the lower mantle: Evidence from mineral inclusions in diamonds. *Doklady Earth Sciences, 453*(2), 1246–1249. doi:10.1134/S1028334X13120155

Ryabchikov I. D., & Kaminsky F. V. (2014). Physicochemical parameters of material in mantle plumes: Evidence from the thermodynamic analysis of mineral inclusions in sublithospheric diamonds. *Geochemistry International, 52*(11), 903–911. doi:10.1134/S001670291411007X

Shen, Y., Wolfe, C. J., & Solomon, S. C. (2003). Seismological evidence for a mid-mantle discontinuity beneath Hawaii and Iceland. *Earth and Planetary Science Letters, 214*(1–2), 143–151. doi:10.1016/S0012-821X(03)00349-2

Souriau, A. (1986). First analyses of broadband records on the geoscope network: Potential for detailed studies of mantle discontinuities. *Geophysical Research Letters, 13* (10), 1011–1014.

Thomas, C., Kendall, J.-M., & Lowman, J. (2004). Lower-mantle seismic discontinuities and the thermal morphology of subducted slabs. *Earth and Planetary Science Letters, 225*, 105–113.

Trampert, J., & Fishtner, A. (2013). Global imaging of the Earth's deep interior seismic constrains on anisotropy, density and attenuation. In: S.-I. Karato, (Ed.), *Physics and Chemistry of Deep Earth*. (pp. 324–351). USA: Wiley.

Tromp, J., & Ishii, M. (1998) Normal mode and free-air gravity constraints on the Large Scale Structure of the Mantle (abstract). Fall AGU Meeting. *Eos, Transactions American Geophysical Union, 79*(45, Suppl.), F598.

Tsuchiya, T. (2011). Elasticity of subducted basaltic crust at the lower mantle pressures: insights on the nature of deep mantle heterogeneity. *Physics of the Earth and Planetary Interiors, 188*, 142–149. doi:10.1016/j.pepi.2011.06.018

Tsuchiya, T. (2013). Elasticity of subducted basaltic crust at the lower mantle pressures: Insights of the nature of deep mantle heterogeneity. *Physics of the Earth and Planetary Interiors, 188*, 142–149. doi:10.1016/j.pepi.2011.06.018

Vanacore, E., Niu, F., & Kawakatsu, H. (2006). Observations of the mid-mantle discontinuity beneath Indonesia from S to P converted waveforms. *Geophysical Research Letters, 33*, L04302. doi:10.1029/2005GL025106

van der Hilst, R. D., & Kárason, H. (1999). Compositional heterogeneity in the bottom 1000 Kilometers of Earth's mantle: Toward a hybrid convection model. *Science, 283*, 1885–1888.

Vinnik, L., Niu, F., & Kawakatsu, H. (1998). Broadband converted phases from midmantle discontinuities. *Earth Planets Space, 50*, 987–99.

Vinnik, L., Kato, M., & Kawakatsu, H. (2001). Search for seismic discontinuities in the lower mantle. *Geophysical Journal International, 147*, 41–56.

Vinnik, L. P., Oreshin, S. I., Speziale, S., & Weber, M. (2010) Mid-mantle layering from SKS receiver functions. *Geophysical Research Letters, 37*, L24302.

Wicks, Jr., C. W., & Richards, M. A. (1993). A detailed map of the 660-kilometer discontinuity beneath the Izu-Bonin subduction zone. *Science, 261*, 1424–1427.

Wicks, C., Weber, M., Le Stunff, Y., & Romanowicz, R. (1996). California broadband array evidence for an upper mantle reflector beneath the West Mariana Ridge (abstract). Fall AGU meeting, San Francisco, *Eos, Transactions American Geophysical Union, 77*(46, Suppl.), F492.

Conclusions

The analysis of natural lower mantle samples demonstrate that the composition of the lower mantle corresponds, in general, to earlier experimental and theoretical constructions: the lower mantle is composed mainly of bridgmanite, ferropericlase and CaSi-perovskite and comprises a zone of *oxide Earth*, in contrast to the silicate Earth (upper mantle). However, the real composition of the lower mantle exhibits some differences from previous models.

1. The lower mantle comprises three major mineral associations: ultramafic, mafic and carbonatitic. The juvenile ultramafic association is predominately represented by the association of four major minerals: bridgmanite + ferropericlase + CaSi-perovskite + silica. The mafic association (bridgmanite + CaSi-perovskite + silica + Al-phase) is formed as a result of the recrystallization of subducted lithospheric slabs. The carbonatitic association is the major metasomatic factor within the lower mantle and the diamond-producing medium. In the ultramafic association, the proportions of major minerals differ from the predicted proportions. Although, to date, the exact proportions of minerals cannot be calculated, ferropericlase plays a greater role than previously suggested, while the proportion of bridgmanite is lower.
2. Bridgmanite is the most abundant mineral in the lower mantle. Two groups of bridgmanite can be distinguished: low-Al (0.25–3.1 wt.% Al_2O_3) and high-Al (8–12.6 wt.% Al_2O_3). Incorporation of Al has either no effect on the thermodynamic properties of bridgmanite or slightly decreases its bulk modulus; in any case, it should not be significant for seismologic interpretation. The addition of Fe in bridgmanite, occurring in the deep part of the lower mantle, causes an increase in its density, bulk modulus, and bulk sound velocity independently of the Al concentration.
3. The composition of ferropericlase is different to that predicted (mg = 0.73–0.88), with the magnesium index mg varying in a wide range from 0.90–0.36, which is most likely caused by magma fractional crystallization in the lower mantle.
4. In addition to the three major minerals comprising the juvenile ultramafic lower mantle (bridgmanite, ferropericlase and CaSi-perovskite), the fourth mineral phase, free silica, permanently occurs in amounts of 2.1–15.0% (average:

F.V. Kaminsky, *The Earth's Lower Mantle*, Springer Geology,
DOI 10.17/978-3-319-55684-0

8.4%). Within the shallow lower mantle this is present as stishovite; at depths of ~1600–1800 km, it gradually transforms into the $CaCl_2$-structured silica, which should cause the reduction of seismic velocities. Deeper, at the D″ layer border, SiO_2 transforms into seifertite with an α-PbO_2 structure. Free silica should be considered to be one of the major initial mineral components of the lower mantle, making it compositionally different to the upper mantle. Its presence in the lower mantle is probably caused by a more silica-rich composition of the lower mantle compared to the upper mantle.

5. Within ultramafic associations, in addition to the four major mineral phases, a set of accessory minerals occurs, such as Mg–Cr–Fe, Ca–Cr and other orthorhombic oxides, jeffbenite, ilmenite (including picroilmenite and manganilmenite), titanite, native Ni and Fe, moissanite and some others.
6. The bulk composition of the juvenile ultramafic lower mantle is not homogeneous. With depth, iron contents in both ferropericlase and bridgmanite increase, reflecting the increasing Fe content in the deep part of the lower mantle.
7. The mafic association does not play such important role in the lower mantle as the in the ultramafic association. Firstly, it is much rarer because of the relatively small volume of subducted lithospheric slabs within the lower mantle. Secondly, the structural changes are not so noticeable compared with that in the ultramafic association. For example, a possible NAL to CF phase transition, accompanied by the increase in density, bulk modulus and share wave anisotropy, could be seismically detectable. However, it is compensated for in the bulk slab composition, by the faster sound velocity of SiO_2 and its high-pressure polymorphic phase(s). The transitions may only be observed locally, within subducting slabs, leading to the conclusion that the mafic association does not play a major role in the general structure of the lower mantle. On the other hand, subducting slabs may cause local seismic anomalies and deliver elements such as Si, Al and water to the deep mantle, forming local water reservoirs within the lower mantle. This may cause local seismic and chemical heterogeneities that can influence deep-mantle geodynamic processes.
8. In addition to the ultramafic and mafic associations, a carbonatitic mineral association exists in the lower mantle. In terms of the mineral composition, it is close to the natrocarbonatitic association. The initial lower mantle carbonatitic melt forms as a result of low-fraction partial melting of lower mantle material at the core-mantle boundary. During ascent to the surface, the initial melt dissociates into two immiscible parts, a carbonate-silicate and a chloride-carbonate melt. The latter melt is parental to the observed natrocarbonatitic lower mantle association. Diamonds with carbonatitic inclusions were formed in carbonatitic melts or high-density natrocarbonatitic fluids.
9. Diamond is a member of all three lower mantle mineral associations. Lower-mantle diamond has a series of features that are different from those in lithospheric diamond. Its morphological and structural peculiarities indicate its origin and long-time residence under high-temperature conditions. The

resorption of crystals in a low total nitrogen content with a great proportion of 'nitrogen-free' stones and almost a 100% aggregation level in nitrogen-containing stones form as a result of a high-temperature diffusion of nitrogen. Most lower-mantle diamonds have a high concentration of hydrogen impurity, which is a specific non-characteristic for astenospheric diamond. Isotopic compositions of carbon and nitrogen in lower-mantle diamond allow average isotopic characteristics for juvenile lower mantle to become established as: $\delta^{13}C_{VPDB} = -4.32‰ \pm 2.45‰(2\sigma)$ and $\delta^{15}N_{atm} = -3.00‰ \pm 2.37‰\ (2\sigma)$. Lower-mantle diamond was formed in carbonate-oxide parental magmas, which experienced fractional crystallization with the decrease of temperature and changes in the melt composition.

10. High-to-low spin state transitions in Fe atoms of iron-containing minerals caused by high pressure–temperature (*P–T*) conditions in the lower mantle create changes in physical, chemical, rheological and transport properties of iron-containing minerals. This process is most prominent in ferropericlase, where the Fe^{2+} ion gradually obtains a low-spin configuration within the depth interval of 1000–1500 km. The high-pressure low-spin state exhibits a much higher bulk modulus and bulk sound velocity than the low-pressure high-spin state; the unit cell volume of ferropericlase decreases. However, the gradual character of the spin transition explains the absence of deviation in the aggregate seismic velocities and, thus, the lack of a seismic signature of the spin crossover, making the spin crossover in ferropericlase a seismologically transparent transition.
11. Under very high-pressure conditions, chemical elements can obtain dramatic new properties in the lower mantle, including the formation of unexpected crystal structures and completely new counter-intuitive compounds. Most of these transformations may occur in specific compositions, which may produce only accessory mineralization. However, they may play a significant role in balancing the Earth's light elements and in the formation of the primordial carbonatitic (or silica-carbonatitic) association, whilst also influencing some major lower mantle phases, such as ferropericlase with the formation of magnesium peroxide MgO_2.
12. The D″ layer occupies the lowermost 200 km part of the lower mantle. The phase transformation from bridgmanite to post-perovskite with 0.9–1.6% density increase, occurring at 124–125 GPa and 2500 K, is the major component involved in distinguishing the D″ layer. It is accompanied by the transition of the orthorhombic $CaCl_2$-structured $SiO_2$to seifertite with a 1.5% density increase. Iron-rich liquid metal from the outer core, containing 5–10% light elements (C, N, O and Si) infiltrates into the lowermost mantle and forms a series of native Fe, iron carbides and nitrides, and silicon carbide. Of particular importance is the presence of Fe_7C_3 and Fe_7N_3, which, when in association with diamond, are the first solidus phases, crystallized from the metallic liquid in the D″ layer. The presence of iron nitrides in the D″ layer is closely related to their suggested presence in the inner core.

13. A series of seismic scatters are recorded within the lower mantle. The most prominent of them are at depths of 800–900, 1100–1200, 1500–1600, 1800–1900, 2000–2100, 2300–2400 and 2600–2700 km. They exist both near subduction zones and beneath plates, and none of them (with the exception of the D″ layer border) can be identified with any mineral phase or spin transition. The stishovite → $CaCl_2$-structured SiO_2 transition, occurring in experiments at 70–80 GPa pressure at the mantle adiabatic conditions (corresponding to depths of ~1600–1800 km) may cause the reduction of seismic wave velocities, particularly V_s (by 12–25%). However, no significant seismic heterogeneities were recorded within this depth range. Another phase transition, wüstite into a rhombohedral structure at the depth of 1700 km, may be realized only at temperatures much lower than mantle adiabatic. The HL iron spin transition in ferropericlase, occurring within the depth interval 1100–1800 km, is very gradual and is also not reflected in seismic heterogeneities. It is therefore highly likely that seismic heterogeneities in the lower mantle are caused by ancient subducted lithospheric plates or from reservoirs of primordial and differentiated material of the lower mantle.
14. The bulk composition of the lower mantle, which is most likely different to that of a 'pyrolite' model, is expected to be the main cause of these features. From the obtained data, the presence of free silica as well as an elevated silica concentration in the lower mantle, favor the Earth's enstatite-chondrite model.

These conclusions require further observations and experiments, which may correct our understandings of the composition and structure of the lower mantle.

Subject Index

F.V. Kaminsky, *The Earth's Lower Mantle*, Springer Geology,
DOI 10.17/978-3-319-55684-0

T

U

W

CPSIA information can be obtained
at www.ICGtesting.com
Printed in the USA
LVOW02*1011110617
537666LV00001B/17/P

9 783319 556833